Primary Geography

Handbook

Edited by Stephen Scoffham

Geographical
Association

Acknowledgements

A handbook of this kind is a team effort involving many authors. The production too is a complex process which depends on the publisher, editor, designer, illustrators and printers. I would like to offer my thanks to all those who have given their time and professional expertise so unstintingly to such good effect. Particular thanks are due to Geographical Association staff, more especially Anna Gunby who served as in-house editor and Fran Royle who has nurtured and supported the handbook since its inception.

This edition of the Primary Geography Handbook is based on an earlier version published in 1998 and edited by Roger Carter. Two years ago, in 2002, representatives from the Geographical Association assembled for a weekend planning meeting in Birmingham at which it was decided to retain as many of the original chapters as possible, but to shift the emphasis away from close adherence to the national curriculum. Not without apprehension, I agreed to take on the role of editor.

Following in Roger's footsteps has been a hard act and much of what appears in this new edition builds on his earlier work. I am also indebted to the team of authors who responded so magnificently to the rather unglamorous task of revising and updating their texts and accommodating my suggestions and alterations. I would also like to welcome the new contributors who have worked closely to their original brief and whose enthusiasm and vision have undoubtedly added a new dimension.

Photographs of children and examples of pupils' work are one of the striking features of this book. These illustrations have been assembled over many years. Where possible we have acknowledged the schools who have allowed us to reproduce original pieces of pupils' work. Where the work has been photographed we have not been able to trace the sources so have acknowledged the photographers. Many thanks to all the schools whose work features.

Finally, I would like to offer thanks to all those friends and colleagues within the Geographical Association and at Canterbury Christ Church University College who have supported me in my role as editor. Being able to draw on their enthusiasm, support and expertise and belonging to a community of geographers has been invaluable.

Stephen Scoffham
October 2004

© Geographical Association, 2004

ISBN 1 84377 103 9
First published 2004
Impression number 10 9 8 7 6 5 4 3 2 1
Year 2007 2006 2005

Published by the Geographical Association, 160 Solly Street, Sheffield S1 4BF.
Website: www.geography.org.uk
E-mail: ga@geography.org.uk
The Geographical Association is a registered charity: no 313129.

The Publications Officer of the GA would be happy to hear from other potential authors who have ideas for geography books. You may contact the Officer via the GA at the address above. The views expressed in this publication are those of the authors and do not necessarily represent those of the Geographical Association.

Editing: Rose Pipes
Design and typesetting: Arkima, Dewsbury
Illustrations: Dan Parry-Jones and Linzi Henry
Cartography: Paul Coles
Printing and binding: In China through Colorcraft Ltd, Hong Kong

Foreword

On the very day I sat down to write this foreword I was somewhat amused to read in the papers of software giant Microsoft's discomfiture at having lost hundreds of millions of dollars simply because so many of its highly paid programmers knew so little about world geography and had made so many mistakes as a result! Better yet, they're all going to be compulsorily re-educated – no doubt covering precisely the kind of territory covered in this invaluable handbook!

Photo | www.JohnBirdsall.co.uk

My take on all this is dead simple: the majority of adults in the rich northern world are living in a state of wilful denial. The hard-edged evidence of accelerating environmental damage is all around us – in our own backyards as much as in the global commons – but we carry on as though nothing was going to have to change.

As a result, I spend my entire life – both with Forum for the Future and the Sustainable Development Commission – trying to transform the mindsets that perpetuate that denial. But the truth of it is that most of the 'decision-makers' I'm dealing with really can't get their heads around 'all this environment stuff' – and for the vast majority of them, the fact that their education was an environment-free zone is a significant contributory factor.

That's now changing – at long last! One of the things that keeps me reasonably optimistic is the fact that every year group leaving the UK educational system has had a little bit more exposure to environmental awareness-raising than the preceding year group. And the contribution that the teaching of geography – both in primary and in secondary schools – makes to that process is enormous.

Improved knowledge about the state of the world and its people is just one part of that. As a Trustee of WWF in the UK, I've seen at first hand the emotional power of the kind of educational materials available to primary school teachers today, providing insights and inspiration that for many young people will stay with them for the rest of their lives. It's all about reinforcing (and validating) a personal relationship with the living world, and gently building that capacity for empathy without which it's so hard truly to understand our role in the world today.

After all, it's not as if we have a choice about learning to live sustainably on Planet Earth. The choices we face are all about when, how, and at what cost? In that respect, tomorrow's politicians will depend on tomorrow's citizens and communities being predisposed to embracing that environmental challenge - and it's difficult to imagine a more critical role in laying the foundations for that shift than today's teachers of primary geography.

Jonathon Porritt
August 2004

Jonathon Porritt is Chairman of the UK Sustainable Development Commission (www.sd-commission.gov.uk) and Programme Director of Forum for the Future (www.forumforthefuture.org.uk).

Contents

Contents

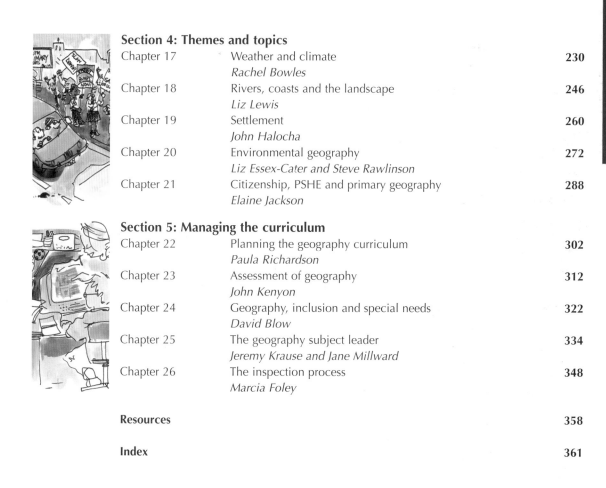

Contributors

Stephen Scoffham is a textbook and atlas author, Honorary Publications Officer for the Geographical Association and a Principal Lecturer at Canterbury Christ Church University College.

Jonathan Barnes is Senior Lecturer in Primary Education at Canterbury Christ Church University College. His main interests outside geography are in cross-curricular and creative thinking, learning and teaching.

Julia Tanner is Principal Lecturer in Education at Leeds Metropolitan University and has a particular interest in children's social and emotional development.

Kathy Alcock is a Principal Lecturer in Education at Canterbury Christ Church University College. Her area of responsibility is Early Professional Development and she teaches primary geography.

Fran Martin is a Senior Lecturer at University College Worcester. Her specialisms are primary geography and the early years. She is also a trustee for The World Studies Trust.

Paula Owens is Deputy Head at Eastchurch Primary School and is completing a PhD about children's early environmental values.

Simon Catling is Professor of Primary Education and Assistant Dean at Oxford Brookes University, Westminster Institute of Education. He teaches and researches in primary geography.

Geoff Dinkele is occasional consultant for geography in education. He was formerly County Advisor/Inspector for geography for Hampshire and then worked part-time in teacher training.

Colin Bridge is a teacher, environmentalist and co-author of the Collins Longman Keystart Atlas Scheme.

Margaret Mackintosh taught in Nigeria and Humberside then at the University of Plymouth. She has been Honorary Editor of *Primary Geographer* since October 1995.

Paula Richardson, former teacher, Advisor and Inspector, is an author, visiting Professor at the University of Minnesota and member of the GA Publications Committee and Fieldwork Group.

Liz Lewis is NE co-ordinator for the GA Valuing Places Project, and teaches Primary Geography education in the Universities of Durham and Sunderland.

Kate Russell is Adviser for Geography with the School Improvement Division in Staffordshire and the new editor of the Staffordshire Learning Net www.sln.org.uk/geography.

Angela Milner is Head of Initial Teacher Training at Edge Hill College of Higher Education, Ormskirk.

Terry Jewson is Education Advisor – Gillingham Partnership Education Action Zone.

Greg Walker is Senior Education Lecturer and Primary Humanities Co-ordinator at Roehampton University. His particular interest is in working with teachers/lecturers to produce primary geography resources.

Maureen Weldon is Head teacher, Nene Valley Primary School, Peterborough; former Advisory Teacher for Geography in Lincolnshire; and author of *Kaptalamwa: A village in Kenya*.

Mary Young promotes the Global Dimension through ITT and CPD in London and the south-east of England for DfID's EES project. Previously she worked for Oxfam.

Rachel Bowles, lately of the University of Greenwich, is co-ordinator of the Register of Research in Primary Geography, a GA Primary Committee and ICT Working Group member.

John Halocha is Reader in Geography Education and Head of Geography at Bishop Grosseteste College, Lincoln; and involved with the Geographical Association and International Geographical Union Commission for Geographical Education.

Liz Essex-Cater is now living in the Champagne-Ardennes region of France after 27 years of teaching geography in the UK.

Steve Rawlinson is Principal Lecturer, Geography and Environmental Education and Senior Admissions Tutor Primary Education, Northumbria University.

Elaine Jackson is Chief Advisor (Primary) for Trafford School Improvement Service and Chair of the Geographical Association's Primary and Middle Schools Section Committee.

John Kenyon is Head teacher, Middlewich Primary School, Cheshire.

David Blow is a Senior Lecturer in Education and has worked in the field of special education, both 'mainstream' and residential, for 30 years.

Jeremy Krause is Senior Regional Director, KS3 National Strategy, formerly Advisor for geography, Cheshire LEA and Past President of the GA.

Jane Millward is Head teacher, Wimboldsley County Primary School, Cheshire, formerly Primary English Consultant, Cheshire LEA.

Marcia Foley, until recently Geography Adviser for Kent LEA, is an experienced independent consultant specialising in primary geography.

Introduction

Geography matters! The things that affect us as individuals do not occur in a vacuum, they happen in a context. Pick up any broadsheet newspaper, watch the television news or listen to people talking to each other and you will quickly discover that places are crucial to our lives. Understanding the world around us, making sense of the way it is inter-related and considering how it might change in the future stand at the heart of geography.

Tracing the term 'geography' back to its linguistic root reveals two key aspects of the subject: 'geo' means 'earth' and 'graphia' means 'writing'. Of course geography as an academic discipline has developed dramatically since the term was first coined in the museum of Alexandria around 300BC. However, the idea that geography is about describing and interpreting our surroundings, both local and distant, is as valid today as it was over 2000 years ago.

The geography national curriculum makes the same point in a different way. The opening paragraph declares: 'geography provokes and answers questions about the natural and human worlds, using different scales of enquiry'. Further on we learn that as pupils study geography they 'encounter different societies and cultures' and learn a range of skills which will equip them for adult life (DfEE/QCA, 1999, p. 108). No other subject can make this claim, and there are few areas of the curriculum which have pioneered such a wide range of teaching techniques as geography.

Although the rationale for teaching geography is no longer included in national curriculum documents, the aims that were identified in the early 1990s still pervade the subject. These were to:

■ stimulate pupils' interest in their surroundings and in the variety of physical and human conditions on the Earth's surface

■ foster their sense of wonder at the beauty of the world

■ develop an informed concern about the environment

■ enhance their sense of responsibility for the care of the Earth and its peoples (DES/WO, 1990, p. 6).

It is immediately apparent that these aims highlight environmental issues. They also include a strong spiritual dimension, focus on values and acknowledge that geography is about the interaction of people and their surroundings.

The knowledge, concepts and skills that geography covers are essential components of a broad and balanced curriculum. Geography also plays an important part in pupils' physical, intellectual, social and emotional development. It has a unique bridging function linking different subject areas and providing a context in which they can be explored (Figure 1). Geographical perspectives are becoming increasingly important as we become aware of the importance of other cultures, the corrosive effects of environmental problems, and the damage caused by

Fieldwork and enquiry
Practical enquiries and investigations are central to geography as they provide pupils with first hand information and direct experience.

Knowing the locality
Studying the locality helps pupils to develop their sense of identity and self-esteem.

Maps and plans
Geography helps pupils to locate themselves in their surroundings and find their way from one place to another.

Photographs, charts and diagrams
Geographers use a great variety of visual devices to communicate findings in non-verbal ways.

Investigating issues
Some of the best work in geography comes when pupils debate issues and propose solutions to problems.

ICT
Interactive whiteboards, digital cameras and the internet now allow pupils to record and interpret the world in new ways.

Cycles, patterns and processes
Geographical concepts such as cycles, patterns, processes and interaction provide a unique way of describing and analysing the world.

Sustainability and the environment
Sustainable development and the care of the environment are key issues for the twenty-first century and form an important part of the geography curriculum.

Critical and creative thinking
The best geographical studies promote critical and creative thinking skills and long-term learning across the curriculum.

Respecting diversity
Geographical studies can tackle biased images and negative stereotypes in a neutral and unemotional way making links to other subjects such as history, citizenship and modern foreign languages.

Global citizenship
Geography recognises that we are all inter-dependent global citizens with a responsibility to the planet and to each other.

Considering the future
Geography helps pupils to develop attitudes and form opinions about current issues, appreciate tensions and uncertainties and consider the future of the world and its peoples.

Figure 1 / How does geography contribute to pupils' learning?
Source: Geographical Association, 2004.

inequalities between nations and peoples. As Professor Andrew Goudie puts it, 'What other subject tells us so much about the great issues of our age?' (cited by DfEE/QCA, p. 108)

About the handbook

This book is about how to teach geography in primary schools. It is written by a volunteer team from around the country and brings together their collected wisdom in a single volume which represents well over five hundred years of teaching experience! It is hardly surprising that enthusiasm and commitment to the subject shines through on every page.

The book is divided into five sections covering learning, skills, places, themes and curriculum management. Some topics such as action research and in-service training are covered throughout the book. As with any manual, it is unlikely you will want to read the text from cover to cover. The navigator opposite is intended to help you locate material on a range of key ideas as quickly and easily as possible. If you want more detailed information, use the index at the back.

Writing about geography at the moment is a complex matter. The pace of change and the speed with which educational initiatives come and go means that the material tends to date. When the first edition of the *Handbook of Primary Geography* (Carter, 1998) appeared the National Literacy and Numeracy Strategies had not even been introduced. Since then PSHE, Citizenship and Education for Sustainable Development have come into their own and ICT has developed beyond all recognition. In the next few years initiatives such as Modern Foreign Languages and creativity and thinking skills may well create new, but as yet unspecified, opportunities for primary geography.

This new *Primary Geography Handbook* seeks to steer away from the details of how to teach a particular government initiative or set of requirements. Instead it provides a bedrock of pedagogical and theoretical ideas which should have lasting credibility. For this reason there are reports on current research into children's ideas and thinking about the world. The references and bibliographies at the end of each chapter will be particularly useful to students and those pursuing higher level degree courses.

Above all, the handbook is intended to inspire. We have taken the opportunity to include new chapters on topics as diverse as emotional geography, creative thinking and global citizenship. Enthusiastic and motivated teachers are the best guarantee for vivid and vibrant classroom practice. Geography has always needed ambassadors who will speak up and argue its case. This is as vital as ever at the moment when there are so many pressures on curriculum time and such a strong focus on measurable achievements.

Some of today's pupils may still be alive at the end of the century. Given the pace of change we can only guess at the conditions and circumstances they will confront as they grow older. What we can try to ensure, however, is that their primary school education gives them the understanding they need for the present and prepares them for a future in which they will continue to learn and develop. Geographical perspectives are an essential part of this foundation. We need to see that children are proficient in basic life skills such as communicating information, way-finding and interpreting the environment. We have a duty to attend to their psychological and emotional well-being and to nurture the formation of positive attitudes to themselves, each other and the environment. In addition we need to help pupils of all abilities and backgrounds – tomorrow's adults – to play their part in a shrinking world. This book is designed to help you achieve these objectives and enable your pupils to live more fulfilling lives.

Stephen Scoffham

References

Carter, R. (ed) (1998) *Handbook of Primary Geography*. Sheffield: Geographical Association.

DES/WO (1990) *Geography for Ages 5 to 16*. London: HMSO.

DfEE/QCA (1999) *The National Curriculum: Handbook for primary teachers in England*. London: DfEE/QCA.

Geographical Association (2004) *Finding Time for Things that Matter* (leaflet) Sheffield: Geographical Association.

Key ideas for navigation

Assessment
Chapters 5, 7, 14, 23, 25

Citizenship
Chapters 3, 4, 6, 16, 20, 21

Creativity
Chapters 2, 4, 22, 24

Concepts
Chapters 1, 5, 11

Differentiation
Chapters 18, 23, 24

Displays
Chapters 11, 13, 15, 17, 26

Distant places
Chapters 4, 14, 15, 16, 21

Emotions
Chapter 3

Enquiries
Chapters 7, 8, 11, 13, 14, 18, 19, 20, 21

Environment
Chapters 2, 12, 13, 16, 19, 20

Europe
Chapters 14, 19

Fieldwork
Chapters 1, 2, 10, 11, 17, 18, 19, 20, 21, 24

Foundation stage
Chapter 5

Games
Chapter 4

Gender differences
Chapters 1, 8, 26

Good practice
Chapters 5, 6, 10, 12, 13, 16, 21, 22, 25, 26

Graphicacy
Chapter 9

ICT
Chapter 12

Inclusion
Chapters 2, 4, 24, 26

Inservice training
Chapters 11, 13, 25

Inspection
Chapters 25, 26

Issues
Chapters 2, 3, 7, 10, 14, 16, 19, 20, 21, 24

Locational knowledge
Chapters1, 8, 10, 15

Mapwork
Chapters 2, 4, 8, 9, 11

Misconceptions
Chapters 1, 9, 14, 15, 16, 17, 19, 20

National Curriculum
Chapters 6, 11, 12, 13, 14, 18, 20, 22, 23, 24, 25, 26

Photographs
Chapters 8, 9, 10

Planning
Chapters 10, 12, 14, 15, 16, 18, 22, 23, 25, 26

Progression
Chapters 1, 6, 7, 8, 9, 11, 14, 17, 18, 21, 22, 23

QCA units
Chapters 7, 14, 15, 17, 18, 19, 22

Questions
Chapters 2, 7, 19

Record keeping
Chapters 10, 17, 23, 25, 26

School journeys
Chapters 5, 8, 10, 13

School linking
Chapter 16

Special educational needs
Chapters 3, 12

Transfer and transition
Chapters 6, 17, 25

Vocabulary
Chapters 11, 17, 19

Photo | Kathy Alcock and John Collar.

Young geographers

'Our present-day knowledge of the child's mind is comparable to a fifteenth-century map of the world – a mixture of truth and error ... vast areas remain to be explored'

**Arnold Gesell,
cited in Fisher, 1995.**

Over the past few years there has been a growing interest in primary geography research and the misconceptions which many children hold about the world around them. The way in which children learn about places is complex and researchers are only slowly piecing together the story. Part of the problem is that there is no entirely logical or inevitable sequence of events. Learning is for many people a surprisingly idiosyncratic process in which ideas are acquired in an apparently random manner. Most people do not become systematic thinkers until they reach adolescence or even later.

It is also remarkably difficult to discover what is actually going on inside a child's head. We were all young once, but it is impossible to recapture the sense of awe and wonder which children experience on doing something for the first time. Adults can often only observe and deduce what seems to be happening from the outside.

Despite these complexities there is, however, one clear message: the skills and competencies of young children appear to have been consistently underestimated. In part this is due to an uncritical acceptance of Piaget's theories. Few people nowadays would contest Piaget's central thesis that children pass through developmental stages as they grow older. However, the age at which this happens, and children's ability to operate at different levels of understanding, is much more variable than was first thought. The more we find out about different aspects of geographical learning the further the roots go back into childhood.

Spatial awareness

Children first begin to discover the location of objects as they play with toys in their cots. Initially, taste and touch provide the main clues, but after about three months their vision improves and they become capable of focusing more sharply on objects. From about ten months onwards children begin to actively explore their environment. As they develop the ability to crawl, and then walk, the scope and range of the places they can visit expands enormously, though parents usually restrict these early forays for fear of dangers. However, as youngsters grow in ability and confidence, so they are allowed to go further afield. This introduces them to the environment beyond the front door.

Maps are essential tools in the discovery process, but how do young children learn to use maps and find their way from place to place? Wiegand (1992) describes an investigation by Bluestein and Acredolo which provides some interesting insights. Sixty 3-5 year-olds were asked to find a toy elephant hidden in a room. They were given a map that showed the layout of the furniture with the position of the toy marked with a cross. The results were impressive. Half the three-year-olds, three-quarters of the four-year-olds and all the five-year-olds were able to find the toy. Generally the children had little difficulty interpreting the map and appeared to understand that the room, furniture and toy were shown symbolically. Subsequent investigations showed the importance of aligning the map. When the map was rotated by 180 degrees only the oldest children were able to use it correctly and even they had difficulty.

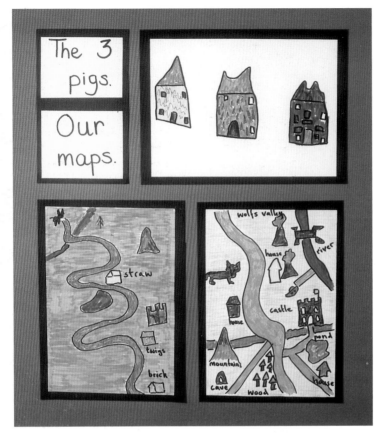

Figure 1 | *Drawings of the route taken by the wolf in 'The Three Little Pigs' by a reception, year 1 and year 2 children.*

Blades and his colleagues (1998) went a stage further and investigated the mapping abilities of four-year-old children in a number of different countries. After discussing the features on a vertical aerial photograph, the children were asked to trace a route from one house to another. Not only were the majority of the children able to complete the task successfully, they all demonstrated an ability to interpret the photograph whatever their cultural background.

Further evidence that spatial reasoning is universal and not related to cultural background comes from a study in rural Kenya conducted by Matthews (1995). Here pupils aged seven to thirteen were asked to draw a map of their village. Despite being almost completely isolated from Western influences and never having seen a formal map, the majority of the pupils completed the exercise successfully. Furthermore they recalled their environment in vivid terms using plan views and other relatively sophisticated representations.

Another issue which has occupied many researchers is whether there is a difference in spatial ability between boys and girls. Simple experiments in which children draw free-hand maps of a familiar journey, such as the route from home to school, have tended to show a clear distinction (Taylor, 1998). Boys are more likely to use plan views and show a larger area than girls who prefer to be precise and include small details. Generally, too, boys appear better at arranging the different elements of a map in their correct relationship and mastering abstract conventions.

These results are interesting because they reveal something about the way children perceive the spaces around them. Whether these differences are innate or the result of social and cultural upbringing is open to question. Matthews (1992), for example, argues that parents allow boys considerable freedom to explore the local surroundings. Girls, by contrast, are expected to help around the house, are allowed out less often and not permitted to go so far.

Research findings also provide compelling reasons for introducing children to mapwork from an early age. Many infant school teachers already take the opportunity to create 'pictures' to show journeys associated with stories and fairy tales. The route taken by the wolf in 'The Three Little Pigs' is a typical example (Figure 1). However, the fact that children seem to draw maps spontaneously long before they learn to read or write suggests that spatial awareness is a fundamental skill which should be developed in nurseries and other pre-school groups as a basic educational entitlement.

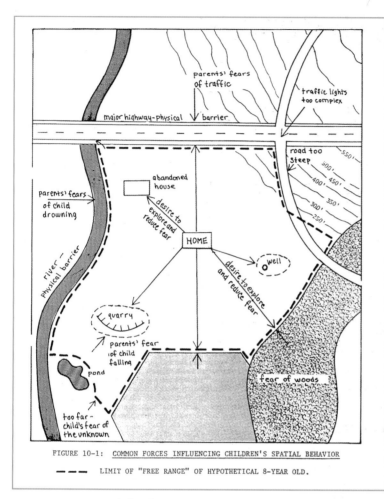

FIGURE 10-1: COMMON FORCES INFLUENCING CHILDREN'S SPATIAL BEHAVIOR

‑ ‑ ‑   LIMIT OF "FREE RANGE" OF HYPOTHETICAL 8-YEAR OLD.

Figure 2 | Children's experience of place – the first ten years. After Hart, 1979.

Exploring the local environment

Most of us remember the place where we were brought up in some considerable detail and often recall it with fondness. These first impressions of the outside world stay with us throughout our lives and provide a rich source of experience. Authors such as Laurie Lee, Virginia Woolf, George Orwell and Marcell Proust recall their early memories with great sensitivity in their novels. Our sense of identity, it seems, derives in some part from the social and physical environment in which we spend our childhood.

The way children interact with their immediate surroundings is important not only for their psychological well-being, it also promotes their educational development. Many play activities involve rehearsing or re-enacting previous events and situations. Through imitation children are able to give full reign to invention and fantasy. Piaget argues that make-believe play allows children to assimilate knowledge and forms the basis of a child's thought even before it can speak. Certainly the opportunity to model and manipulate experiences seems to be an essential part of the learning process.

Several researchers have attempted to find out more about children's private geographies. The classic study was undertaken by Hart (1979) who made a detailed investigation of a New England township over a two-year period. Hart discovered that the children put a particularly high value on water features such as rivers, lakes and ponds. They also favoured trees for climbing and hiding games. The places they feared matched the archetypal scary places of children's literature – attics, cellars and abandoned buildings, and bedrooms and garages at night. Very few of the children selected places for their aesthetic qualities alone. Hart comments on the way children treasure informal routes and pathways which they often use as 'short-cuts' even when they are actually longer. Other researchers, too, remark on children's affinity for secret routes and alleyways. As they explore their surroundings children construct private geographies which meet their physical and emotional needs (Figure 2).

Further insight into children's thinking is provided by the names they invent for their favourite places. Matthews (1992) reports how, when drawing maps of suburban Coventry, children labelled a variety of local features. Examples included the 'Moth-hawk tree', the

Figure 3 | *Exploring their environment helps children develop their sense of attachment and identity. Photo: Diane Wright.*

'dump', 'Charlie's field' and the 'back alley'. Sometimes these personal names denote the activities that can be done in a place rather than its appearance, e.g. 'Roller-coaster place'. In her study of young children's environmental preferences Owens (2003) too found that many children prized the activities they could do in a place above all its other qualities. In one example she describes how a metal bar used for fastening a door was variously identified by reception class pupils as a meeting place, something you can lean and swing on and something you must not touch.

In thinking about pupils' perceptions of the environment it is important to recognise that their experience may be limited and not to make assumptions. For example, Owens (2003) discovered through discussion with reception class pupils that many of them had no idea that there was a field attached to their school. Similarly, there are adolescents living in London who have never travelled by bus or underground or ventured across the River Thames and those from the Channel ports have never been to mainland Europe.

At the same time children are naturally curious and inventive. They invest their surroundings with personal meaning and interpret it according to their needs. It is worth remembering in this context that the most intensively used play areas are often small patches of dirt rather than designated facilities. Children also need places where they simply loiter or day-dream (Figure 3). Edmund Gosse, the Victorian naturalist, is one of many authors who have left us with a description of their childhood pleasures:

> *By the side of the road, between the school and my home, there was a large horse pond. Here I created a maritime empire – islands, a seaboard with harbours, lighthouses, fortifications. My geographical inventiveness had its full swing. Sometimes, while I was creating, a cart would be driven roughly into the pond, shattering my ports with what was worse than a typhoon. But I immediately set to work, as soon as the cart was gone and the mud had settled, to tidy up my coastline again and scoop out anew my harbours* (Gosse, 1965, p. 136).

It is through transactions of this kind that children come to invest their environment with meaning. The attachment to places which we develop as adults is derived from these childhood interactions. We identify with our home area in lots of different ways. Some people support their local football team, others become involved with local history or trace their family tree. Historically people used to believe they belonged to the soil of a particular place in an organic and religious relationship. The Romans recognised the spirit or essence of a locality (its *genius loci*) by setting up shrines to local deities. Anthropologists have recorded similar beliefs among the original inhabitants of Australia and North and South America.

These studies remind us that the quality of an environment is a very complex issue. As well as quantifying physical and human processes, geographers seek to take account of subjective and personal responses. What a place is like is not simply a matter of fact. It depends equally on how we perceive it and what we feel about it. Today, concerns over personal security and the relentless growth of road traffic are serving to erode children's personal freedom and their links with the environment. In an authoritative and far-reaching study Hillman (1993) found that the number of unaccompanied activities undertaken by junior school pupils at weekends had halved between 1971 and 1990, may affect emotional and social development. When children explore the environment they have opportunities to take initiatives, learn survival skills, develop a sense of adventure, gain self-esteem and accept responsibility for their actions. The physical activity also helps to keep them fit and healthy.

Restoring this richness to children's lives is a challenge which schools cannot expect to meet on their own. However, they can at least promote pupils' awareness and provide some form of environmental experience. Geographical fieldwork has a unique contribution to make to this process. On one level fieldwork can consist of environmental walks and simple data collection activities in and around the school buildings. It can also involve work in local streets and journeys to nearby places. In addition many schools organise some form of residential experience or school journey.

It may also be possible to make better use of the daily journey from home to school. In a study of 150 key stage 1 and Foundation Stage pupils, Large (2003) found that those who talked to an adult about their journey were much more aware of the route than those who did not, even if they travelled by car. Parental involvement was the crucial factor in turning children from passive participants to active learners. Equally, teachers can encourage pupils to engage with their surroundings by setting them challenges such as identifying the nearest post box or making a survey of front door colours.

Distant places

Children's ideas about distant places develop alongside their knowledge of the local area. Fairytales and picture books often provide settings which are way beyond a child's immediate experience. Some stories involve journeys to the other side of the world. Others describe adventures in mountains, forests or deserts. Birds and animals provide much of the interest in young children's literature. Pictures of lions, tigers and other creatures frequently decorate the walls of toddlers' bedrooms and many children first learn about Africa and Australia through wildlife.

At the same time, images of the wider world are beamed into our homes through the media. Satellite images and scenes from foreign lands appear regularly on our televisions. Soap operas, with their focus on people and strong storylines, are a particularly powerful influence, although they give a surprisingly limited impression of places. Advertisements and magazines are another source of images and information. Children's natural curiosity about distant places feeds off these stimuli. At first they may not be too sure about the difference between real and fictitious places. They will probably also be very confused about distances. However, as they grow older they gradually begin to sort out their ideas into a more logical framework.

As well as being exotic and exciting, some psychologists have postulated a deeper reason for children's interest in faraway lands. Egan (1990) suggests that infants try to cut loose from the restrictions of family and neighbourhoods by thinking about imaginary or distant locations. In this way they begin to acquire detailed information about specific places such as villages, harbours and small islands. As well as being reassuring, a knowledge of the boundaries of the

Photo | *Paula Richardson.*

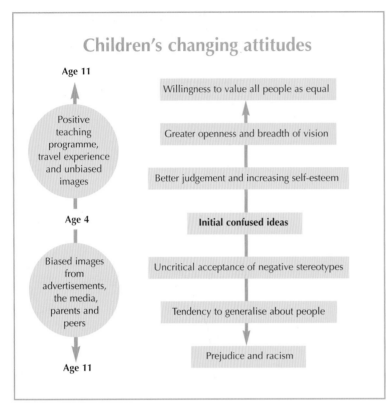

Figure 4 | *Positive images of the developing world are best developed in the primary years before attitudinal rigidity sets in.*

world builds up children's self confidence and helps to develop their sense of individuality. What then are the images of the Earth which young children carry around in their heads? Wiegand (1992) began at a fundamental level by asking 222 pupils from different Yorkshire primary schools to write down the names of all the countries that they knew. The seven-year-olds typically named about five countries, focusing especially on the larger land masses such as America, Africa, India and Australia. By age eleven the pupils' knowledge had expanded and they were on average able to name 15 countries. Generally they appeared to have a good knowledge of Western Europe but the countries of Africa and many other parts of the developing world still did not feature significantly on their mental maps.

What children think about a country is of course more important than the ability to name it. In recent years numerous activities have sought to reveal their images and ideas. Generally it appears that infants and lower juniors tend to associate countries with food and animals but are likely to be confused. Older children exhibit a wider range of responses but also include an increasing number of negative images such as war, famine and poverty.

A detailed study of children's perceptions undertaken by Graham and Lynn (1989) carried these investigations further. Working with top infant and top junior classes they interviewed pupils in groups of six to eight. Each group was shown photographs portraying scenes in the developing world and invited to discuss them using an open-ended schedule of questions. They reported that: '... a large proportion of infants and many less mature upper juniors associated the scenes in Bengal and Bangladesh with a hunter-gatherer lifestyle. One infant group, in no way an exception, thought there would be no roads and people would rub sticks to make fire, "cos there's no matches"' (Graham and Lynn, 1989, pp. 29-30).

Another junior group discussing daily life in Bangladesh declared, 'They sleep on skins from the bears they've killed with spears'. Speculating on their findings, Graham and Lynn suggest that children go through a number of stages as their ideas develop. Young children appear to relish stories about prehistoric life and the simple hunting life which pictures of the developing world can easily reinforce. Aboud (1988) identified the same pattern but contends that children over the age of seven are cognitively capable of making their own judgements on racial issues if they have access to appropriate information.

Teachers need to be aware of these findings so that they can provide classroom activities

Figure 5 | *We can help children to relate to people in other parts of the world by stressing common human needs and emotions.*
Photo | *Stephen Scoffham.*

which will broaden pupils' thinking. If left unchallenged, crude stereotypes can easily harden into prejudices, especially when reinforced by peer or group pressure. Infants, it appears, learn attitudes in the same way that they learn facts. Stereotypes are thus fairly easy to dislodge at this age. By the time pupils reach secondary school their attitudes are much more entrenched and difficult to modify (see Figures 4 and 5).

Teaching strategies

The fact that children develop their geographical understanding from an early age has significant implications for teachers. To begin with, it is important to discover what pupils already know before teaching them something new and to engage with their personal experiences, which are often highly individual. Not only does this serve to eliminate repetition, it also helps with lesson planning. New knowledge, as Bruner (1960) has pointed out, is much more secure if it is keyed into existing patterns of understanding. Conversely, misconceptions can obstruct even the best-planned lesson, confusing pupils as they seek to grasp new ideas.

Figure 6 | *Primary geography involves studies of both the local area and distant places from the earliest years.*

Regarding curriculum content, there are strong reasons for introducing children to geography from the earliest ages upwards. Children are naturally curious about their surroundings and we can harness this energy to enhance their knowledge and understanding. One of the aims of curriculum planning is to ensure that pupils have a reasonable balance of topics and are introduced to new ideas in a logical and progressive manner. Blyth and Krause (1995) have proposed a model of progression in geography which is particularly compelling.

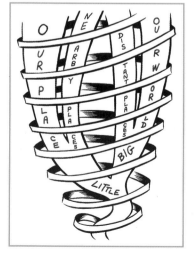

They see the curriculum in terms of four main strands:
1) the local area
2) nearby places
3) distant places
4) the wider world

Basing their ideas loosely on Bruner's spiral curriculum, they argue that all four strands need to represented at each stage of development. However, as children grow older they will revisit the strands in increasing breadth and complexity. One of the strengths of this model is that it acknowledges how the local and the distant are intertwined. Another advantage is that it acknowledges that pupils need to consolidate their learning by revisiting it in a number of different ways (Figure 6).

In recent years there have been a number of new developments which have further illuminated our thinking about primary geography. Perhaps the most profound comes from research into the workings of the brain.

Using radioactive trackers, scientists are now able to identify which areas of the brain 'light up' when we do particular tasks. Not only has this shown that very simple activities can involve complex brain operations, but also that almost everything that we do involves both an emotional and intellectual component. It seems our 'thinking' and 'emotional' brains work in tandem and cannot easily be separated.

One of the reasons why the emotions are so crucial is because they focus our attention which in turn drives our learning and memory systems. Deep and meaningful learning can only occur when we are emotionally engaged. This in turn depends on a secure and affirmative classroom environment.

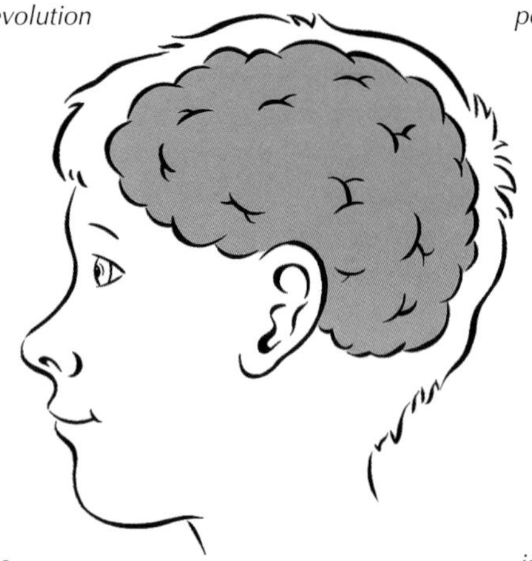

Fieldwork provides a rich, multi-modal environment from which the brain can extract many different meanings

Mapwork develops the key survival skill of spatial awareness which appears to have been built into human capabilities through evolution

Thinking about real life issues is a powerful motivator because the brain responds best to problems that it perceives as meaningful

Asking questions and raising doubts can create rich and sustained brain activity

Enquiries and investigations draw children into a problem so that they relate new learning to previous understanding and reinforce neural pathways

Role plays and simulations involve children emotionally, make learning relevant and engage the brain's natural capacities

Over-arching concepts such as change, pattern and process provide a framework that replicates the brain's natural structures and helps children organise and transfer ideas

Graphical techniques involve modes of thinking which are quite different from literacy and numeracy and which are deeply embedded in human thinking

Figure 7 | *Many of the strategies used in primary geography teaching are compatible with our new understanding about the working of the brain.*

Fortunately geography, because it is such a rich and varied subject, provides young children with a great number of entry points, any one of which can harness their enthusiasm. In addition, geography teachers have a long tradition of linking classroom learning with the outside world which makes it more personal and meaningful. They have also promoted methods of learning that encourage pupils to ask questions, test hypotheses and transfer ideas from one situation to another. These different approaches, which we know are effective from long experience, can now be matched against the evidence from neurology (Figure 7).

Investigating the quality and character of different places is an essential component of any worthwhile geography curriculum. Children are uniquely equipped to undertake these studies as they have a freshness of vision and a strong natural desire to explore their surroundings. As the nineteenth-century educationalist Friedrich Froebel put it, children 'seek adventure high and low' in their 'desire to control the diversity of things' and 'see individual things in their connection with the whole' (Lilley, 1967). We need to encourage them in this endeavour from an early age.

References

Aboud, E. (1998) *Children and Prejudice*. Oxford: Blackwell.

Blades, M., Blaut, J., Darviseh, Z., Elguea, S., Soni, D., Sowden, S., Spencer, S., Stea, S., Surajpaul, R. and Utall, D. (1998) 'A cross-cultural study of young children's mapping ability', *The Transactions of the Institute of British Geographers*, NS, 23, pp. 269-77.

Blyth, A. and Krause, J. (1995) *Primary Geography*. London: Hodder & Stoughton.

Bruner, J. (1960) *The Process of Education*. Cambridge, MA: Harvard University Press.

DES (1990) *Geography for Ages 5-16*. London, Cardiff: DES/WO.

Egan, K. (1990) *Romantic Understanding: The development of rationality and imagination, ages 8-15*. London: Routledge.

Fisher, S. (1995) *Teaching Children to Think*. Cheltenham: Nelson Thornes.

Gosse, E. (1965) *Father and Son*. London: Heinemann.

Graham, L. and Lynn, S. (1989) 'Mud huts and flints: children's images of the Third World', *Education 3-13*, 17, 2, pp. 29-32.

Hart, R. (1979) *Children's Experience of Place*. New York, NY: Irvington Press.

Hillman, M. (1993) *Children, Transport and the Quality of Life*. London: Policy Studies Institute.

Large, J. (2004) 'Watch how you go', *Primary Geographer*, 53, pp. 26-7.

Lee, L. (1959) *Cider With Rosie*. Harmondsworth: Penguin.

Lilley, I. (ed) (1967) *Friedrich Froebel: A selection from his writings*. Cambridge: Cambridge University Press.

Matthews, H. (1995) 'Cultural, environmental experience and environmental awareness: Making sense of young Kenyan children's views of place', *The Geographical Journal*, 161, 3, pp. 285-95.

Matthews, M.H. (1992) *Making Sense of Place*. Hemel Hempstead: Harvester Wheatsheaf.

Owens, P. (2003) *'Fields of Meaning – The Construction of Environmental Meaning and Value in the Early Years'*, Unpublished PhD thesis, Canterbury: Canterbury Christ Church University College.

Scoffham, S. (1998) 'Young children's perceptions of the world' in David, T. (ed) *Changing Minds 1*. London: Paul Chapman, pp. 125-38.

Taylor, S. (1998) 'Progression and gender differences in mapwork' in Scoffham, S. (ed) *Primary Sources: Research findings in primary geography*. Sheffield: Geographical Association, pp. 14-15.

Wiegand, P. (1992) *Places in the Primary School*. London: Falmer.

IN THIS CHAPTER YOU WILL FIND KEY IDEAS ON
CREATIVITY • FIELDWORK • INCLUSION • ISSUES • MAPWORK • QUESTIONS

Geography, creativity and place

Figure 1 | *Ed makes a detailed and poetic description of the physical properties of an old wall. Photo by Rhian Langford.*

Geography and creativity are not words that are usually associated with each other. Etymologically speaking, 'geography' comes from the Greek word *Geographia* meaning 'Earth writing'; a concrete and clear-cut activity if ever there was one. Over the last century, geography has been considered to be a subject dealing with hard facts about our world, such as the names of continents, capitals and countries, and descriptions of landscapes, rivers and economies. Today's geography teachers know, however, that definitions of the subject have been significantly widened by the inclusion of broad concepts which help us understand the world. These ideas include sustainability, environment, change, character, interdependence, fieldwork (Figure 1), environmental improvement and futures education. We might say that the best in traditional geography has always sought to understand the global, find the facts, learn some kind of objective truth, use patterns and processes to make generalisations and analyse places. As a result geography has always been imbued with creative potential.

Creativity involves imagination and activity to produce an outcome which is both original and of some value. While creativity varies in degree and impact, it always involves making connections between two previously unconnected items or ideas. Creativity may range from exceptional insights which permanently change the thinking and experience of a field of knowledge, to everyday flashes of inventiveness that are valuable to us only for the moment. However, it is important to recognise that creativity is much more than just inspiration. It also involves the rigorous testing and sifting of ideas.

Contrary to popular belief, one of the other features of creativity is that it can be taught. The key point here is that creativity does not automatically imply talent, nor is it something that is fixed. Everyone is capable of creative activity and achievement and everyone can improve their performance. It follows that as teachers we can provide the conditions for everyday creativity, by providing multiple opportunities for lateral thinking, reflection, imagination and the sharing of ideas. A creative education might be expected to be personal, imaginative,

Figure 2 | *Rainforest work by Reception pupils at Chilham Primary School. Photo: Chris Johnson.*

value-rich, individually configured and emotional. It is no coincidence that these are precisely the qualities which Britain will depend on both economically and socially as the twenty-first century develops (Robinson, 2001, Claxton, 1998, 2003).

A creative view of geography

Teachers today have the opportunity to adopt an enlightened approach to teaching geography, following the lead of the QCA, DfES and Geographical Association, all of which have prompted teachers to encourage in pupils a curiosity about places, sensitivity towards the environment, open-mindedness towards peoples or cultures and a questioning, reflective mind. By adopting a more personal, emotional and creative approach to their subject, teachers of primary geography can ensure that it survives the 'information revolution', offering a good deal more to pupils than the diet of facts and simple data that is now available to them via the internet. One way of considering this approach is in terms of integrating the following:

- The factual and the imaginative
- Objective truth and personal values
- The general and the unique
- The analytical and the emotional
- The global and the local

What follow are examples of actual lessons which illustrate a 'creative approach' to geography in the context of the above pairings.

Lesson 1: The factual and the imaginative

Harvey's nursery class is listening to a story, but the words do not seem as important to the pupils as the bold and interesting illustrations in the book. The book, *Where the Forest Meets the Sea* (Baker, 1988), contains a series of illustrations of the rainforest in Australia. The pupils are eagerly telling Harvey what they can see in the pictures. As Harvey observed afterwards, 'The influence of the illustrations was such that the storyline took a backseat in their learning'.

Through this activity pupils of three and four are engaging in some of the core activities of geography. Among other things they are:

- Describing environments
- Observing and discussing change in the environment
- Coming to conclusions about what counts for quality in the environment
- Thinking about damaged environments and how we might improve them.

After the focused discussions Harvey suggests making a painting of the rainforest. The pupils begin the task with real gusto, eager to put down in picture form their thoughts and new understandings (Figure 2).

Figure 3 | *The thinking behind the artwork.*

Objectives	Transcribed conversation
Understand key vocabulary associated with landscape	Teacher: Wow that looks great! Is that the sea? Pupil: Yeah and I got sand too. And lots of monkeys. Teacher: What else have you painted? Pupil: Birds and flowers and lots of nice birds.
Compare and contrast with home area	Pupil: I've got a tree in my garden. Teacher: Have you? Does it look like the one in the picture? Pupil: Yeah, it does, but mine's smaller; those are massive. I like my tree though. They look nice too. I can climb my tree to the top. Those are really big!
Interpreting pictures	Teacher: What are you painting? It looks like a rainforest. Pupil: Yes. Teacher: Can you tell me what's in your picture? Pupil: It's a forest with creepers and vines and lots of trees. Teacher: What's the brown for? Pupil: That's the mud.
Knowledge of distant places	Teacher: What do you think the climate will be like in the rainforest? Pupil: Climate ...? Teacher: Will it be cold? Pupil: No! It's hot, wet and hot, really hot and sweaty!

Their paintings look simple, but conversations with each pupil reveal real and deep geographical learning (Figure 3).

Harvey has taken the pupils on a journey that has led to a demonstrable sensitivity to a world quite distant from them. At least in the short term, attitudes seem to have changed. He reports: 'By the end of the scheme of work the pupils' misconceptions (e.g. about "people living in trees", about "Spiderman living there", or about people living there being "dirty and smelly because of the mud") seem to have disappeared and been replaced by positive, relevant and geographically valid ideas'. This journey started with the use of image and imagination to engage and motivate pupils to deeper thought and more focused thinking. Essentially, pupils were given time to dream and reflect though the painting exercise; creative activity was the door to deeper engagement and transferable geographical learning.

Lesson 2: Objective truth and personal values

Cherry wants to teach her class about a place in India. The pupils in her year 3 class have severe learning difficulties. She knows that they learn better through relationship (touch, sound, taste, smell, sight and movement) than they do by listening to her speak. She knows too that they concentrate particularly hard when they are involved in their favourite activities, such as singing, making music and cooking.

Cherry has visited India and has collected a wide range of teaching resources. She decides to start her lesson by singing a gentle Indian melody against a silent video of the street market in Dindigul, a town in southern India. The pupils participate by tapping small bells and drums (Figure 4) that instantly give the song (now sung by several pupils and support staff too) an Indian flavour. Cherry tells the pupils they are going to find out about living in a far away place and helps them dress in local costumes such as saris, dhotis and headscarves. As they are dressing, Indian tabla and sitar music plays from a CD. Once they are in their Indian clothes, the pupils are assisted in climbing on an imaginary plane, holding passports and luggage. They fly (with an appropriate flying song) on the long journey to Chennai, the capital of Tamil Nadu.

The school's sensory room represents Chennai for the pupils. The room is lit with twinkly colours and hung with oriental fabrics. The atmosphere is warmed with hot-air blowers and made damp with fine sprays of water. It feels very humid. Cherry asks the pupils what it feels like now they have

Figure 4 | *Music used as a way in to a distant locality. Photo: Cherry Barnes.*

Photo | *Paula Richardson.*

arrived in Chennai. Some are able to answer 'hot and sweaty'; others are engrossed in the video which is now showing pictures of traffic and congestion in a busy Indian city.

Meanwhile, Bandana, a school volunteer, has arrived in the classroom to cook some Indian food. When the pupils return, their room is already filled with different smells and the sound of spices being crushed in a granite mortar. The pupils help to crush cinnamon, cardamom and anise and pass the powders around to each other to smell. While they are helping with the cooking, Cherry and a teaching assistant have laid out a market stall behind the group. Fruit, vegetables, brushes and sacks of rice are arranged on the floor to represent a street market and the pupils are invited to come and shop. They want to ask their friends from the next door classroom to come too. Tomorrow they will learn an Indian dance.

Activities like these would be exciting for any pupils, but Cherry feels that they also promote the learning of information and teach positive values. Pupils in her class are learning that difference is not necessarily threatening, that unusual sounds, tastes and smells can be good, that people from other cultures are the same as them and that diversity is interesting. Cherry can tell instantly through facial expressions and individual sounds that the pupils are enjoying experiences like these; she knows the lesson has enriched their lives. The fact that the pupils wanted to share the experience with other classes was strong evidence that the session had affected their values.

This lesson highlights some key issues for teachers of primary geography. As children (and adults) become increasingly aware of the interdependence of people around the world, it becomes clear to them that what they regard as 'true' or 'right' may not be the same for all other people. This is so within complex multicultural societies such as our own, and by recognising it we can also appreciate why it is so difficult for multinational organisations such as the UN and EU to come to unanimous decisions. The government has addressed this issue by including including a statement of values in the latest version of the national curriculum. Some of these statements are particularly relevant to geography – for example:
We should:

■ Refuse to support values or actions that may be harmful to individuals or communities
■ Respect religious and cultural diversity
■ Contribute to, as well as benefit fairly from, economic and cultural resources
■ Accept our responsibility to maintain a sustainable environment for future generations
■ Understand the place of human beings within nature
■ Ensure that development can be justified
■ Preserve balance and diversity in nature wherever possible
■ Preserve areas of beauty and interest for future generations
■ Repair, wherever possible, habitats damaged by human development and other means
(DfEE/QCA, 1999, pp 147-49).

Lesson 3: The general and the unique

Looking for patterns in the weather, landscape, human activity and environment has been a key geographical activity for generations, but Richard wanted to help his pupils understand

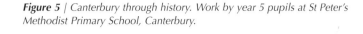

Figure 5 | *Canterbury through history. Work by year 5 pupils at St Peter's Methodist Primary School, Canterbury.*

local geography in a more personal and meaningful way. Since part of the town centre was being redeveloped as a shopping centre, Richard invited his year 5 class to think about continuity and change in their locality (Rogers and Barnes, 2003). The pupils had already learnt about the history of the area by visiting the archaeological excavation being undertaken at the redevelopment site. They had met executives from the development company and seen films and projections of what the site would look like in the future. Richard set about collecting all the maps he could find which showed this area of the town at different periods. He found copies of maps made in 1640, 1750, 1801, the 1870s, 1912 and the 1950s to add to the school collection of modern maps. He then used the photocopier to reduce or enlarge each map to a similar scale. His lesson used the map reading skills the pupils had been taught earlier in the term.

Richard divided the pupils into groups of four. Each group was given a map of the redevelopment area, but from a different historical period. The first task was to note down the main features shown on the map. Bringing this information together after ten minutes it became obvious that certain features were the same whatever the date. For example, the city wall, the burial mound, the cathedral, the flint church, the pub and piece of open land all remained constants. The class was then briefly reminded about the orientation exercises they had done while on a weekend adventure holiday. In their groups they were asked to work together to imagine what the views would have been like to the north, south, east and west from a central point at the time their map was made. This activity generated a great deal of talk as the groups formed their ideas using direction, map symbols, scale and historical knowledge. The views were drawn as borders around the edge of their photocopied maps (Figure 5) and displayed. Change and continuity were depicted strongly in the different scenes shown in the frames. The work made a strong impression on each of the participants and they remembered the detail and implications very precisely when they were interviewed months later.

The experience of provoking imagination through studying maps was to become even more personal. In the following lesson Richard asked the class to imagine the future, after the new shopping centre was built, after the saplings had matured, even perhaps after the motorcar had been superseded. Each pupil used a base map of familiar streets to demonstrate their own understanding of what might be visible from the same place in 2050. The results were startlingly individual, but also displayed the internalisation of the idea of change and continuity and feelings about what it was important to preserve in the context of a specific place.

The world is full of patterns and processes. Learning in geography is very much about identifying, analysing and evaluating them. We should never lose sight of the fact that the patterns geography exposes and the processes it describes help us to make sense of a world that might otherwise seem chaotic. Geographical patterns and processes are nonetheless thoughtful generalisations and teachers need to help pupils construct such generalisations, starting by helping pupils to see the patterns and processes in the place where they are.

Photo | Paula Richardson.

Lesson 4: The analytical and the emotional

A year 4 class is working with their teacher, Kirsty, beside a river and an old building behind the school. They have been given a diagram and asked to label it. The river bank slopes gently where the pupils are standing, but on the opposite side, where the water is travelling faster, some pupils have noticed that the bank is steep. They are excited by this observation and are discussing whether it might be more dangerous to stand on the far side:

Pupil A: 'If you stood there you'd fall right in.'
Pupil B: 'No you wouldn't, it's safe.'
Pupil A: 'The bank would fall down, 'cos it's just soil.'
Teacher: 'That's interesting, do you think the steep bank has got anything to do with the fast water?'
Pupil B: 'It might be wearing it away. Look.' (points to a large twig scraping the river bank)

This discussion is typical of those that happen repeatedly when pupils learn in an educational setting which truly promotes thinking. Fieldwork instantly puts pupils in a context that provokes and interests them because natural and built environments beyond school are automatically richer and more generative than even the most stimulating classrooms.

The lesson continues and the pupils go on collecting words and making descriptions of the river. The little drawings and diagrams they produce involve a relatively low level or 'everyday' creativity, but the atmosphere changes when Kirsty asks them to sit on the grass with their eyes closed and listen to the water. After a full four minutes of silent listening with (she hopes) open minds, Kirsty very quietly asks the pupils to say the words which have come into their heads. She writes them down – rustling, wind, tweeting, brrr, swirling, silent, swishing, quiet, click, crunch of twigs, shimmering, lapping, tinkling. As the responses are beginning to slow down Kirsty reminds the pupils of the Japanese poems they had been reading and learning about the day before. She asks the pupils to find their own space to try to write a simple poem in the same style. Here are two examples:

Photo | Diane Wright.

Walls with small bumps
Plaster pointed lumps
River flowing at a lazy pace
Water ripples and weaves around columns
Swirling and whirling as if fine lace.
(Jonathan)

Photo | *Diane Wright.*

Water swirling across stones sticking up
Very silent swishing
The water bed covered with little pebbles
A dark browny colour the water is
Bright green leaves shining in the water
(Dorothy)

This may be described as a literacy activity, but it has arisen out of a geography field trip, so is it now geography? It does not matter of course. What Kirsty and her pupils have discovered is that real experience cuts across subject disciplines all the time. The poems which Dorothy and Jonathan wrote employ skills from both literacy and geography. Their work exhibits imagination, intuition, analysis and evaluation in full measure and it is clear that lasting learning has taken place. Geographical truths came via physical and emotional engagement. The mind, spirit, body and senses combined to effect attitude change. These pupils will not feel the same about this river again.

Lesson 5: The global and the local

Stephen is teaching a year 6 class about Tanzania as part of the school's Africa Week. He has explained how he visited Tanzania the previous summer and visited the village of Vunta in the Pare Hills where aid dependency is a serious problem. He has also provided pupils with information about the climate, landscape and people who live in the area. Now the pupils are working in small groups on a number of detailed (and real) farm plans. They have already learned how to interpret a map and are used to reading climate statistics. They have also examined real artefacts from Tanzania in a previous session.

It is clear that the pupils have made sense of and identified with the four hectares of steeply sloping and fragmented farmland which has been allocated to them. Each group has also been allocated 5000 Tanzanian shillings (about £50) to spend for the year ahead and is responsible for feeding, caring for and educating the rest of their 'family' (eight people). In addition to these meagre funds, they can plan to use their own physical resources to improve their particular plot of land and make it profitable. The conversations around the farm plans are revealing:

- 'We can't build a fishpond here, it would just dry up. There's no shade.'
- 'Can't we just sell off that extra bit of land over there, I'm sure Njogu's family would buy it, and then we could buy one of those big cows that give lots of milk. You know, those Fresian ones.'
- 'Yeah, then we could sell it at the market.'
- 'I wish we had even a little stream. Couldn't we all work together and build a channel which could water three farms?'
- 'How are we going to pay the workers?'

The situation seems very real to the pupils, and their individual contributions are focused and thoughtful. However, this imaginative activity is more than just emotionally engaging, it introduces pupils to development issues in a meaningful way. The problems they are discussing are similar to those which face poor farmers in all parts of the developing world. Pupils in this class are learning from the inside about the complexity of development, about

the need for small-scale appropriate solutions and about the difference working in teams can make. At the same time the work gives pupils the opportunity to apply some of the key skills of geography and to show the degree to which they understand and can use tables, maps, graphs and climate statistics. Finally, they are relating the problems of the developing world to their own situation and circumstances. We all need food, water and shelter: the question is, how can people best meet these needs in different parts of the world?

In his assessment of individual responses to this task Stephen is convinced that 80% of his pupils can readily transfer these skills to other situations. Metacognitive learning of this kind has enduring value. Such life skills are best taught in a shared context, and pupils may develop individual insight and attitudes through their part in powerful and informed role play.

What do these examples show?

Each of the above examples illustrates how primary geography lessons can be creative. The examples also show how to teach or extend key geographical skills. These skills include map reading and making, fieldwork, use of geographical vocabulary, use of pictures and photos, and recording, observing or expressing geographical views.

All the lessons have built up a knowledge and understanding of places and have identified patterns and processes. Most have helped pupils to understand the notion of environmental change and consider sustainable development. Such activities deal in facts and specialised understandings that would be familiar to geographers anywhere. These lessons also illustrate another important principle; that traditional and creative approaches in geography are not necessarily opposed.

■ An imaginative stimulus generated an atmosphere in which geographical vocabulary and facts were easily internalised by nursery-aged children

■ Positive images of India resulted from an imaginary visit linked to video footage of a market and supported by food and smells

■ Discussing a map of a well-known area in a group setting resulted in a highly individualised but informed view of both the past and the future of the area

■ Feelings about places and environment arose from a fieldwork trip to a river

■ Role-playing an African farmer gave rise to personal responses and led to an understanding of global issues in development education, and transferable geographical skills

Why is place so important?

The skills, knowledge and understandings of geography cannot be fully understood unless applied to actual places. Place is a central concept in geography, but why should we use a place to stimulate creativity? Here are a number of reasons:

- The place where we work is ours. We have an inbuilt emotional involvement with these unique environments and since learning requires an emotional involvement this makes a good starting point.
- Our place of work is, or should be, safe, and a feeling of emotional and physical safety is essential because our brains have evolved so that emotional or physical survival over-rides all other considerations.
- The school grounds and buildings round about us are multi-faceted. They represent a fusion of art, design, technology, science, philosophy, geography, history, mathematics, language, culture and creativity.
- The local area, whether inner city, suburban or rural, provides a rich environment; authentic, multi-sensory, not imagined and easily related to other knowledge.
- Our places are (sometimes literally) concrete, but our pupils spend much of their time being asked to consider abstract answers to abstract problems when we know that many of them have real difficulty in abstract thought. Our 'survival oriented' brains are good at fine discrimination, categorising, handling and sensing but need metaphors and external prompts to help them develop and handle new thoughts.

We know environments, like our minds, are plastic; they are always changing, but we can have some say over those changes. If either our places or minds are not supple, they become ineffective, inappropriate and ultimately useless. Personal and community inventiveness has a big part to play in keeping mind and place flexible. There is a strong metaphorical and practical link between our minds and our places. All places shaped by people are repositories of what Perkins calls 'distributed intelligence' (Perkins, 1992, p. 186), where history, culture, community, technologies and the environment itself can help us see patterns, note processes and extend our thinking by inspiring new connections. In a modern and fast-changing world where the total information available to us doubles every two years (Robinson, 2001, p. 46), we can no longer teach geography as a rigid body of knowledge, but must consider it as a disciplined but imaginative process which deepens our understanding of the places we share.

References and further reading

Baker, J. (1988) *Where the Forest Meets the Sea*. London: Walker Books.

Claxton, G. (1998) *Hare Brain Tortoise Mind*. London: Fourth Estate.

DfEE/QCA (1999) *The National Curriculum: Handbook for primary teachers in England*. London: DfEE/QCA.

Gardner, H. (1999) *The Disciplined Mind*. New York, NY: Simon and Schuster.

Mithin, S. (1996) *The Prehistory of the Mind*. London: Thames and Hudson.

Perkins, D. (1992) *Smart Schools*. New York, NY: Free Press.

Pinker, S. (2002) *The Blank Slate*. London: Penguin.

Robinson, K. (2001) *Out of our Minds*. London: Capstone.

Rogers, R. and Barnes, J. (2003) 'Past, present, future', *Primary Geographer*, 50, pp. 22-3.

Standish, A. (2002) *'Constructing a value map'* (http://www.spiked-online.com/Articles/00000 006DB23.htm)

Sternberg, R. (ed) (1999) *The Handbook of Creativity*. Cambridge: Cambridge University Press.

Photo | Kathy Alcock and John Collar

Geography and the emotions

G eography is about places, and places can provoke powerful emotional reactions. Dramatically beautiful places evoke feelings of awe and wonder and harsh environments a sense of fear or alienation. People feel affection for otherwise ordinary places which have personal significance. Environmental issues, too, can arouse strong passions, as witnessed by any proposal to build a new by-pass, demolish an attractive building, limit car usage or cut carbon emissions.

The term emotional literacy was first used by Claude Steiner. He suggests emotional literacy has three elements: 'the ability to understand your emotions, the ability to listen to others and empathise with their emotions, and the ability to express emotions productively' (Steiner and Perry, 1997, p. 11).

The term emotional intelligence has been popularised by Daniel Goleman, who argues that emotional intelligence may be more important than intelligence quotient (IQ) for success in life (Goleman, 1996). Building on Howard Gardner's work on multiple intelligences (Gardner, 1984), Goleman suggests that emotional intelligence involves five main domains. The first three of these relate to what Gardner refers to as intrapersonal intelligence, whereas the last two relate to the interpersonal aspects of emotional quotient (EQ):

1. Knowing one's emotions involves self awareness and recognising feelings as they happen, and is the keystone of emotional intelligence.
2. Managing emotions is concerned with self regulation and handling feelings in an appropriate way.
3. Motivating oneself is concerned with goal achievement and so is essential for maintaining attention, self-motivation, and creativity.

4. Recognising emotions in others involves understanding others and developing empathy, and is a fundamental interpersonal skill.
5. Handling relationships is concerned with social competence and is therefore related to the ability to influence others, communicate well, manage conflict and work collaboratively.

Goleman has created world-wide interest in the concept of emotional intelligence, arguing that teaching pupils the skills of emotional literacy is as important as teaching other forms of literacy:

'In navigating our lives, it is our fears and envies, our rages and depressions, our worries and anxieties that steer us day to day. Even the most academically brilliant amongst us are vulnerable to being undone by unruly emotions. The price we paid for emotional illiteracy is in failed marriages and troubled families, in stunted social and work lives, in deteriorating physical health and mental anguish' (Goleman, 1996, p. 43).

The two concepts are closely related, and are often used interchangeably. Emotional literacy appears to be a broader term, increasingly used in education, while emotional intelligence is more associated with business and leadership.

To nurture emotional literacy and emotional intelligence we need to attend to children's feelings and reactions. Emotionally literate people are able to recognise and manage their own feelings and build constructive and effective relationships with others. It is now widely recognised that emotional intelligence is crucial for success in personal and professional life (Goleman, 1996). The difference between emotional literacy and emotional intelligence is explored in Figure 1.

Geography has a particular contribution to make to the development of emotional literacy because the study of real places, real people and real-life issues is at its core. In investigating places, people and issues, primary age pupils can explore the significance of the affective domain in relation to place, space and the environment.

Figure 1 | *Emotional literacy and emotional intelligence.*

This chapter shows how studying places and issues in primary geography can provide a context within which emotional literacy or intelligence may be fostered and nurtured. Although some useful general guidance for teachers on fostering children's emotional development (e.g. Greenhaugh, 1994) and on promoting emotional literacy (e.g. Sharp, 2001; Antidote, 2003; Weare, 2003) has been published, little has yet been written about the development of emotional literacy in specific subject areas.

The chapter begins with a brief discussion of the importance of the affective domain in learning, and outlines the arguments for developing children's personal and interpersonal skills. This involves exploring how the concepts of emotional literacy and emotional intelligence relate to geography and the wider primary curriculum. There then follows a review of the evidence concerning children's feelings about places, and discussion about how a focus on this can enhance and enliven the study of places and environmental issues. The chapter then considers the role of geography in promoting the interpersonal skills associated with emotional intelligence, and concludes with the characteristics of emotionally literate geography.

Educating the emotions

There is a long tradition of interest in the role of education in promoting emotional development which may be traced back to the work of John Dewey and other progressive educators. But, as Suzie Orbach, a psychoanalyst and co-founder of Antidote (a national charity set up in 1995 to promote emotional literacy) has argued, the recent focus on raising standards and on targets, tests and league tables has meant that 'it has been difficult for arguments about the importance of emotional development in schooling to find a place on the agenda' (Orbach, 1998). Indeed the philosophical underpinning for the national curriculum derives ultimately from the work of thinkers such as Kant and Descartes who specifically separated the affective and the intellectual.

However, recent developments in neuroscience suggest that there are strong links between emotion and thinking, and that feeling and understanding are deeply interconnected. We now know that the mid-brain, the amydala and limbic system control our emotions, and that the upper-brain, or neo-cortex, is the 'thinking' part of the brain. It seems that learning requires activity in both the limbic system and neo-cortex. Carter (1999) suggests that the neuro-connections between the two parts of the brain mean that the cortex can control emotions

originating in the limbic system, but that the connections take time to develop. This helps to explain young children's relative lack of control of their emotions. 'The young brain is essentially unbalanced – the immature cortex [is] no match for the powerful amydala' (Carter, 1999, p. 90).

Our new understanding of the significance of emotions in learning suggests that paying more attention to the affective domain may help to raise standards of achievement across the curriculum. Pupils with special educational needs are particularly likely to benefit whether they are slow learners or gifted and talented. McCarthy and Park (1998) argue that emotional learning matters because:

Photo | *Kathy Alcock and John Collar*

- understanding emotions is directly connected with motivation and with cognitive achievement;
- dealing with emotions helps to develop better relationships and a sense of psychological and mental well-being;
- emotionally developed young people are better equipped to live with difference;
- our moral outlook and value systems are deeply shaped by our attitudes and feelings; and
- our sense of meaning and purpose is derived as much from feeling as from understanding.

Emotional literacy in the school curriculum

In terms of the wider curriculum framework, emotional literacy and intelligence have particularly strong links with personal, social and health education (PSHE). PSHE is concerned with preparing children for life now and in the future. It is about developing their self-knowledge, their ability to understand and manage their feelings, to build relationships with other pupils and adults, and to understand their local community and national society. As Elaine Jackson has comprehensively demonstrated, there are strong natural connections between primary geography and PSHE. Both provide opportunities to express opinions; to develop an understanding of the place of the individual in society, and of collective rights and responsibilities; to develop empathy, and an awareness of the points of view of others, and cultural understanding of other societies. In addition, active learning in geography requires purposeful pupil interaction, group activities, co-operative working, collaboration, reasoned debate, negotiation and informed decision-making. (Jackson, 2000, p. 34).

A decade ago, Frances Slater suggested that geography education should seek to develop both reason and feeling (Slater, 1994), echoing a persistent argument that really meaningful learning in any subject engages both the intellect and emotions. Geography offers three major opportunities for developing emotional literacy:

- It helps children to recognise and express emotions associated with places and environmental issues;
- It provides opportunities to develop empathetic understanding of others' feelings and views; and
- It develops interpersonal skills through the active learning approaches required by meaningful geographical enquiry.

Attachment to place

It is well known that secure attachment to significant people is important for young children's psychological growth and health. It is also argued that attachment to place can be equally significant, for 'to be attached to places and have profound ties with them is an important human need' (Relph, 1976). Identification with and attachment to place is expressed in many ways, such as supporting a local football team, identifying oneself as a Yorkshire woman, campaigning to preserve a locally significant building, or studying local history. Most people appear to be interested in and care about the place they live, and this attachment to place seems to have its roots in childhood experiences.

For many adults some of their strongest memories and recollections of childhood relate to places. Most of us recall in some detail, and often with affection, the homes where we were

raised, the buildings in which we were educated, and the places we played in. While I was writing this chapter I drove every day past the site where a gracious but outdated Edwardian school was being replaced. Several weeks before the site was due to be vacated, a large sign appeared outside inviting people to 'Say goodbye to the old buildings'. This invitation represents an acknowledgement that former pupils, teachers and local residents may have feelings of sadness when a locally significant landmark is demolished. In her study of adults' memories of childhood places, Louise Chawla found that affectionate memories were associated with 'places to which we trace our roots, which are associated with happiness and security ... there is a parallel between the warmth of feeling for the place and people in it' (Chawla, 1986, quoted in Titman, 1994, p. 7).

Simon Catling argues that children's experience of places is a vital part of their lives, contributing to their sense of self, identity and self-esteem (Catling, 2003). It is also often said that young children are natural geographers (Scoffham, 1998). From early infancy, they start to learn about their surroundings through direct first-hand experience, and so become familiar with 'their place'. Through secondary sources of information such as television, books, photographs and pictures, they also begin to form perceptions about 'other places' beyond their experience (Goodey, 1971). Both known places and (as yet) unvisited places evoke feelings in, and have meaning for, children and inform their developing sense of place.

> A sense of place describes a particular kind of relationship between individuals and localities. For individuals different places are imbued with different meanings (Matthews, 1992).

By the time they enter school, children will already have developed knowledge of their familiar environments, and will also have feelings about them. Their environments will be, as Catling points out, not only physical places but also 'affective places, places for fun and enjoyment with freedom of movement or places of limits, exclusions and, perhaps, fears' (Catling, 2003, p. 173). During their primary years, children's experience of known places extends as they are given or take greater freedom to move unsupervised around the local area.

Furthermore, these early childhood experiences have an impact which carries though into adult life. Many of today's environmental activists, for example, can point to first-hand experiences in the natural world that nurtured their later enthusiasm and interests (Palmer, 1998). The importance of 'significant life experiences' of this kind is also recognised in many biographies.

Titman (1994) too notes that children consistently express strong positive reactions to relatively 'natural' areas. She found that although they were keenly aware of the intrinsic value of the environment and of the need to care for the natural world as the home of all living things, children also valued natural places because of the way they made them feel. Natural places were judged to be peaceful, but were also associated with freedom, adventure and challenge. Environmental elements seen positively included natural colour, trees, woods, shady areas, big grassy areas, places with different levels, places where you can climb/hide/explore/make a den, places that have 'millions of bits', and places with wildlife. Dirt, pollution, rubbish, litter, damaged things, unnatural colour, tarmac, places where you

Special places

Pupils in a year 5 class were asked by their teacher to think about places which were special to them. Many produced carefully executed pictures to show their special place, and wrote about why it mattered to them, and how they felt about it when in it.

Of the 30 pupils, 16 (both boys and girls) nominated their own bedroom as their special place. Many described their bedroom in great detail and explained what they enjoyed doing in it. Bedrooms were referred to as places of safety, where they could relax and *'be themselves'*.

'I can have a happy time reading. I am relaxed and calm, especially when I read one of my books or magazines. My bed is very cosy and comfortable and warm. It is blue with daisies all over. I have lots of cuddly toys in my bed and I feel safe.'

'Because nobody shares it me with me – it's always there to turn to when I am sad and weary.'

Several others reported that they valued their bedroom as a place where they could withdraw from family life, especially contact with siblings: *'What I like best is that my brother can't come in'*. Friends, however, were welcome in many bedrooms, as their presence was associated with fun. One pupil offered a particularly detailed picture of her bedroom and what it meant to her:

'My special place is a corner of my room. It has got a bean bag, books, a CD player, games, a cuddly toy, cushions and a lamp. It's special to me because I can have some space and just chill out. I normally just sit down, put my music on low and read a book. It really helps when I do that. If I have just fallen out with my mum and dad, I can relax and get it all out of my system. What I like best is the space to do whatever I want. It really annoys me when people break the atmosphere by knocking on my door, just to ask a stupid question'.

A wide range of other types of places were nominated by the pupils as particularly special to them. These included the local park, a tent, a shed in the garden, a shop owned by parents, a grandma's house, Disneyland Paris, the London Eye, an Indian beach, a fast-food restaurant in India and an aqua park in Corfu.

In many cases these places were associated with enjoyable activities, happy memories, or with people to whom the pupils were attached. The girl who chose the local park said:

'I like this place because I made my first best friend there. I also meet my friends there sometimes, we have lots of fun. It also brings back memories like the time my brother was playing football with his friends and I was the commentator. I feel very relaxed, especially with my friends, and cheerful and glad too. The thing I like best is that you feel free and have a lot of fresh air. The park will always be my special place even if I move away'.

Animals were significant for two girls, one who enjoyed going to her grandma's to play with the dog, and the other to the shed where she kept guinea pigs:

'My special place is in my shed because my guinea pigs comfort me when I'm down, and when my parents tell me off I feel safe and secure when I'm around my guinea pigs and I enjoy stroking them'.

Only one of the year 5 pupils, a girl, nominated a fantasy place; *'a secret hidden away beach'* which she can imagine *'really easily'*, and where she goes (in her imagination) when she feels down or bored, because *'it feels safe'*.

Figure 2 | *Special places – a year 5 case study.*

can't climb/hide/explore/make a den, and places that were 'boring' or 'too open' were all viewed negatively by children in the study.

Working with feelings

One way of helping children to recognise and acknowledge their feelings about their surroundings is by asking them to think about their favourite places. What parts of their home, school and immediate locality do they value and why? Key stage 1 pupils can conduct a survey of the most liked and disliked places in the school grounds or local area, and discuss the results. This will enable them to explore why people tend to feel more positive about some features or areas than others. The study of a contrasting locality, whether in the UK or overseas, should provide opportunities for children to consider the questions 'How do I feel about this place?', and 'How do other people, including those who live there, feel about it?'. As a broader perspective is developed in key stage 2, pupils can consider their responses to different types of environment – urban, rural, wild and pockets of one within another, e.g. open spaces in an urban area, traditional buildings in rural landscapes. They should be encouraged to attend more deeply to their feelings for the places studied, and to try to understand, articulate and explain them (Figure 2).

At any age, children can be encouraged to express and communicate their affective response to places, spaces and environments in a variety of ways. Many of the oral strategies used in circle time to explore feelings can be successfully adapted to have a geographical focus. For example, the activity 'Someone else's shoes' (Mosley, 1996, p. 163-4) provides a

Photo | *Paula Richardson.*

structure for developing empathy, and would be a useful way of explaining differing views and opinions about an environmental issue. Mosley also suggests that creative visualisations can focus on different places, such as a beach or part of the Amazon rainforest (Mosley, 1998, pp. 109-15).

In addition to communicating through language, teachers can make meaningful cross-curricular links by using the creative and expressive arts. Drama, dance and movement, music and art are all traditionally associated with personal and social development. The way that children, who may be otherwise reticent and withdrawn, take on new persona when placed in a role is well attested. This can be an excellent way of exploring the impact of changes and local issues. Such approaches encourage children to express their feelings about places and issues creatively, and helps them to develop empathy for others' emotions and opinions.

Places as contexts for learning

The school grounds

In recent years, a great deal of attention has been paid to the way school grounds and the environment can be developed for learning and play. Many schools have sought to improve their grounds by setting up ponds, wildlife areas, fitness trails, vegetable and herb gardens, murals, areas for different types of play activity, hedge and tree planting. School grounds improvement projects provide an excellent opportunity for pupils to participate in real-life decision making and environmental enhancement, and experience the roller-coaster of emotions that often accompanies such initiatives.

A fascinating study undertaken by Wendy Titman (1994) confirms that the school grounds are very significant places to primary age children. Titman found that pupils had a well-developed concept of their ideal school grounds which would offer plenty of opportunities for them to 'feel' and 'be' as well as 'do' and 'think'. In Titman's words they wanted the school grounds to offer:

> *A place for feeling – which presented colour and beauty and interest, which engendered a sense of ownership and pride and belonging, in which they could be 'small' without feeling vulnerable, when they could care for the place and people in it and feel cared for themselves ... [and] ... a place for being – which allowed them to be themselves, which recognised their individuality, their need to have a private persona in a public place, privacy, being alone and with friends, for being quiet in noise, for being a child* (Titman, 1994, p. 58).

The local area

The study of the local area is the bedrock of primary geography. It is important to study the school grounds and the immediate locality partly because, in geography, as in other subject areas, we should build on what pupils already know. However, it is also important because through this work, we demonstrate to pupils that where they come from matters and is worthy of study. This in turn helps to create a sense of personal worth. The investigation of local streets, buildings, landmarks and open spaces can enable pupils to really 'see' them for the first time and to appreciate their inherent interest and qualities.

In all key stages, pupils should have opportunities to notice and think about buildings or features of the local area which they or other people value, and which are worthy of care and preservation. Some of these may be of purely local significance, and valued for their amenity value, historical origins, or aesthetic qualities. In every locality there will also be areas or buildings which are in need of improvement or repair or could be better cared for. In addition, many schools will have easy access to an officially recognised 'special' place. There are over 500,000 buildings in the UK listed for their architectural or historical merit, nearly 30,000 monuments and 10,000 conservation areas. Identifying these and considering how they can be used in your geography could be a stimulating early evening INSET session involving the whole staff.

To nurture emotional literacy through local study, teachers can encourage pupils to consider questions such as:

■ Where is special in our local area?

■ What makes it special?

■ How do I feel about this place, building or feature?

■ How do other people feel about it?

■ How can we encourage everyone to look after it and preserve and enhance its special qualities?

Finding out about the character of different places is central to geography. The best way to do it is through first-hand enquiries and observations.

Volcano Terror

Molton lava flows through the town of Goma in Africa. 14 villages have been destroyed. Lots of people have died and people have lost their homes. Entire neighbourhoods have been burnt down.

Volcano

Nyirangongo volcano is almost 4,000 meters high. It is round about 1,000 meters wide and over 1,000 metres deep. The lava has been flowing at 40 mph. The town is full of billowing smoke.

The people have got no other clothes and they are so scared. People are walking along the roads covering there mouths and noses. The lava is 6 feet deep. 50 people have died at a petrol station.

All of the pollution is flowing into lake Kivu. Over 5,000 people have escaped to Rwanda.

We hope you can help it will make a lot of difference.

Reported by Sarah Lochhead

***Work** | Y4, Perton First School, Staffordshire.*

Distant places

The study of more distant places, within and beyond the UK, also offers opportunities for pupils to attend to their feelings about places and develop empathy for the people who inhabit them. They can be asked to consider a parallel set of questions to those about the local area noted above, for example:

■ What is special about this place?

■ How do I feel about it?

■ How do the people (children/adults, men/women, richer/poorer people) who live there feel about it?

■ What could be done to look after this place, and enhance its special qualities?

Learning about a range of places around the world can also help to allay children's natural fears and uncertainties. We all tend to be wary of the unknown and approach it with a mixture of suspicion and excitement. By teaching children about distant places we can help them to develop more complete and balanced images which are based on fact rather than the incomplete, stereotypical and limited images they may pick up from elsewhere.

Photo | *Anna Gunby.*

Environmental issues

The study of local and more distant environmental issues also offers pupils the opportunity to express their own feelings and opinions. As they get older and develop a less egocentric view of the world, children are increasingly able to understand that other people may feel differently from themselves, and that there might be a range of legitimate perspectives on environmental issues or problems.

Topical, real-life, local environmental issues can arouse strong feelings, as reading the letters page of any local paper will testify! The study of local issues is very motivating, particularly if it engages pupils in debate with local residents or officials, or leads to action for change (Chambers, 1995). In most communities there is a wide range of issues which could be tackled:

■ Poor public transport provision

■ Lack of safe places for children to play

■ Problems with traffic congestion and parking

■ Pedestrian safety

■ Protection of open spaces, attractive buildings or significant trees

■ Provision of recycling facilities

■ Vandalism and graffiti

■ Footpath maintenance

■ Planning applications for demolition, extensions, or change of use

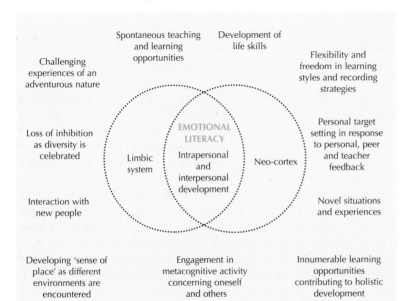

Figure 3 | *Visual framework for the development of emotional skills through residential fieldtrips. After Chantler, 2004.*

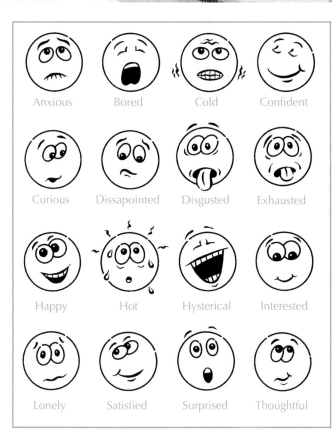

Figure 4 | *Fieldwork can stimulate a range of emotions. After FSC Castle Head Centre.*

Anxious   Bored   Cold   Confident

Curious   Dissapointed   Disgusted   Exhausted

Happy   Hot   Hysterical   Interested

Lonely   Satisfied   Surprised   Thoughtful

At one school in a village on the edge of Leeds, pupils investigated traffic problems and then wrote to the City Planning Department about their concerns. In an inner city school, year 2 pupils lobbied local residents and Councillors about the problem of dog mess in the adjacent park – the only open space in the area. In a third school the parking problems caused by parents driving their children to school led to the establishment of a Walk to School campaign committee. In all these examples, pupils' interest and motivation was high, and very high quality work was completed because both their emotions and intellect were fully engaged. This point has been noted in many HMI and Ofsted reports. In their annual report for 2001/2, for example, the Inspectors noted that 'some of the best work involved pupils in practical activities linked to relevant and real issues' (Ofsted, 2003).

Promoting empathy and interpersonal skills in different contexts

Geographical learning in the primary school provides a valuable context for developing the interpersonal aspects of emotional intelligence. The enquiry approach in particular requires collaborative and co-operative work, through which pupils develop their ability to handle relationships, exercise responsibility and initiative, and work in teams. Fieldwork too usually demands group work with peers and may involve interaction with adults including members of the public.

Fieldwork

Fieldwork is a core element in primary geography. It brings pupils into contact with the natural and built environment, helps to develop their sense of place and encourages their notion of stewardship and care for the environment. For many pupils, it offers the chance to experience places where they would not otherwise go. Fieldwork entails many opportunities for pupils to notice, attend to and communicate their emotions and feelings. Figure 3 suggests ways in which fieldtrips may develop children's emotional literacy.

An excellent resource developed and produced by the staff of the Field Studies Council is a sheet of cartoon faces humorously illustrating seventy alphabetically organised emotions that could be experienced during fieldwork (Figure 4). These range from anxious, bored and exhausted to curious, satisfied and ecstatic. One innovative use of the sheet involves asking pupils to draw the faces to complete a 'feelings graph' to show how they feel at various times during a day visit to Box Hill. This approach could easily be adapted for field work in the local area.

Figure 5 | *Road with Cypresses, 1890 (oil on canvas). Gogh, Vincent van (1853-90)/Rijksmuseum Kroller-Muller, Otterlo, Netherlands.*

Paintings and photographs

Many great artists have been moved to interpret the places they cared about through their art, and numerous well-known works of art are representations of places. Painters such as Constable, Van Gogh and Cézanne have left us with enduring images and transformed our perception of the landscape (Figure 5). You can encourage pupils to talk about their feelings using slides or high quality reproductions of these scenes. Useful starting points for discussion include:

- How do you feel when you look at this picture?

- Why do you think that artist chose to paint/draw this place?

- How do you think s/he felt about the place?

- What feelings did s/he want to provoke in people looking at the picture?

- How does s/he encourage these feelings?

Questions like these will provide an excellent focus for discussing the feelings associated with places, and for recognising that different people may respond differently to the same place. Using similar materials, techniques or style to the artist to create a representation of the local area or other place will also help develop their sensitivity to the affective impact of visual images.

Photographs of known or unfamiliar places can also provide a successful stimulus for exploring emotional response.

- Pupils can be asked to consider how selected photographs make them feel, using a feeling word list as a prompt if appropriate.

- A set of striking photographs can be displayed and the class provided with small coloured stickers to represent different emotions, e.g. red for angry, blue for sad, yellow for happy. Pupils can then be invited to place their stickers against any photographs which make them feel that emotion. The pattern of response can then be discussed, e.g. 'Why did this photograph make so many of us feel sad?', and individuals invited to explain their response: 'I felt happy when I looked at this photograph because …'.

- Pupils can draw themselves on a post-it, and then 'put themselves in the picture' by placing the post-it on a photograph. Ask the children to imagine themselves in the photograph and to say what they can see, hear, touch etc., and how they feel about being in that place.

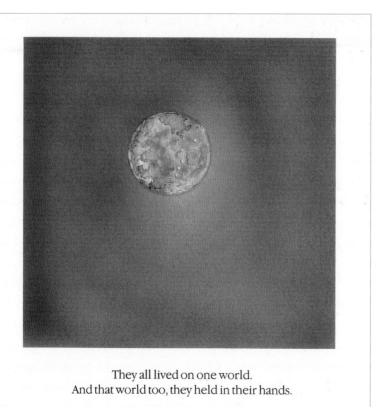

Figure 6 | *One World © 1990 by Michael Foreman. First published by Andersen Press Ltd, London, ISBN 1 84270 334 X*

They all lived on one world.
And that world too, they held in their hands.

- High-quality photographs of children in the distant localities studied can be used to develop empathy by considering questions such as 'What was he/she doing when the photograph was taken?' 'What do you think he/she was thinking?' 'How do you think he/she was feeling?'.

- Carefully selected photographs can also provide a stimulus for role play or drama, e.g. by role playing an interview with someone in a photograph, or acting out what happened the moment after the photograph was taken.

Story and fiction

Picture and story books and fiction offer many opportunities for developing a sense of place, for exploring feelings about places and issues, and for developing empathy for characters. Many primary teachers are skilled in using fiction in this way, and many of the ideas in this chapter can be developed in relation to picture and story books. Fiction offers children the opportunity vicariously to experience otherwise inaccessible experiences, places, events and situations, and to explore and articulate the emotions associated with them (Figure 6). Two publications from the Development Education Centre (DEC) in Birmingham – *Start with a Story* (1991) and *Long Ago and Far Away* (1994) suggest many excellent ideas and approaches. *Hadithi Nzuri* (1998) from Action Aid is another valuable resource that reviews almost 200 books from Africa, Asia and Latin America.

Conclusion

An emotionally literate geography education provides the opportunity for primary age children to:

- Recognise and acknowledge attachment to personally significant places
- Understand the power of place to provoke affective responses
- Recognise and express emotions about known and unknown places
- Listen to, understand and acknowledge others' feelings about places
- Develop empathy for others' feelings about places
- Know that affective response to places may be represented through different media, e.g., 3D and textile art forms, music, photographs, dance and movement
- Represent and communicate feelings about places in a variety of ways
- Understand that environmental issues may arouse strong feelings
- Express their own emotions associated with environmental issues appropriately

Many children growing up today have fractured families and disrupted home lives and are thus likely to have experienced greater change and be more rootless than their predecessors. The fear of strangers and exaggerated worries about safety are also causing parents to be more vigilant. The result is that more and more children are restricted in where they can go independently and are ferried from place to place by car. Not only are children going out less, they also now have televisions in their bedrooms or spend hours staring at a screen as they surf the internet, with well-documented adverse consequences for their physical and social development (Hillman *et al.*, 1990).

Geography can make a significant contribution to the development of emotional literacy. Fully rounded studies of the local area and other localities will provide children with opportunities to attend to, acknowledge and express their feelings about places. At a time when education is dominated by standards, accountability and measurable outcomes, it is particularly important to keep sight of the ultimate goal. If we can help our pupils to lead more complete, fulfilling and balanced lives we will have achieved something really worthwhile. If we can lead them to reflect on the needs of others and contribute positively to the world, both locally and globally, the benefits will be even greater. Geography can be and should be part of this vision.

References and further reading

Birmingham DEC (1989) *Get the Picture*. Birmingham: Birmingham DEC.

Birmingham DEC (1991) *Start with a Story*. Birmingham: Birmingham DEC.

Birmingham DEC (1994) *Long Ago and Far Away*. Birmingham: Birmingham DEC.

Carter, R. (1999) *Mapping the Mind*. London: Seven Dials.

Catling, S. (2003) 'Curriculum Contested: Primary geography and social justice', *Geography*, 88, 3, pp. 164-210.

Chambers, B. (1995) *Awareness into Action: Environmental education in the primary school*. Sheffield: Geographical Association.

Chantler, L. (2004) *What Effect does Residential Fieldwork in Geography Have on Children's Emotional Development?* Unpublished dissertation. Canterbury: Canterbury Christ Church University College.

Gardner, H. (1983) *Frames of Mind: The theory of multiple intelligences.* New York, NY: Basic Books.

Goleman, D. (1996) *Emotional Intelligence: Why it can matter more than IQ.* London: Bloomsbury.

Goodey, B. (1971) *Perceptions of the Environment.* Birmingham: Centre of Urban and Regional Studies, University of Birmingham.

Greenhaugh, P. (1994) *Emotional Growth and Learning.* London: Routledge.

Hart, R. (1979) *Children's Experience of Place.* New York, NY: Irvington Press.

Hillman, M., Adams, J. and Whitelegg, J. (1990) *One False Move: A study of children's independent mobility.* London: The Policy Studies Institute.

Jackson, E. (2000) 'Personal, Social and Health Education' in Grimwade, K. (ed) *Geography and the New Agenda.* Sheffield: Geographical Association, pp 33-43.

Matthews, M.H. (1992) *Making Sense of Place.* Hemel Hempstead: Harvester Wheatsheaf.

McCarthy, K. and Park, J. (1998) *Learning by Heart: The role of emotional education in raising school achievement.* London: Calouste Gulbenkian Foundation.

Mosley, J. (1996) *Quality Circle Time in the Primary Classroom.* Cambridge: LDA Learning.

Mosley, J. (1998) *More Quality Circle Time.* Cambridge: LDA Learning.

Ofsted (2003) *Geography in Primary Schools: Ofsted subject reports 2001/02.* London: Ofsted

Orbach, S. (1998) 'Emotional literacy', *Young Minds*, 3, pp. 12-13.

Palmer, J. (1998) *Environmental Education in the 21st Century.* London, Routledge.

Park, J., Haddon, A. and Goodman, H. (2003) *The Emotional Literacy Handbook: Promoting whole-school strategies.* London: David Fulton.

Relph, E. (1976) *Place and Placelessness.* London: Pion.

Scoffham, S. (1998) 'Places, attachment and identity' in Scoffham, S. (ed) *Primary Sources: Research findings in primary geography.* Sheffield: Geographical Association, pp. 26-7.

Sharp, P. (2001) *Nurturing Emotional Literacy: A practical guide for teachers, parents and those in the caring professions.* London: David Fulton.

Slater, F. (1994) *Learning Through Geography.* Oxford: Heinemann Education.

Steiner, C. and Perry, P. (1997) *Achieving Emotional Literacy.* London: Bloomsbury.

Titman, W. (1994) *Special People, Special Places.* Godalming: WWF UK/Learning Through Landscapes.

Weare, K. (2003) *Developing the Emotionally Literate School.* London: Paul Chapman Publications.

Acknowledgement

I would like to thank all the colleagues, teachers and children who have contributed ideas and work to illustrate this chapter, and especially Rosemary Stirk of Calverley Church of England Primary School, Leeds, for the case study material.

Photo | John Halocha.

IN THIS CHAPTER YOU WILL FIND KEY IDEAS ON
CITIZENSHIP • CREATIVITY • DISTANT PLACES • GAMES • INCLUSION • MAPWORK

Making geography fun

Look around the world we inhabit, glance out of the window, switch on the television, look at a newspaper. There are places which are flat, mountainous, dry, wet, barren, cultivated, empty, covered with buildings, teaming with human life, totally desolate, safe, dangerous, dull, interesting … the list could go on. Bring to mind some of the people you know and their characteristics, the jobs they do, the languages they speak, the colour of their skin, the leisure activities they engage in. Now extend that to the people you meet and interact with indirectly through the media, through travelling, through surfing the internet and so on. Reflect on some of the issues with which geography engages. They affect almost every aspect of our lives and those of the people living across the world now and in the future.

Have you ever wondered why, with all this diversity and variety, geography is not top of the popularity list for school subjects? How is it that pupils and adults both within education and outside it fail to find geography riveting, captivating, mind-blowing and all the other superlatives people associate with subjects which inspire them? We cannot blame the raw material with which we operate. We could suggest that the pupils have some part to play, but it is my belief that it is the way the material has been, and often still is, presented within our schools which is the root cause of any disaffection. As Kirchberg writes:

> The didactics of teaching do not need to be redesigned, but new paths have to be followed, with new contents, methods and lesson procedures that are better adapted to today's children and adolescents.
> (Kirchberg, 2000, pp. 5-16)

If geography educators are to become more effective they need to learn lessons from the plethora of recent research about learning and apply appropriate strategies in their teaching. To use words attributed to Mahatma Gandhi: 'Be the change you want to see in the world' (Gilbert, 2002). It is my passion, and one I know is shared by many

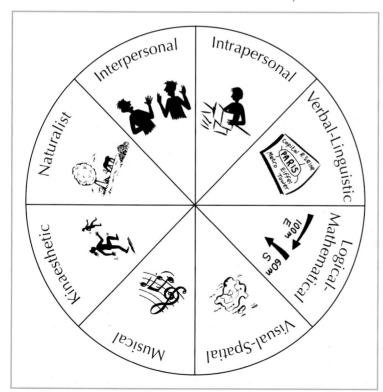

Figure 1 | *Diagram to show multiple intelligences. Based on material from Gardner (2000).*

others, that young people should be introduced to geography in ways that will motivate them to discover more for themselves: to enquire, investigate and search for answers; to enjoy their learning and pass on positive rather than negative messages. We can use the enthusiasm and interests of the pupils. We can find out how they learn, what they know, want to know, enjoy doing, question, don't understand, what is going to be useful to them. We can vary the teaching strategies employed.

If teachers wish to be truly inclusive, they must ensure they take account of differences, be aware of the need for versatility and design activities which take account of specific needs – it is my premise that there are few geographical experiences which cannot be enjoyed and provide learning opportunities for all. Many of the ideas offered in this chapter are open ended, offering potential for gifted and talented pupils to be adequately challenged. They can also be used, easily adapted or modified to cater for pupils with specific needs.

If we, as teachers can engage with geography in a positive way and develop our own vision of what the subject encompasses it is likely that our teaching will be enriched, leading to improvement in the learning of the pupils with whom we work. If we are having fun, they are likely to have fun too.

Different ways of thinking

The way in which a topic is introduced can determine the attitude pupils have to the rest of the work. Howard Gardner's multiple intelligence theory (Gardner, 1983) suggests that the human mind, rather than being a unified entity, is better considered as a community of separate intelligences. In daily life these intelligences usually work in harmony so their individual characteristics are not apparent. In teaching, however, we need to be aware of these different modes so we can develop each one to its full potential. Howard Gardner argues that the human mind consists of different forms of intelligence each of which is equally valid (Figure 1).

We need to take account of these various 'intelligences' when we begin any new topic, as well as throughout a unit of work. For example, if you are introducing work on the local area to your class, whether through one of the QCA schemes or a school-specific plan, you could use one of the activities in Figure 2 to 'get pupils interested and alert' each lesson. It is envisaged that these activities will last for 10 minutes maximum. Other 'starter' activities are suggested later in this chapter.

Photo | *John Collar and Kathy Alcock.*

Figure 2 | *Suggestions for 'starter' activities about the local area that tap into different forms of intelligence.*

Interpersonal

Divide pupils into groups of two or three and ask them to talk for about three minutes about the places they like going to in their local area. Then get the pupils to change groups and discuss people they know who work in the area, again for three minutes. The final discussion, in another group, could relate to an issue in the local area which the pupils believe is important.

Intrapersonal

Present the class with a number of statements about the place they are studying. They have to rank the statements in order depending on how important they are to them, thus making value judgements. Examples might include: 'There are too many new houses being built here' or 'The proposed waste disposal site should be allowed'. In this exercise it should be stressed that all opinions are valued and it may be better not to share results.

Verbal/linguistic

Ask pupils to work with a partner and to put the name of the place in the centre of a large sheet of paper. They then write around it all the words and phrases they associate with that place. Alternatively, give the pupils an initial letter and tell them to think of all the words associated with the place beginning with that letter. After about one minute change the letter, and so on.

Logical/mathematical

Give pupils a pie-chart showing the different shops and services available within the chosen area. What can they deduce from studying the chart?

Visual/spatial

Give pupils a map of the local area drawn to an appropriate scale. Display a series of pictures around the room. Ask the pupils what they think the pictures show, where they are found on a map, and to identify each location with a mark. You could also ask them to devise a key.

Musical

As the lesson begins, play music associated with the local area. This could be music from the local place of worship, sounds which might be heard in the streets, or music which evokes a sense of the place, e.g. traditional sea shanties or part of Beethoven's 'Pastoral Symphony'.

Kinaesthetic

Allow pupils to walk around the school grounds (or the school building if more appropriate) and look at what they can see of the local area from a variety of viewing points. If possible the class should be divided into small groups (two or three pupils) to facilitate discussion.

Naturalist

Set up a display of wildlife specimens (leaves, grasses, plants) that are found in the area in the classroom. Ask pupils to identify them and say where they might be found.

Fun and games in geography

'Research has shown that mental limbering up makes for more effective learning' (Gilbert, 2002, p. 119).

Start a geography lesson by saying 'we'll start off with a game' and you will have many pupils engaged immediately. Try this in the middle of a lesson, when attention may be flagging, and the strategy is equally successful. If you finish off with a game, they might not want the lesson to finish. The following websites include geographical games:

Puzzlemaker: (http://www.puzzlemaker.com) – Children devise their own map games and puzzles.

BBC Weather: (http://www.bbc.co.uk/weather) – The weatherwise section contains some stimulating and enjoyable weather games.

Mapzone: (http://www.mapzone.co.uk) – The Ordnance Survey children's website which includes map games.

Volcano World: (http://volcano.und.nodak.edu) – Games and simulations to with volcanoes.

Sea and Sky: (http://seasky.org) – Contains some excellent animations of life in air and water.

Global Gang: (http://www.globalgang.org.uk) – The Christian Aid site includes games like 'Milk the Goat'.

Quizzes and puzzles

The numerous quiz shows on TV and radio have great potential as fun ways of learning, or assessing knowledge and understanding. Try these:

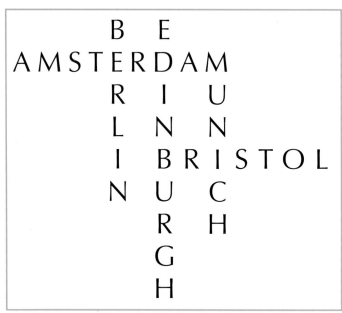

Present geographical words or phrases as anagrams – a good way of warming up the brain:

IVRSER AYM DOLFO NI AVHEY NRIA

Put one word on the board and see how many other words of three letters or more can be made from it – find 20 in 5 minutes:

AFGHANISTAN

Write the name, for example, of a European country or city on the board or a large sheet of paper and challenge the class to create a crossword as quickly as possible. How quickly could you get to 15 European cities? Keep the crossword accessible so that pupils can devise the clues during the following week, possibly in free time.

Globe games

Use an inflated or soft globe and let the class throw and catch it. Have a letter spinner or set of alphabet cards and a set of geographical 'category' cards. One pupil at a time chooses the initial letter and another the category so that when the globe is caught the pupil (or pair of pupils if that is more appropriate) has to say the name of a place or feature beginning with that initial letter. For example letter 'E', category 'mountain'. What would you say? These games can be adapted by adding 'Ask a question about ...' or 'Say anything you know about ...' or 'Find this place on the globe ...'.

Map cards and games

Pupils will undoubtedly learn from and enjoy playing the games on the Mapzone website (www.mapzone. co.uk), but you can devise your own. Here are some ideas to get you started:

Make or buy a set of individual Ordnance Survey map symbols flashcards. Now make a second set of cards with the corresponding names of the features from the symbol cards. Divide the class into groups and take it in turns to 'say the symbol' or 'sketch it'. Share the cards out and make up a story which includes the features on the cards. As you say the word the pupil with that symbol stands up and holds up the card. Make it really fun by increasing the frequency with which these words appear. As pupils get more proficient get them to tell their own story for the rest of the class to react to. A further variation can involve the pupils creating a class set of geographical feature cards (Figure 3).

Figure 3 | *River Snap. Work by Y5/6 pupils, Shelley Primary School, Horsham.*

Using maps

If you have access to a range of 1:50,000 OS maps, ask pupils to select a map, then a place on the map they would like to visit for a holiday. Give them a set time (say three minutes) to use the map evidence – contours, symbols, place names and so on – to 'pack' a rucksack or a case with clothes and equipment which they might need during their holiday. Pupils can record their answers visually in a pre-drawn rucksack or case outline, or verbally in pairs. (This idea also appears in 'Making use of schemes of work' see page 58).

Initial letter, word and guessing games

In small groups, ask pupils to think of all the features beginning with a certain letter they might find, for example, in the countryside, a town, or a named area. For example, how many features can you think of that you might see at the seaside beginning with s? Then ask pupils to make up a story using the words identified. With younger pupils you might ask them for just one or two things then change the letter. Alternatively, play a version of 'I spy' with the teacher or pupils taking it in turn to identify the themes, e.g. building, mode of transport, physical feature. A similar game involves the pupils in asking questions. You say, for example, 'I am thinking of something I can see at the seaside' and write the word or draw the picture on a whiteboard which the pupils can't see. They ask questions until they find the right answer. This could be made more challenging by restricting the pupils to asking questions that can only be answered 'yes' or 'no'.

Picture postcards

Picture postcards are useful for sorting activities with younger pupils. Make sets of category labels, enough for one set per group, such as: Local area, UK, Europe, Distant Places; or seaside, transport, buildings, people. Put a variety of postcards in sorting bags and give one bag to each group. Pupils take turns to draw out a card and put it in the appropriate category, explaining briefly their reasons for the choice.

If the cards have been used, the reverse side will also contain information which pupils can use, such as the postmark, stamp and message. This can become a game if they work in pairs or small groups and give clues from the information so that others can work out the source of the card.

You could also cut up old postcards to create jigsaws. Pupils should then re-assemble them using the clues from the picture. Cut the cards up into different size and shape pieces to differentiate the activity for appropriate levels of ability.

Photo | *Kathy Alcock and John Collar.*

Developing key geography skills through having fun

Back-to-back: Using geographical vocabulary and maps

In groups of three, give each pupil a label – A, B or C – and a pencil or washable pen. Pupils A and B must sit back-to-back and not look over their shoulder. Pupil C needs blank paper. Give pupils A and B a photocopied or laminated map each that you have prepared earlier at an appropriate scale. (Laminating the maps and using washable pens means you can use the same resource many times.) Pupil B decides on a route to get from one place to another on the map, and marks this on their copy. Pupil B then gives instructions to pupil A so that they are able to replicate the route on their map. Pupil A can ask questions to help them. Pupil C remains silent but makes written comments on what is happening. After a pre-determined time period, the drawing stops and the two routes are compared. Pupil C reports back on what was successful or not during the activity. Using this information, the activity is repeated, with roles rotated and a different route selected, until all three pupils have completed each part.

This could be used as an assessment strategy. The pupils will have fun and learn – so will you. Younger pupils can engage in this activity providing the map you give them is appropriate. Do ensure the pupils think about what they have learnt and the knowledge and understanding they have used, perhaps as a plenary session.

Sing a song

Sing songs with a geographical theme. 'The Continents Song' (Education through Music, 1994) is an excellent way of helping pupils to remember the names of all of them. This accommodates those pupils with 'musical intelligence' and also acts as a method of injecting variety.

Hold it up!

Create 'story sticks' (Figure 4) to help pupils explain and re-tell geographical processes. For younger pupils teachers will need to make this resource themselves. Older pupils can make these themselves by researching, for example, 'The story of a banana from tree to tummy' or 'The journey of the River Rhine from source to mouth'. They select five or six stages in the story which are presented visually and mounted on canes or pencils. Groups of pupils use these prompts

Figure 4 | *Story sticks. Photo: Stephen Scoffham.*

to re-tell the process in an imaginative way. You will be amazed how this can release pupils to excel when they can hide behind a role and present knowledge and understanding orally and creatively. Another use for this type of resource is as a 'put in the right order' activity. This makes better use of time than colouring, cutting out or sticking in books which is so often seen in classrooms and is not good geography, even though the learning that it is promoting may be.

Find a furry friend to help!
Puppets and toys are also useful in helping pupils to display their understanding. Pupils can hide behind the character, and if they get it wrong it is the puppet not them! Role play can also help pupils escape from the fear of making mistakes. They can re-tell familiar stories, or create their own, using appropriate geographical vocabulary, and recount journeys and experiences undertaken. Barnaby Bear has many a story to tell, and indeed, so will the toys belonging to your pupils.

Having fun learning in sand and water
Early Years pupils can have great fun in the sand pit creating landscapes – tunnels, mountains, bridges, settlements. While engaging in water play they are learning about how water moves, carries things, transports and deposits. In the role-play area they engage in real-life activity given the right environment for learning. Talk to the pupils as they play, show them pictures, play with them, let them ask you questions, listen to them. You will learn as much about what they know and understand by just being there. This should be seen as a vital part of your role as a teacher and not left to the supporting adults.

Maths in mind
Mental maths starters should be set in a real-life context. Try this idea: A bus leaves Canterbury bus station with 23 people on board. At Safeway's it picks up 4 more. Outside the College, 5 get off and 2 get on. Once it reaches the Park and Ride 10 get off and 5 get on. The bus travels on the A2 towards Dover. It stops at the Woolage turn to pick up one more passenger. How many people are now on the bus?

This idea can be developed in numerous ways. You can use a map together with the mental maths, find the route taken, ask questions, or suggest enquiries. Pupils can take turns in preparing the challenges.

Trails as treasures for teachers
One of the features of geography lessons is that they can easily provide a context for learning in a number of other subject areas as well. A trail around the local area or a visit to a contrasting locality could involve the following:

- writing words to describe your feelings about a place which develops into a poem, letter or piece of descriptive writing;
- noting or recording sounds which can lead to making music that evokes the character of the locality;
- investigating change within the buildings in a community can be linked to historical enquiry;
- field sketches can be drawn and annotated to cover a range of themes.

Picture this: virtual visits

You can use geography resources such as photopacks, slide shows and video clips for other curriculum areas. For example, use a picture showing life in an Indian city as the context for a story. You begin the story then divide pupils into groups and they take turns to continue to make up the story using the characters, buildings and features portrayed in the picture. For recording of evidence purposes the story telling can be taped.

Problem solving

Problem solving can lead to disaffected children being stimulated to learn and discover for themselves. I witnessed an example of this in a school in Kent. The teacher began the lesson by saying: 'You have a problem to solve'. This was clearly the first time this approach had been used in this way. Without exception the pupils sat up higher in their seats and looked engaged. I recorded the exchanges that followed, some of which are given here ('pupil' refers to different pupils):

Pupil: 'What's the problem?'
Teacher: 'The village needs a new by-pass.'
Pupil: 'How do you know?'
Teacher: 'Well what do you think?'
Pupils began to discuss quietly amongst themselves.
Teacher: 'OK. You spend a couple of minutes thinking about that.'

After a few minutes the consensus was in agreement with the statement but they decided they needed some evidence to prove they were right. They also went on to decide what other information they needed in order to solve the problem.

What followed was one of the 'magic moments' of my career in education as I witnessed one of the best geography lessons I had seen. Maps and aerial photographs were found, the internet was searched, pupils went to ask questions of others in the school, telephone calls were made, faxes and e-mails sent, plans for visits and further investigations were made. I returned to the school several weeks' later and the pupils couldn't wait to tell me what they were doing 'in geography' now. Magic!

Lesson planning

Motivation through choice

All too often, differentiation is not apparent within geography lessons and when it does occur it generally means a different level of difficulty in terms of what is recorded. Differentiation may be by task, by text, by interest, by ability, by choice. In taking account of differences, pupils themselves may be able to select an appropriate activity to motivate them to progress their learning. Alternatively, teachers may need to guide and direct choice in order to address the gaps in knowledge and understanding as well as issues relating to learning styles. Having a range of possibilities on offer allows for this to occur. Here are two examples:

Option 1: Reading the landscape

Use either the view from the window, an overhead projector transparency or large picture (for example, of an area which is due to be developed in some way or a distant locality view). Ask the pupils to select from the five activities listed below. They should do one (or more if they wish) but not all of them.

Task	**Follow-up activity**
▨ Generate a list of questions	▨ Devise a quiz for others to complete
▨ Create a list of words or phrases	▨ Compose a rap, song or poem
▨ Describe the scene in words or pictures	▨ Create a collage from photographs, drawings and descriptions
▨ Explain the activity within the picture	▨ Act out the activity as 'playlets'
▨ Create a survey based on the scene	▨ Undertake the survey and present results in a variety of ways

Photo | *Paula Richardson.*

Option 2: Create a travel guide for the teacher

The aim is to make a class book or resource pack which relates to significant places around the world, such as those referred to in the framework of locational knowledge for key stage 2 (DFEE/QCA, 1999, p.115). Research undertaken by primary geography specialists at Canterbury Christ Church University College in 2001-02 showed that place knowledge is an area of weakness for adults as well as children.

Tell the class you want to go on holiday, spending a few days in several different places. (The range will depend on the pupils. It may be within your local area, county or region, the country, or be worldwide.) You will plot the route but need their help in order to plan your journey, visiting interesting places. Ask them to create a double-page spread of interactive activities and information about a place in your chosen area they have visited or know well. Give them some guidance as to suitable material without limiting the scope of their creativity. A brief class discussion before beginning the task could ascertain that they may include maps, pictures, letters in envelopes written from the place, lift-the-flap activities, quizzes, games, pictures of artefacts, information panels, websites and so on. This activity could be linked to ICT depending on the resources available and age and ICT competency of the pupils (and teacher).

Making use of schemes of work

Many schools are currently teaching geography through the QCA schemes of work (DfEE/QCA, 2000). These provide a valuable framework but it is important to realise that you are expected to personalise the materials to make them applicable to your own situation. If you are responsible for leading INSET, clearly this is a message which should be passed on to colleagues along with the benefits of using personal experiences, enthusiasm and themes which are of particular interest to them as teachers. This is likely to lead to a higher degree of engagement.

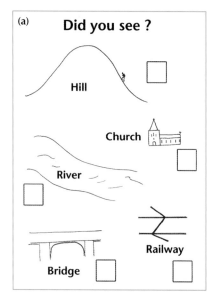

(a) Did you see ?

Hill

Church

River

Railway

Bridge

(b) Which of these did you see during the journey?

Feature	
boats	
bridge	
castle	
church	
factory	
fields	
harbour	
hill	
houses	
offices	
petrol station	
railway	
river	
school	
shops	

Figure 5 | *Using a journey to promote geography with (a) Foundation Stage and (b) key stage 1 pupils.* © *Kathy Alcock.*

Where the QCA schemes are used creatively, pupils become engaged in the associated activities as is clear from the success of Barnaby Bear. Many articles in *Primary Geographer* have exemplified this modification approach, e.g. Martin (2003) and Martin and Matthews (2002). Here are a few more suggestions to inject extra fun into units 3, 13, 15 and 24 in particular:

■ Pack a case or rucksack with the clothes and items you would take if you were visiting an island, a contrasting locality by the sea or a mountain environment, or going on a round-the-world trip. Discuss with your pupils what you have selected.

■ Give pupils a map of the area you are studying and ask them to suggest what they would need to take with them if going on holiday there, using only the evidence from the map. This is an excellent way of assessing pupils' ability to read and use maps in a meaningful context.

■ Listen to and make music related to the themes covered. The *Exploring Geography* CD (Education through Music, 1994) provides a range of songs which are useful here, such as 'The Continents song', 'I like looking at maps' and 'Where does a river come from?'.

■ Show video clips from travel programmes and other appropriate programmes without the sound track and ask pupils to create their own accompanying commentary. (Tip: keep a dedicated video tape ready at home to record on whenever you see something relevant to geography. You may need a 'keep off' note for the rest of the family!)

■ You might also show the class extracts from commercial videos. The 1996 film *Alaska* (directed by Fraser Heston and available from Columbia Tristar Home videos) has some stunning river scenes which have been used with great success to support work on rivers in Units 11 and 23. Further support material for using video in teaching is available from Film Education (www.filmeducation.org).

Journeys and visits

Fieldwork and first-hand outdoors experiences are an important component of geographical education. For many of us, it was precisely these practical adventures in learning which fired our enthusiasm for the subject. The academic geographer, Peter Gould, decided to become a geographer while bird-watching on the cliffs of Devon as an adolescent; I committed myself to a future career in geography teaching when involved in a frightening, wet, climb down Gordale Scar in the Malham area of Yorkshire. Sadly, residential fieldwork is not possible for many pupils today. However, you can compensate to some extent by using journeys to support and develop geographical activities. Here are some suggestions:

■ When taking pupils on a journey (whatever the curriculum subject) ask them to look at the map and follow the route beforehand, then to complete an 'I have seen' checksheet as they go along (Figure 5). This could become part of your school policy.

Stuck in the Mud

How to play Stuck in the Mud

1) Choose 4 players to be the chasers.

2) Everyone runs around. The chasers wait for a minute.

3) If the chasers catch you then you have to stand still. You have to stand in a star shape.

4) To be back in the game someone has to crawl under the persons legs who is stuck.

5) You are free to play until a whistle has blown at the end of the game. Keep a look out for your friends who are " stuck in the mud "

This was made up by Charlie Cole

Figure 7 | *Teaching how to play. Work by Y1 pupils.*

■ When visits are planned, encourage the staff involved to include some geographical games such as 'I spy to the north/south/east/west ...'. For younger pupils this may need simplifying to in front/behind/above/below.

■ During fieldwork ensure you include activities which relate to people as well as places. One idea is to ask pupils to survey what people are doing or carrying as they walk past. Prior to the visit get pupils to create their own categories such as walking, communicating, riding, listening, using a mobile phone, walking a dog, etc.

Holidays

Holidays and out-of-school excursions are vastly under-used resources. Create a school 'holiday pack' of ideas for pupils to undertake when going away on holiday. Figure 6 lists suggestions for making the most of these valuable, first-hand experiences of other localities.

Learning about distant places

Engaging pupils with the wider world to create balance and understanding is essential if we are to help them to become global citizens. This area is developed more fully in Chapters 16 and 21, but here is one idea which was used successfully.

Year 1 pupils wanted to know about a project which our church was supporting to help children in a region of Northern Ghana who have severe eye problems. I had agreed to show a video and share my knowledge of the area and the project. During the discussion, I asked the pupils what games they liked to play and what games they thought the children in Lingbinsi would want to play once they could see again. They made some suggestions and together we decided it would be good if they could send information to the children in Lingbinsi about the games they liked to play. Figure 7 shows an example of what the pupils produced. They laminated them and sent them together with crayons and small toys which they thought the children might enjoy. In return, the school received photographs and information from the Ghanaian teachers about what their pupils liked to do, to eat and games they played in their village. The UK pupils had their vision widened and positive images were created.

Photo | *Paula Richardson.*

Figure 6 | *Holiday geography.*
After Alcock, 2001.

Pupil activities for undertaking in the classroom before, during, or after a holiday

- Using a map at an appropriate scale find the place of origin of cards or letters you receive.
- Display a map showing the route to the holiday destination and the itinerary if visiting more than one place.
- Plan the detail of the journey – the stops, meal breaks, where and when fuel might be needed, overnight stops.
- Search for information about the places to be visited while on holiday.
- Make a list of questions to find the answers to while you are away.
- Write a letter or send an e-mail request for information from the local Tourist Information office.
- Compile a display showing what life is like in the place visited.
- Run a question and answer session in class for other pupils to ask questions about the place visited.
- Find the cost of buying several different products from a local shop on holiday and compare these with buying the same or similar items from a local shop at home.

Activities adults could help children undertake while on holiday

- Send a postcard to the class with a picture which shows some significant landmark, feature or aspect of life in the place you are staying. Use stamps of different values if possible.
- Find out what is grown or produced in the local area. Are there any regional or local specialities?
- Eat a meal made from locally-produced products or cooked in a traditional way for the area, e.g. Cornish pasty, Spanish paella, Italian pizza, French cheese, Devon cream tea, Scottish shortbread, Bakewell Tart.
- Buy an article using local currency if appropriate. For young children, unused to dealing with money themselves, holidays are often a good time to introduce this idea.
- Devise a list of products which can be bought in most towns in Britain and abroad, for example, one litre of milk, a loaf of bread, six eggs, a bottle of fizzy drink, one kilo of tomatoes, four oranges, one ice cream. Find the cost of these things in a shop in your holiday destination. Convert the prices into pounds sterling if appropriate.
- Discover how many different forms of transport are used in the place visited.
- Take a journey, even if only short, on local public transport.
- Visit and makes notes about a place of interest which can be used in a talk to the rest of the class.
- Collect information about the area from, for example, the local Tourist Information office.

- Try to discover the meaning and origin of local place names.
- Discover how the place you are visiting is changing. What are the current issues affecting the lives of local people?
- If possible, read a local newspaper to discover some things which are similar to and different from the place where you live. Consider, for example, the sports played, the goods for sale in shops, and the entertainment available.
- Draw pictures, write notes or make a sketch map of what you can see from where you are staying, to show what it is really like to people who have never been there.
- Take photographs of people and places which you can talk about back at school.
- Talk to people who live or work in the area to find out what they like or dislike about living in that place and about the work they do. How are their lives changing?
- Look for signs and symbols and discover what they mean.
- Make sketches of interesting buildings and label them to help you to remember what they were really like.
- Make a list of the names of the people who you meet during your holiday and the jobs they do.
- Collect artefacts such as coins, stamps, tickets, leaflets, carrier bags and wrapping paper, local produce, leaves and flowers, shells and stones. Make sure you note where you found/got them and do not take away anything from conservation areas.
- Find a quiet place and write down all the words which come into your head as you look around. Do the same in a noisy place.
- If staying in a foreign country, learn a few words or phrases in the local language, e.g. good morning, thank you, please, my name is.
- Find out if any famous people came from the area where you are staying.

Activities for pupils to undertake on the journey to their holiday destination

- Make animals from pipe cleaners.
- Make up games using car registration plates. For example, find words which use the letters in the same order, or devise sums which involve adding or multiplying the numerals together.
- Complete an 'I have seen ...' tick sheet with words or pictures of things you see out of the window.
- Make surveys for about five minutes of, for example, colours of cars, types of vehicles, shapes passed, number of traffic lights.
- Play games such as 'I'm thinking of ... (see page 52)
- Go through the alphabet thinking of objects or names beginning with each letter in turn.

Note

Many of the suggestions and ideas in this chapter involve activity and experiences which will not lead to anything being recorded in the traditional way in geography books or folders. This should not prevent you from using them. Be creative about the way you record the evidence. Use audio and video recording techniques and take photographs. Use technology to support what you believe in and be able to justify your practice by demonstrating improved knowledge and understanding from the pupils.

References and further reading

Alcock, K. (2000) *Our World*. Dunstable: Folens/Belair.

Alcock, K. (2001) 'Distance learning', *Primary Geographer,* 44, pp. 32-3.

DfEE/QCA (1999) *The National Curriculum: Handbook for primary teachers in England.* London: DfEE/QCA.

DfEE/QCA (2000) *Geography: A scheme of work for key stages 1 and 2 (Update).* London: DfEE/QCA.

Education through Music (1994) *Exploring Geography* CD. Coventry: ETM.

Gardner, H. (1983) *Frames of Mind.* New York, NY: Basic Books.

Gardner, H. (2000) *Intelligence Reframed: Multiple intelligences for the 21st century.* New York, NY: Basic Books.

Geographical Association (2003) *Primary Geographer 51: Focus on Early Years.* Sheffield: GA.

Gilbert, I. (2002) *Essential Motivation in the Classroom.* London: Falmer.

Kirchberg, G. (2000) 'Changes in Youth: No Changes in Teaching Geography? Aspects of a neglected problem in the didactics of geography', *International Research in Geographical and Environmental Education*, 9, 1, p. 5-16.

Martin, F. and Matthews, C. (2002) 'Thinking through units', *Primary Geographer,* 47, pp. 12-14.

Martin, F. (2003) 'QCA Creativity Project: Creative work with rivers', *Primary Geographer,* 50, pp. 14-15.

Blow wind blow

Photo | Kathy Alcock and John Collar.

Young children making sense of their place in the world

The Foundation Stage applies to children aged between three and five years in a variety of settings such as nurseries, playgroups and schools. In these important early years, children are forming values and attitudes that will colour their perception of the world around them and predispose them to learning about it. This is a crucial period in which educators need to ply their craft appropriately if they are to lay secure foundations for future development. One of the main aims of the Foundation Stage curriculum is to make children aware of their place in the world and help them to make sense of it. As a result geography is at the very heart of early learning, encompassing as it does aspects of social, physical and cultural worlds that are necessary for meaningful understanding of our surroundings.

This chapter outlines the principles, provision and processes that facilitate the learning and teaching of geography in the Early Years. The Foundation Stage curriculum is organised into six areas of learning. While we recognise that geography is subsumed under the area entitled 'Knowledge and Understanding of the World', we use the term 'geography' throughout this chapter for two reasons. First, geography can be found in all of the six areas of learning, and second, practitioners need to know what experiences and skills are essentially geographical within all areas of the curriculum in order to promote a balanced approach to geographical learning.

Principles underpinning early geographical learning experiences

A principle is an important underlying assumption or belief that shapes the way we act. The Geographical Association has identified three core principles that underpin all learning in the Foundation Stage, namely:

Illustration | *From 'Handa's Surprise' by Eileen Browne. © 1994, Eileen Browne. Reproduced by permission of Walker Books Ltd, London SE11 5HJ.*

Photo | *Kathy Alcock and John Collar.*

- Young children learn best through self-initiated and supported play.
- Links should be made between the 'areas of learning' to reflect a holistic approach in the education of young children.
- Young children learn best when they feel valued, secure and are able to build positive relationships with peers and other adults.
(Geographical Association, 2003).

Research findings strongly suggest that young children learn best in situations that promote self-initiated and self-directed activities, in other words, play (David, 1999). Young children need time to freely explore the world around them and try out a range of responses in order to investigate, discover, reflect on and evaluate a range of learning experiences. First-hand experience and sensory interaction are vital components in play, as is appropriate intervention to reinforce, extend or challenge newly emerging concepts. The skill of the practitioner lies in the provision of stimulating and appropriate materials, resources and contexts and the ability to intervene at the optimum moment for learning. This skill is necessary for both child-initiated and supported play.

Play is a rich and varied activity that knows no subject boundaries, but rather encompasses a range of different learning categories in which common and specific skills will need to be learnt and practised. Skills are often subject transferable. For example, the enquiry skills used by geographers are also appropriate tools for many other areas of learning. We believe that it is beneficial to children's learning that the value of skills acquisition and their role in learning is flagged up by practitioners. Thus one key part of early geographical learning is to teach essential skills in holistic learning contexts, thereby enabling children to reflect on what they have done and how they have done it.

Young children are enabled to make sense of their place in the world best when:
- They are inspired to think about their own place in the world, their values, and their rights and responsibilities to other people and the environment (DfEE, 1999, p. 14) and given opportunities to actively participate;
- Their wealth of knowledge, understandings and feelings about people and places are drawn on by practitioners;
- They are supported and encouraged to construct their own meanings about people and places;

- They are encouraged to identify with other people and places in positive ways;
- Practitioners recognise that children's sense of identity is rooted in the places where they live and play;
- Their home areas, families and communities are valued; and
- They actively experience a range of high quality, stimulating environments.

We recognise the specialised skills that parents and practitioners have that allow them to intervene sensitively and appropriately, as partners in children's learning.

The best practitioners impart the message that the child is valued for who he or she is and what she or he already knows. Children who feel valued in this way are more confident and will be more inclined to try new and different things and more willing to share their findings with others. We can, and should, model the value that we place on the lives and backgrounds of others. As recent research by Siraj-Blatchford *et al.* (2002) has shown, Early Years practitioners are powerful role models for the young children in their care.

Figure 1 | *Principles underpinning the Geographical Association Position statement on geography in the Foundation Stage. Source: Geographical Association, 2003.*

Geography starts when we begin to value our place, and that of others, in the world and acknowledge the importance of our actions within it. The specific principles that we believe underpin this process are outlined in Figure 1.

The importance of allowing children to participate actively in real learning contexts has been shown by Hart (1997). He argued that children's concern for the environment was born out of their affective unmediated contact with it and that the best outcomes were achieved when children's opinions were genuinely valued by adults. Indeed, it seems that even very young children are capable of making and enacting decisions about their environment with appropriate support. This of course stresses the value of first-hand experience in the outdoors. It is a requirement of the Foundation Stage that children have regular and structured outdoor experiences and desirable that the outdoors is used whenever appropriate to stimulate and facilitate geographical learning. Outdoor provision should include a range of stimulating materials and settings in which pupils are encouraged to actively enquire, reflect and participate in desired change. Such learning is all the more valuable because of the pupils' role within it and renders it meaningful and relevant. Catling (2003) rightly questions the lack of voice that young children often have within our society and calls for this to be corrected.

Photo | Paula Owens.

Children's early experiences of place may vary enormously but they all bring some awareness and understanding of geographical concepts to early learning contexts. Research by Blades and Spencer (1987, cited in Matthews, 1992) has shown that children as young as three were able to use simple maps in a rudimentary way, while Spencer and Darvizeh (1981, cited in Matthews, 1992) demonstrated that children of this age were capable of recalling routes travelled just once, using given photographs. The practitioner can build on this burgeoning understanding by actively teaching relevant skills inherent in observation, enquiry, communication, reflection and evaluation.

It is also important, when building on children's knowledge, to take opportunities to reinforce positive concepts and challenge negative ones. In fact the early years are the best time to challenge stereotypical thinking. As Scoffham (1999) observes, children's attitudes become entrenched as they get older and are harder to change. For this reason it is important that children have access to positive images of other peoples and cultures in global and local contexts. These images may be gleaned through a range of resources and first-hand experiences. Appropriate examples include:

- Experiencing the story *Handa's Surprise* (Browne, 1997) coupled with the opportunity to taste a range of delicious fruits.

- A visit from a parent who may show artefacts and talk about their culture and beliefs.

- An e-mail link with a school or institution from another country in which photographs and/or drawings can be exchanged.

While geographers tend to draw out similarities and differences between our own and other cultures and places, with very young children it is more fruitful to dwell on aspects of similarity. As practitioners, we have a duty to encourage tolerance, understanding, responsibility and respect, with a view to developing global citizens who will have the confidence, knowledge and skills to act fairly, capably and equably in the world they will inherit. These attitudes for learning need to be started early.

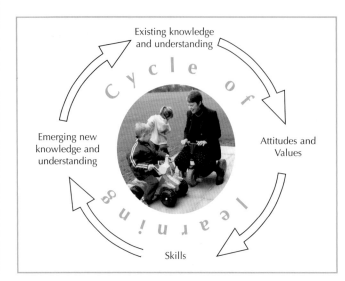

Existing knowledge
and understanding

Cycle of
learning

Emerging new
knowledge and
understanding

Attitudes and
Values

Skills

*Figure 2 | The cycle of learning.
Source: Geographical Association,
2003. Photo: Kathy Alcock and
John Collar.*

The process of learning

We have already stated our belief that play is central to young children's learning. The Geographical Association (2003) sets out a model for learning that has play at the core (Figure 2). The model suggests that learning can be seen as a cycle that uses children's present levels of knowledge, understanding, skills and attitudes as a starting point.

We see the model as a flexible planning tool. A cycle of learning may be of different lengths, ranging from a single activity to a sequence undertaken over several weeks or more. We also recognise that within this process there is a constant cycle of learning that reflects the enquiry process (see Chapter 7), as new knowledge becomes existing knowledge and the process starts again. In addition, children need to revisit stages in the cycle in order to consolidate what they have learned. So while it may be possible to progress through the stages in a linear fashion such a progression is likely to lead to shallow learning that lacks meaningful roots. The rationale for this model is as follows:

1. Why start with prior knowledge and understanding?

All teachers are aware of the importance of establishing what their pupils already know, understand and can do before embarking on new learning. This helps to ensure that, whether activities are pupil-initiated or teacher-led, the work that you plan is at a suitable level and pupils have a cognitive framework into which the new learning can be accommodated.

2. Why is it important to develop attitudes as a base for new learning?

The Foundation Stage stresses the importance of engaging young children's curiosity by drawing on their existing knowledge and fostering their motivation to learn new things. It sees children as eager, willing learners and has a number of aims, all of which involve attitudes and values. These include:

■ Encouraging and supporting children's natural curiosity in the world around them

■ Developing respect for and positive attitudes towards self, other peoples and cultures

■ Developing a sense of wonder at the beauty of natural and built environments

■ Encouraging and developing a sense of responsibility for the environment and their actions within it

In establishing the values base for learning it is important to consider the ethos of the school, the provision that is made available, and the example provided by teachers and other role models. You also need to make the most of opportunities as they arise, whether stemming from pupils' current interests and ideas, or from an opportunity identified by a practitioner. For example, take your pupils on a walk to observe the changes in autumn, giving them the opportunity to play in the leaves and to select some to take back to school to use as they wish.

Figure 3 | *Opportunities for geographical experiences can be found in all aspects of the Foundation Stage curriculum.*

Opportunities can be found here...

Knowledge and understanding of the world (KUW)
Use a range of first-hand experiences to:
- Explore a range of real and created environments – school grounds, local facilities, model farm
- Observe similarities and differences in environmental and social contexts – compare our home to a place in a story
- Record the world around us in a variety of ways – models, maps, sounds, pictures
- Develop enquiry skills – develop curiosity through asking questions, make predictions, make informed choices, model research techniques using non-fiction books, CD-Roms

Physical development (PD)
- Enjoy and take part in physical challenges – follow routes, climb up/down hills, up/down climbing frame, negotiate obstacles
- Use relevant vocabulary – movement, direction, distance, location, speed
- Use spaces around them – control physical movements (forward/backward, left/right, up/down), find personal space in which to work, play or sit
- Take part in a range of activities in places outdoors – dig and plant, experience weather, visit wildlife area/other place, and indoors – use small world toys to create new or improve existing small worlds and environments; role play travel agents, garden centre staff, airport staff, weather reporters

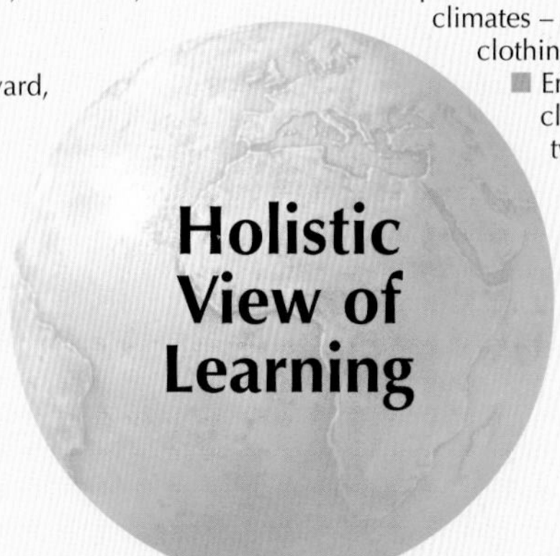

And everywhere...

Communication, language and literacy (CLL)
- Look at, and listen to, stories, non-fiction or rhymes about weather, places, people, cultures, journeys
- Make and read signs and symbols – label places (quiet area, messy area), label features in the room, look at symbols and signs in the environment, make personal pictorial representations, become familiar with other symbolic languages (e.g. Braille)
- Use a range of vocabulary to communicate knowledge and understanding of the world – naming, describing, positional vocabulary
- Talk about the world for a variety of purposes – discuss likes and dislikes, plan a journey, give instructions for a route to follow, describe pictures/features, investigate/enquire/ask questions about the environment
- Communicate own views and listen to those of others – Where shall we put the bird table?

Personal, social and emotional development (PSE)
- Develop positive attitudes towards, and a respect for, people, places and cultures
- Embrace ethnic and cultural diversity through positive images – displays, stories, pictures, non-fiction texts, visits and visitors
- Show awe and wonder in response to places and events in the environment – rainbows, waterfalls, mountains, sunsets, waves, skyscrapers, bridges, places of worship
- Become curious about the world around them – Where do rivers begin and end? Where does rain come from? Why does it get dark?

There...

Creative development (CD)
- Explore environments using all their senses – street, playground, supermarket, station, park, countryside, beach
- Explore resources from a variety of cultures and climates – artefacts, pictures, music, foods, clothing, poetry, story
- Employ natural materials creatively – clay, straw, wool, wood, shells, sand, twigs, leaves, water
- Appreciate the creativity of others' religions and cultures through visits, festivals, architecture, dance
- Use natural events as a stimulus for creative responses – waves, wind, rain, thunder

Holistic View of Learning

Mathematical development (MD)
- Observe, recognise and sort number, patterns and symbols in a range of environments – shapes in buildings, brick patterns, sorting and making symbols for outdoor play area
- Measure, order and record the world around us – distances in footsteps, pebbles and rocks by size, weight, types of transport
- Use positional vocabulary – on, in, under, next to, near, far away, through
- Use and give a range of directions – using control technology, e.g. roamer to programme left, right, forwards, backwards

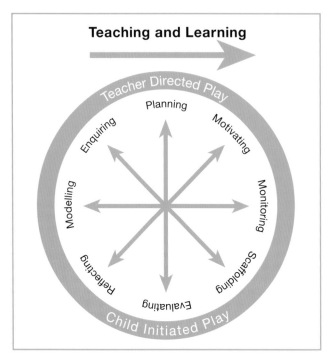

Teaching and Learning

Teacher Directed Play

Planning

Enquiring

Motivating

Modelling

Monitoring

Reflecting

Scaffolding

Evaluating

Child Initiated Play

Figure 4 | Enablers for learning.

3. Why skills next?

Investigating the world using a multi-sensory approach, developing an enquiring mind, making connections and exploring varied modes of communication are key elements in the Foundation Stage curriculum. If children are to accommodate new learning they need to refine and develop new skills. In working towards the overall goal of helping children make sense of their place in the world, they will be learning to input information about the world (asking questions, observing, recording, etc.) and learning to output information about the world (communication).

Try to develop awareness of skills and how they are used (metacognition) through teaching, practising, sharing and valuing. For example, a pupil making a painting as a means of exploring different colours and shapes of leaves is also encouraged to talk about the painting – colours chosen, why and how it was painted.

4. Why put new and emerging knowledge at the end of the cycle?

The Foundation Stage aims to:

■ Develop knowledge and understanding about places and peoples, near and far.

■ Provide a range of positive experiences in varied and challenging environments.

We believe that children acquire new knowledge as a result of motivation and skills acquisition. Recognising new learning, and having existing/uncertain knowledge challenged in supportive ways are important elements in this process.

The teacher's role will be to challenge existing knowledge, to reinforce and to extend knowledge as appropriate. This must go together with applying knowledge. For example, ask the pupil 'If you were making a book on autumn what might you include and what materials might you use?'.

Geographical learning in the Foundation Stage curriculum

The way in which geographical experiences can be developed in all of the six areas of learning in the Foundation Stage curriculum is shown in Figure 3. Recognising what 'counts' as geography will not only help practitioners, it will also ease pupils in their transition from one stage to the next.

While it is important to plan for pupil-initiated and teacher-directed activities, ideally these two modes of experience should mesh seamlessly into a unified learning process. We have identified eight key enablers for learning (Figure 4), that describe different approaches a practitioner might take to intercept periods

Photo | Kathy Alcock and John Collar.

Teacher and child role	Enabling strategy	Stage in learning cycle
1. Teacher directed Take the children for a walk around the outside area and gather interesting items, e.g. autumn would be a good time to gather different coloured leaves, and different sized twigs and sticks, although seasonal and landscape variations will pose different collection opportunities. Use stopping points en route to discuss what can be seen, heard and so on. Find out what children know and think about their outdoor area.	Planning Motivating Enquiring Reflecting	Building on existing knowledge. Beginning to foster attitudes, e.g. curiosity, wonder.
2. Teacher directed Model some of the creative possibilities using found materials, e.g. create simple radiating patterns using twigs. Encourage language of shape, colour, texture.	Planning Motivating Modelling Reflecting	Teaching skills of observation, sorting, describing using existing and new vocabulary. Developing new knowledge (lower order level).
3. Child initiated Provide a table with a range of natural materials, including found objects from the walk and other media such as masking tape, glue, and paper. Teacher and peer scaffolding may take place as children explore, discuss and create patterns and sculptures.	Scaffolding Enquiring Monitoring Evaluating	Encourage and support creative exploration, curiosity, vocabulary and technique as children make 'environmental sculptures'. Reinforce new language.
4. Teacher directed Discuss with children about the patterns, colours, shapes, textures used. Ask children where in the outdoors they could place finished work and introduce concept of a route.	Reflecting Enquiring Planning	Reinforcing skills and new vocabulary. Modelling value, of work created and children's choice in locating work.
5a. Teacher directed Show children different sites outside, model locational language. **5b. Child initiated** Explore outdoors and discuss with peers where work should go.	Modelling Scaffolding Enquiring Reflecting Planning	Building on value of work created to foster critical skills of evaluation and extending new knowledge of location and vocabulary.
6. Child initiated Children play freely outside, walking the trail and having opportunities to look at and discuss their displayed work with peers. Teacher to monitor activities and talk. Freedom to add to/change display and exchange new ideas about outdoor area.	Reflecting Evaluating Monitoring	Building on existing knowledge, reinforcing and extending value and participatory skills.
7. Teacher directed Discuss the outcomes of the trail. Encourage children to say how/why the trail has improved their environment.	Reflection Evaluation	Emerging new knowledge and understanding and values.

Figure 5 | *Cycle of learning: Creating a sculpture trail.*

Photo | *Paula Owens.*

of pupil-initiated and teacher-directed play. In practice there is no ideal prescribed order or combination in which enablers might be utilised. Rather we see them as flexible and sensitive tools in a vibrant learning environment. The subsequent examples illustrate how they might be used and demonstrate their wider relevance.

Cycle of learning: example 1
Evaluating and developing the outdoor area by creating a sculpture trail

We have chosen this as our first example of a cycle of learning because it utilises first-hand experience of the outdoors, builds on prior knowledge and encourages pupil participation.

Teacher and child role	Enabling strategy	Stage in learning cycle
1. Teacher directed Gather all the waste generated in the nursery/reception classroom over the course of a single day. Next day, with children, look at the waste (health and safety regulations important here) and consider the implications for what happens to it/ where it goes/possible impact on environment.	Planning Motivating Reflecting	Building on existing knowledge. Beginning to foster attitudes.
2. Teacher directed With maximum child involvement consider what needs to be done to change home corner area so that it becomes a 'waste management' area, encouraging recycling, reusing and repairing of waste from the home.	Planning Scaffolding Enquiring Reflecting	Using design-and-make skills to create waste management area Developing new knowledge (lower order).
3. Child-initiated 'Free' play in home waste management area. Children talk and play out some ideas generated during the planning and making phase. Peer scaffolding may be taking place here with the teacher acting as an observer	Modelling Enquiring Monitoring Evaluating	Using recently gained new knowledge. Using skills of sorting. Fostering attitudes and values.
4. Teacher directed Circle time discussion to gain children's ideas about what they have been doing in the waste management area, how they feel about it, whether any changes are needed, does it 'work', roles generated.	Reflecting Evaluating Planning	Reinforcing skills, values, knowledge and understanding (key concepts).
5a. Teacher directed Teacher role plays with children in the home corner, modelling roles/questions/activities. **5b. Child-initiated** Child role plays and teacher monitors whether play has changed as a result of circle time discussion.	Modelling Scaffolding Monitoring Evaluating Planning	Building on existing knowledge. Extending skills and understanding of key concepts (higher order).
6. Teacher directed Discuss and evaluate the use of the waste management area over the last week(s) and consider why it is important and relate to real life and waste centres.	Evaluating Assessing Planning	Emerging new knowledge and understanding and values.

Figure 6 | *Cycle of learning: Waste management.*

Overall aims of the sequence of activities

To develop skills of observation, communication and evaluation and enable the acquisition of relevant vocabulary and, as a result, to encourage critical decision making in the design of part of the outdoor area.

Learning objectives

- Develop enquiry and investigation skills (KUW)
- Develop co-operation and decision-making skills (PSE)
- Acquire a range of relevant vocabulary (CLL)
- Express responses to outdoor settings in a range of media (CD)
- Plan a simple route (KUW)
- Be able to take action to improve the environment (KUW/PSE)
- Express views about aspects of the environment (KUW/ CLL)

This activity (Figure 5) could take place at different times of the year using different found and gathered materials.

Cycle of learning: example 2

Using the home corner area as an example of waste management in the home

We have chosen this as our second example because there is encouragement from the government for citizenship and sustainable development to be integrated into learning at all levels of education. We believe that geographical learning provides an excellent context for these cross-curricular themes and that their emphasis on positive action has the potential to lead to some really purposeful and deep learning.

Key geographical concepts	
Sense of place	What is this place like?
Similarity and difference	How is it like your kitchen? How is it different to your kitchen at home?
Location and spatial pattern	Where are the waste bins? Are they easy to get to?
Change and continuity	Is there anything you would like to change in our area? What would you like to keep the same?
Cause and effect	Why did we change our home corner area? What difference has it made? If we didn't sort all our rubbish like this what would happen to it?
Appreciation	How does this help us look after our place?
Roles, responsibility and participation	Who decided to change our home corner area? Why did we decide to do it? Do any of you do this at home?

Figure 7 | Developing key geographical concepts.

Overall aims of the sequence of activities

To involve pupils in redesigning the home corner area in order to promote ideas about recycling and the re-use of materials as necessary for waste reduction, thus developing the knowledge, understanding and skills for them to take positive action to improve the environment.

Learning objectives

- Develop enquiry and investigation skills (KUW)
- Understand what a home is like – inputs and outputs (KUW)
- Understand the effects human activity can have on places (positive and negative) (KUW)
- Be able to sort materials into categories – recycle, re-use and repair (KUW/MD)
- Be able to take action to improve the environment (KUW/PSE)
- Plan out and then make the waste management area in home corner (CD/MD)
- Be able to re-use materials creatively (CD)
- Use vocabulary related to home and waste management (CLL)
- Ask and respond to questions about place (adaptation of home corner) (KUW/CLL)
- Talk about the world they are 'manipulating' using increasingly complex sentences (CLL)
- Understand symbols for recycling/re-using/repairing – standard and ones generated by themselves (CLL)

Photo | Kathy Alcock and John Collar.

Resources

- Usual home-corner kitchen items, e.g. sink, cooker, washing machine
- Waste area – different bins labelled for recycling packaging, re-using packaging and recycling food waste with one bin for anything that cannot be recycled, repaired or re-used
- Posters/symbols to put around the area in appropriate places

Activities

Figure 6 sets out a sequence of activities which may take place over a week or more. As well as indicating the relationship between learning that is teacher-led and that which is child-initiated, it also shows which enabling strategies can be used (see Figure 4) and which stage in the learning cycle each activity relates to (see Figure 2).

Photo | *Kathy Alcock and John Collar.*

The teacher's role

These two examples of developing a sculpture trail and a waste management project show how the teacher's role shifts from directing and leading tasks to observing and monitoring pupil-initiated activity and back again. The enabling strategies used at any point will depend on the stage of learning.

In the waste management project the teacher draws on pupils' existing knowledge, using it to help them think in a way that they may not have considered before. They all know that we throw rubbish away during the school day, but they may not have thought about the amount we dispose of and the effect this is having on the environment. The discussion and activity aims to stimulate the pupils' interest in what is to follow. Depending on how the pupils respond, the teacher will adapt the loose plans for the next activity to take account of general and individual needs. For example, some pupils may need to develop subject-specific vocabulary, others may have misconceptions about the names and types of materials involved.

As far as planning is concerned the teacher's role will clearly shift from week to week. At the start of a project there needs to be an overall plan that sets out some clear learning objectives and indicates the range of activities that might take place, including a possible sequence. It is also important to relate the work to the six areas of learning in the Foundation Stage curriculum, identifying the resources required and the points at which pupils' learning can be monitored (for example, through recorded observations that have a clear focus). Short-term planning is also needed, on a day-to-day and weekly basis, which takes account of observations and evaluation of the pupils' activity and learning. In any learning experience pupils will be operating at a number of different stages in terms of their development, skills, knowledge and understanding.

In the examples cited in this chapter it is not necessary to plan different activities to match pupils' abilities. The activities are open-ended and it is possible for each pupil to act out roles, interact with others and use vocabulary that is appropriate for their level of understanding. The key role of the practitioner, therefore, will be to differentiate through intervention and scaffolding – varying the types of questions asked of each pupil, and making suggestions for further play based on careful observation of the previous activity.

Using targeted intervention to scaffold learning means that assessment has a critical part to play in the learning process. It also requires a certain level of knowledge on the part of the practitioner – i.e. the knowledge to be able to recognise what is taking place, to link it with frameworks for knowledge, understanding and attitudes, and therefore to identify what to do or say next that will best support each pupil.

One way of prompting pupils and helping to direct their thoughts is by asking appropriate questions. This approach is particularly compatible with geography which makes extensive use of enquiry questions to structure learning. Figure 7 identifies some key geographical concepts and shows how they can be developed. Further advice on asking questions is given throughout this handbook (see index).

Conclusion

The teaching and learning of very young children can be likened to a wheel that is forever turning, albeit at different speeds, and which rotates through periods of pupil-initiated and teacher-directed play. Although these periods will be planned for, it is essential that the settings chosen for Foundation Stage geographical studies are flexible enough to allow for unexpected opportunities and demands as they arise. For example, a pupil may return after a holiday and bring in postcards and objects that prompt curiosity in the class, or an unusual weather event may provoke a raft of questions.

Good provision for learning in the Foundation Stage will also involve a range of indoor and outdoor settings, a variety of resources and a clear underpinning ethic. Young children need to be exposed to a range of experiences that encourage curiosity, invite physical and sensory exploration, and provide opportunities for them to evaluate and change their environment. Research shows that even very young children are capable of making decisions about their surroundings and enacting change when supported by peers and adults, (Owens, 2003). Personal involvement is one of the best ways of developing values and is a powerful motivator for learning. Geography, because it taps into a range of skills and different ways of thinking, is uniquely placed to develop this. When children begin to value themselves and others and make meaningful connections about their actions within physical and social spheres both near and far, then they truly begin to make sense of their place within the world.

References

Browne, E. (1997) *Handa's Surprise*. London: Walker Books.

Catling, S. (2003) 'Curriculum Contested: Primary Geography and Social Justice', *Geography*, 88, 3 pp. 164-210.

David, T. (1999) (ed) *Teaching Young Children*. London: Paul Chapman.

DfEE/QCA (2000) *Curriculum Guidance for the Foundation Stage*. London: DfEE/QCA.

DfEE/QCA (1999) *Geography: The national curriculum for England key stages 1-3*. London: DfEE/QCA.

Geographical Association (2003) *Making connections: Geography in the Foundation Stage – a position statement from the Geographical Association*. Sheffield: Geographical Association.

Hart, R. (1997) *Children's Participation: The theory and practice of involving young citizens in community development and environmental care*. London: UNICEF/Earthscan.

Matthews, M. H. (1992) *Making Sense of Place: Children's understanding of large-scale environments*. Hemel Hempstead: Harvester Wheatsheaf.

Owens, P. (2003) 'Environmental Education and the Eco-school Experience: Practice and evaluation' in Hills, P. and Man C.S. (eds) *New Directions in Environmental Education*. Hong Kong: Centre for Urban Planning and Environmental Management, pp. 76-92.

Scoffham, S. (1999) 'Young Children's Perceptions of the World' in David, T. (ed) *Teaching Young Children*. London: Paul Chapman, pp 125-138.

Siraj-Blatchford, I., Sylva, K., Muttock, S., Gilden, R. and Bell, D. (2002) *Researching Effective Pedagogy In The Early Years*. DfES Brief 356. London: DfES.

PASSPORT

I look like this

Photo | Kathy Alcock and John Collar.

IN THIS CHAPTER YOU WILL FIND KEY IDEAS ON
CITIZENSHIP • GOOD PRACTICE • NATIONAL CURRICULUM • PROGRESSION • TRANSFER AND TRANSITION

Understanding and developing primary geography

'[Geography] is learning about the world and the environment.'
[10 year old]

'[Geography] is education about map work and learning how our world is getting destroyed, and understanding how big the world is.'
[10 year old]

What is geography?

Geography is part and parcel of our personal survival kit. It is a way of looking at the world that focuses our learning on what places and the environment are like, why they are important to us, how they are changing and how they might develop in the future. To make sense of the features and layout of our immediate and the wider world, we map it, both to see where things are and to help us understand how it is organised. Recognising spatial patterns helps us to understand the variety of natural and human processes at work in the environment. It also enables us to plan ahead, whether to devise routes for travel or to reorganise parts of the locality.

But such knowledge and understanding are only one part of geographical awareness. Of equal importance is how we feel about different places. Places have meaning for us; they are where we are, not just where we reside or go to school, to play or to work, but where we feel 'at home' or 'out of place'. We relate to places, and this relationship is a key element of our personal identity.

We also have a wider concern for the world around us. We have a clearer understanding today about how we are affecting the environment than ever before. In our concern to improve our own lives and the lives of others, we have begun to realise that using resources wisely, managing the natural environment and repairing inadvertent damage is essential for our future. Geography encompasses all of this. The idea of the global citizen who realises the interplay and interdependence of the local and the wider world and who argues for responsible action in day-to-day life lies at its heart.

Traditionally geography has been defined as 'describing the Earth'. For each of us, today, geography:

- helps us to know where we are and know what is there, giving us a sense of location;
- develops our understanding of the environment, the natural world, modified landscapes and the social environment;
- provides a vocabulary to describe the features of the environment and the processes that shape it;
- enables us to understand the spatial layout and organisation of the world about us and recognise the spatial distributions, patterns and relationships in the environment;
- takes forward our understanding and appreciation of places, of their importance to us, of their impact on us and of how we can manage and develop them;
- introduces us to the nature, role and value of maps in 'seeing', understanding, interpreting and valuing the world, from the local to the global;
- help us to recognise how changes to places and the environment happen and affect us, both as a result of natural processes and through human activity;

Attitudes and values	Knowledge, concepts and understanding
The aims of geography in primary education are to:	The aims of geography in primary education are to:

Attitudes and values

The aims of geography in primary education are to:

- Stimulate and develop pupils' interest in their own surroundings and in the variety of natural, modified and social environments and conditions around the Earth
- Engage pupils' sense of wonder at the beauty and variety of the world
- Inspire in pupils an awareness of their own place in the world
- Develop in pupils informed concern about human impact on the quality of the environment and places in order to enhance their sense of responsibility for the care of the Earth and its peoples in the future

Knowledge, concepts and understanding

The aims of geography in primary education are to:

- Enable pupils to extend their knowledge about the features, nature and character of their immediate surroundings and more distant environments and places, to support a sense of what it means to live in one place rather than another and as the basis for a framework in which to place information appropriately in a local and global context
- Develop pupils' appreciation of geographical location, links and interdependence and their understanding of social and natural processes in creating and changing geographical patterns and relationships in different landscapes and human activities
- Build pupils' competence in using the methods of and skills in undertaking geographical enquiry and in analysing and communicating geographical information

Figure 1 | *Aims for geography in the national curriculum. Adapted from Catling, 1998; DES, 1986; DES/WO, 1990; DfEE/QCA, 1999a.*

- develops our sense of stewardship for our world, at a local and global scale;
- helps us to recognise and value our interdependence with other people in our own area and the wider world and to appreciate the diversity evident in the world among peoples and environments.

Inevitably, then, primary geography teaching is concerned with knowing about and understanding the Earth, with developing the skills to do this well, and with fostering attitudes and values that enhance people's lives, places and the environment. All of this is natural for younger children, who are curious about their own and the wider world, who want to explore and investigate people, places and the environment, who wish to develop and enhance the skills to do so, and who enjoy and value the variety they see about them and have concerns for fairness and their future.

Including geography in the national curriculum

The importance and value of geographical learning was recognised in the debate leading towards the introduction of the national curriculum in the 1980s. As a result geography was one of the subjects included when the national curriculum was introduced through the Education Reform Act of 1989. The first national curriculum geography programmes of study and attainment targets were introduced in 1991, followed by revisions in 1995 and 1999. From 1996 onwards the Foundation Stage curriculum was also developed for younger children. This included a number of areas of learning one of which, 'knowledge and understanding of the world', relates directly to geography. As a result geography, in one context or another, has become a required component of the school experience of all 3-14 year-old children since 2000.

It is worth noting, then, that the period of formal geographical learning extends for eight years during nursery and primary school, with only three more years at secondary level (though a sizeable minority of pupils might take it for another four years). This gives early years and primary school teachers considerable opportunities to enhance pupils' geographical learning. Central government has provided two particularly useful documents on how to plan and resource units of work at this level: *Curriculum Guidance for the Foundation Stage* (DfEE/QCA, 2000a) and *Geography: A scheme of work for key stages 1 and 2* (DfEE/QCA, 2000b).

The aims of early years and primary geography

From the outline above, it can be seen that the study of geography in the early and primary years:

- helps pupils to make sense of their own experience and personal geographies;
- introduces pupils to and extends their awareness, knowledge and understanding of the wider world;
- develops pupils' spatial awareness and understanding;
- develops pupils' locational knowledge and understanding;
- develops pupils' knowledge and understanding of what places and environments are like, why they are like they are and how they are changing;
- fosters pupils' appreciation of the environment and of the Earth as their home, and helps them to understand why sustainability is important;
- encourages pupils to be thoughtful about how their own decisions affect their lives and the lives of others, including those they will never know.

These intentions provide the basis for the aims for primary geography set out in Figure 1. They emphasise the importance of undertaking rigorous geographical investigations and enquiries. They also indicate clearly that national curriculum geography is not only about being informed and having understanding about the world; it is also about fostering particular values and attitudes to the ways we treat our place, our planet and the lives of others.

The focus of national curriculum geography

Three key elements stand at the heart of geography in the early years and primary curriculum:
1. an understanding of places;
2. an understanding of the natural and social (or human) environment; and
3. an understanding of spatial organisation.

These different elements interact with each other and are underpinned by a fourth component – geographical skills and study methods. Without the gathering, examination and evaluation of data there can be no understanding of the interactions between place, environment and spatial organisation.

Understanding places

The study of places is concerned with what places are like, how they came to be as they are, how we use them, how they might change and what we feel about and think of them. Developing the idea of place is not simply about exploring one place – our own – but involves comparing a number of places, from within the UK and across the world. For younger pupils the focus is on small areas which they can examine in detail. Through their studies, particularly of their own places, pupils will begin to develop their sense of place and personal identity. In addition, they will start constructing cognitive or mental maps of the locality, country, Europe and the world as a basic locational framework to which to relate other places they encounter throughout life.

Understanding natural and social environments

We cannot look at places without looking at their environment. The word environment is frequently used simply to refer to the natural world but for geographers it has a wider meaning. As well as looking at natural features such as rivers, mountains, seas and coasts, geographers seek to understand the physical processes that created them. It is also important to recognise that human activity has modified the environment to a considerable extent. For example, farming, industry and settlements now cover much of the natural habitat and landscape of the UK. Examining social forces is another aspect of geographical work. People have an impact on places through work and leisure activities. What they think about the environment – their perceptions and feelings – affect the decisions that they take. Geography seeks to examine why we might want to preserve natural environments or revive derelict sites and how we might use natural resources more sparingly and sustainably. Inevitably, this will involve studying controversial issues and recognising and weighing up different viewpoints. Pupils will also need to explore their own attitudes and values, considering where they stand on some issues and looking at what they might do to bring about improvements. Essential to this is the recognition that people do not always do things in the same way in different parts of the world. Culture plays a role. Where you live, what access you have to resources, how you are treated, and what your society values will all play a part in how you respond. Fundamentally, geography examines the interplay within and between the social, modified and natural environments to help children learn how to contribute to their own community and to the world at large.

Photo | *Tina Horler.*

Understanding geographical space

To understand our environment we need to understand its spatial organisation. Knowing where something is, either in relation to other things or on an abstract grid, is a key geographical idea. This is why maps are so important. Modern photographic and computer technologies use satellite-based global positioning systems (GPS) to pinpoint just about anywhere on the Earth's surface. By mapping locations and features geographers can identify distribution patterns which provide clues about the character of different places. As well as identifying land use patterns in cities, towns and agricultural areas, space satellites provide information about weather patterns as they develop and move and so help us to make predictions. In all these ways, spatial and temporal information about the environment helps us to understand how the world works.

Using methods of study in geography

Pupils need to learn how to read, interpret, use and make maps, not least as a life skill but also as a key tool in appreciating the nature of the environment and the processes that shape it. At its heart geography is about the world 'out there'; it does not exist without a wide variety of information gathered from the environment. This can only be obtained by going outside. Fieldwork provides the raw data, the observations, the sampling, the measurements, the photographs, the sketch maps and the conversations that provide the basis for geographical understanding.

Photo | *John Halocha.*

Pupils need a clear sense of purpose and the capacity to ask focused geographical questions if they are to use techniques well and know when the evidence is useful. Thus, at the core of environmental observation and data gathering is effective enquiry. For example:

- What is it?
- Where is it?
- What is it like?
- How did it come to be like this?
- How is it changing?
- How might it change and what are the alternative possibilities?
- What impact is change having or might it have?
- What different viewpoints and opinions are there, and what do I think and feel about this?
- What, realistically, might, could or should be done next?
- What do I think, feel and/or do?
- How does it compare to other similar or different examples?

To use and respond to such questions does not just require gathering information at first-hand; it will often be essential to use secondary sources as well. Maps, photographs, drawings, charts, tables and diagrams are some of the resources available. But however the information is collected it will need evaluation. As well as describing what environments and places are like geography also involves making considered judgements about current processes and causes and indicating what might happen next based on the balance of probabilities. Furthermore, geography also looks at what it might be best to do and, thus, how the environment and places might be managed.

The requirements of national curriculum geography

The early learning goals and the national curriculum identify the range of geography to be studied across the early years and primary curriculum (DfEE/QCA, 1999a, 1999b). To illustrate the connections between, the continuity in and the potential for progression across the varied elements of pupils' geographical experience in the Foundation Stage and key stages 1 and 2, the geography requirements are outlined in Figure 2.

In the Foundation Stage pupils should be encouraged to develop their observational skills and to find out more about features in the school and the immediate area as part of their growing awareness of the world. They need to find out about people and physical features. Their observations and enquiries should be built around their personal experience and they should be encouraged to express their own views and opinions. The information that they encounter about other peoples and places will form the basis for discussion and examination of the wider world beyond direct experience in the nursery and reception class. Learning an initial vocabulary about the environment is a key aspect of this experience.

In key stage 1, pupils' understanding will be developed through more structured investigations of the local area and a contrasting place in or beyond the UK. Learning about people and the way the environment is changing will continue to be important. Pupils will also develop their awareness of the world beyond the local area using atlases and maps to

Aspect of geography	Foundation Stage	Key Stage 1	Key Stage 2
Knowledge, Skills and Understanding			
Geographical Enquiry	Ask questions, observe, find out about and identify Talk about those features they like and dislike	Ask geographical questions Observe and record evidence Communicate in different ways Express personal views about people, places and the environment	Ask geographical questions Collect and record evidence Analyse evidence and draw conclusions Communicate in ways appropriate to the task and audience Identify and explain different views that people, including themselves, hold about topical geographical issues
Geographical Skills	Use everyday words to describe shape, size, position Observe, find out about and identify features in the place they live	Use geographical vocabulary Use fieldwork skills Use globes at a variety of scales Use plans and maps at a variety of scales Make plans and maps Use secondary sources of information	Use appropriate geographical vocabulary Use appropriate fieldwork techniques and instruments Use globes and atlases at a variety of scales Use plans and maps at a variety of scales Draw plans and maps at a range of scales Use secondary sources of evidence, including aerial photographs Use ICT to help in geographical investigations Develop decision-making skills
Knowledge and Understanding of Places	Find out about their environment Begin to know about their own cultures and beliefs and those of other people	Identify and describe what places are like Locate and describe where places are Find out how places have become the way they are Identify how places are changing Describe how places compare with other places Identify how places are linked to other places	Identify and describe what places are like Describe where places are Locate places and environments studied and other significant places and environments. Locate named features on maps of the British Isles, Europe and the World Explain why places are like they are Identify how and why places are changing and how they may change in the future Describe and explain how and why places are similar to/different from other places in the same country and elsewhere in the world Recognise how places fit within a wider geographical context and are interdependent
Knowledge and Understanding of Patterns and Processes		Make observations about where things are located and about other features	Recognise and explain patterns of individual physical and human features
Knowledge and Understanding of Environmental Change and Sustainable Development		Recognise changes in physical and human features Recognise changes in the environment Recognise how the environment might be improved and sustained	Recognise some physical and human processes and explain how these can cause changes in places and environments Recognise how people can improve or damage the environment Recognise how decisions about places and environments affect the future quality of people's lives Recognise how and why people may seek to manage environments sustainably, and identify opportunities for their own involvement

Aspect of geography	Foundation Stage	Key Stage 1	Key Stage 2
Breadth of study			
Scale and Range of Study of Localities and Themes	Look at places and environments at an immediate and local scale	Find out about places and environments at a local scale	Investigate places and environments at a range of scales: local, regional and national, in a range of places in the UK, European Union and other parts of the world
Fieldwork	Observe outdoors	Carry out fieldwork investigations outside the classroom	Include fieldwork investigation outside the classroom
Localities to Study	The school grounds and the school's immediate vicinity	The locality of the school	A locality in the UK
		A UK or overseas locality that has physical and/or human features that contrast with those in the locality of the school	A locality in a country that is less economically developed
Geographical Themes to Study			Include a range of places and environments in different parts of the world
			Water and its effects on landscapes and people, and the physical features of rivers or coasts and the processes of erosion and deposition that affect them
			How settlements differ and change, why they differ in size and character, and an issue arising from changes in land use
			An environmental issue, caused by change in an environment, and attempts to manage the environment sustainably

Figure 2 | Curriculum 2000: Geography in the Foundation Stage and KS1 and 2. Source: DfEE/QCA, 1999a, 1999b.

Photo | Tina Horler.

Figure 3 | Progression in geographical enquiry from the Foundation Stage to year 6.

Foundation Stage	Years 1 & 2	Years 3 & 4	Years 5 & 6
Talk about observations of features in the immediate environment and in photographs	Make observations about features and activities in the local environment and in secondary sources	Give reasons for some observations about places and the environment from direct experience and secondary sources	Explain reasons for making observations about places and the environment from direct experience and secondary sources
Answer questions about features	Respond to questions about places and environmental topics	Respond to geographical questions, for instance, why might we find shops grouped together in towns?	Respond to geographical questions, for example, about the relationship within river processes
Ask questions about what they see around and in pictures to gain information	Ask questions about places and environmental topics	Ask geographical questions, such as why are there more vehicles using this street?	Suggest suitable geographical questions for investigation, such as, Where would a new shop best be situated?
Talk about how they find out information	Discuss ways to carry out an enquiry	Offer own ideas for planning the enquiry, e.g. for a fieldwork investigation	Plan a geographical enquiry, for instance, about a particular locality
Sometimes collect and record observations and information	Collect and record evidence	Collect and record evidence, for example through a survey	Collect, sift and record appropriate evidence
Talk about what they have observed about the place or environment	Use evidence to describe what there is or what happens in a place or environment	Begin to analyse evidence and draw conclusions	Analyse and evaluate evidence to draw conclusions about the topic of and questions in the enquiry
Talk about what they think about the place or environment	Express views on features in the environment	Identify their own views about geographical topics Begin to recognise that others hold views about places and the environment	Identify and explain the various viewpoints held by different people about places and the environment. Begin to appreciate the arguments put forward for viewpoints and that different people may hold opposing views sincerely
Extend vocabulary to describe features and activities in places and the environment	Extend vocabulary to describe features and activities in places and the environment Begin to use appropriate geographical vocabulary	Extend further their use of appropriate geographical vocabulary	Extend further their use of appropriate geographical vocabulary, including the use of more technical terms
Talk about what they have seen and found out	Communicate findings in different ways	Communicate findings as if to an audience and appropriate to the task	Communicate findings in ways appropriate to the topic of the enquiry and as if to an audience
Talk about what they liked or did not like about finding out	Discuss what they have learnt from the enquiry and if they liked doing it	Evaluate the outcome(s) of enquiries and investigations and what they have learnt about undertaking an enquiry	Evaluate and reflect upon the outcome(s) of the enquiries and investigations, to consider both the skills, knowledge and understanding that they have gained and what has been learnt about undertaking an enquiry that can be applied in their next geographical enquiry

find out about places. In observing the environment pupils should consider ways to look after and to improve it. This will develop their sense of environmental care and initiate an early understanding of sustainability. In undertaking these studies, younger pupils will also begin to think about why some parts of the environment change and others do not, and about the locations of different features and activities they see. Pupils should be encouraged to use and extend their growing 'geographical' vocabulary as they explain their ideas.

Encountering Geography: The Foundation Stage

Pupils should:

Observe and find out about features and activities in the nursery/school and local streets;

Talk about what they like and dislike about their immediate environment;

Use pictures and discussion to develop awareness of other peoples, environments and places;

Talk about the lives of other people;

Play with environmental toys, and listen to stories about what people do in a variety of places and environments.

Geographical Awareness: Years 1 & 2

Pupils should:

Describe features and activities in the local area and other environments and what places are like;

Use enquiry questions and a variety of sources to find out about localities near and far;

Examine people's lives in places, and their cultures;

Look at similarities and differences between places;

Express views about places and environmental matters;

Observe that environments change and find out how they can be looked after.

Geographical Engagement: Years 3 & 4

Pupils should:

Contribute to planning their geographical enquiries and how they can carry them out;

Examine the location of features and activities and patterns they create;

Consider how some changes in places and the environment are caused, and look at the impact on people's lives;

Look at what settlements are like, using a locality as one example among others;

Study how people are affected by and can affect the environment;

Consider how people can improve the environment.

Geographical Involvement: Years 5 & 6

Pupils should:

Plan, undertake and share enquiries using a variety of resources;

Analyse ways in which human and natural processes create and shape localities and aspects of the environment, such as rivers or coasts, and affect locations and geographic patterns;

Compare how places are changing, and consider the effects of change;

Explain the importance of the interdependence of peoples and places;

Study how decisions affect places and environmental issues, and appreciate the reasons why different people hold different views on issues;

Identify and justify ways in which they can be involved in sustaining environments.

Geographical skills

Geographical skills are to be threaded through all years. Children need to develop experience and skills that support their development of:

- the use of photographs of all sorts - elevation, oblique and vertical - and at a range of scales;

- the use of maps of all sorts and at a range of scales, from playmat maps and large scale plans to atlas and wall maps;

- the use of various secondary sources including information books and texts of all sorts, the visual media, and ICT (eg CD-roms, websites);

- the development of vocabulary, from everyday words and phrases to geographical terms;

- the use of fieldwork, to draw on the key geographical primary source.

In key stage 2, these burgeoning understandings are taken further. Pupils' ability to initiate and organise enquiries grows as they begin to gain a deeper understanding of geographical questions. Examining localities, now perhaps at the scale of the school catchment area, in the UK, Europe or the less economically developed world, provides a basis for comparisons. Pupils should also look at specific aspects of the human and physical environment, developing ideas about settlements and water and their role in people's lives. This will provide the opportunity to examine the location and distribution of features and events and to look at patterns in the environment, as well as at the processes at work shaping settlements and rivers or coasts. In doing so, pupils will discover how people affect and manage the environment, damaging, sustaining or improving it through the decisions they make. Through their encounters and studies of places and environments around the world pupils will build up their locational knowledge of places. Building on their descriptions, pupils will develop their analytic and evaluative skills and increasingly use more precise 'geographical' vocabulary in their studies.

This outline indicates a potential progression from the nursery to the end of the primary school. It involves moving from the local outwards in increasing depth

By the end of key stage 2 most pupils should be able to:

■ 'explain the physical and human characteristics of places, and their similarities and differences;

■ know the location of key places in the United Kingdom, Europe and the world;

■ explain patterns of physical and human features;

■ recognise how selected physical and human processes cause changes in the character of places and environments;

■ describe how people can affect the environment and explain the different views held by people about environmental change; and

■ undertake geographical investigations by asking and responding to questions and using a range of geographical enquiry skills, resources and their own observations'.

Figure 5 | *Expectations for achievement in geography by the end of year 6. Source: QCA, 1998, p. 13.*

from the earliest years. From asking rather general and, indeed, random questions about places and the environment, pupils build up a sense of effective geographical questions and, with good guidance, become more selective and thorough in their enquiries. Figure 3 illustrates such a progression in the development of geographical enquiry from the Foundation Stage to the end of key stage 2. This involves using increasingly precise geographical vocabulary, as well as moving from specific examples to more abstract ideas and concepts. Over time, pupils extend their information base, and can better make comparisons and identify possible ways to improve the environment and consider reasonably how they might act to do so. Figure 4 presents one way in which this progression in geographical learning could be structured. It outlines the way in which the focus of geographical studies might evolve from the Foundation Stage to the last two years of key stage 2. It also provides a basis for identifying the emerging geographical understanding which most primary pupils should have when they move on to secondary school and to study geography at key stage 3. This is outlined in Figure 5, as a summative statement for pupils' geographical learning in primary education.

Interconnectedness in geography teaching

No aspect of geography can really be taught in isolation. In planning geography teaching, it is important to consider geographical methods at the same time as geographical content. When pupils study their locality, for example, they will talk about what people do there and what the place means to them, examine different physical features and decide on suitable enquiry questions and the mapwork and fieldwork skills they will employ. The working group that devised the national curriculum always envisaged that schools would integrate skills, places and themes in their geographical studies (DES/WO, 1990). This inter-relationship can be expressed visually using a cube to represent the curriculum (Figure 6). One of the advantages of this model is that it provides a unifying framework. At the same time it is flexible enough to match different physical environments and a range of planning strategies.

It is also important to recognise that geography makes strong links with other curriculum areas. The study of water, soils and rocks in science, for instance, is integral to an appreciation of how rivers shape the landscape. Equally, recording the lie of the land in art or representing the seashore in music provides another dimension to geographical investigations. More than any other subject, geography serves to bridge and unify different curriculum areas (Figure 7). You can use these overlaps to your advantage when it comes to finding time for geography in an over-crowded timetable.

Moving nursery and primary geography forward

Reflections on the current state of geography in schools

During the lifetime of the national curriculum there has been a steady improvement in the quality of teaching of geography in the early and primary years. By the beginning of the twenty-first century geography teaching was at least satisfactory in over 90% of schools, and

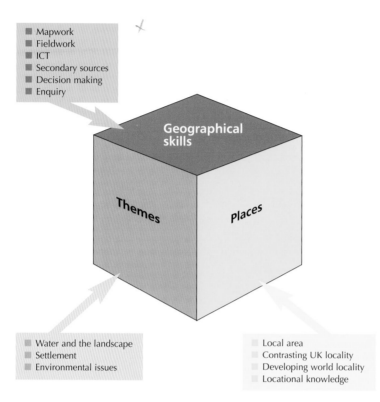

- Mapwork
- Fieldwork
- ICT
- Secondary sources
- Decision making
- Enquiry

Geographical skills

Themes

Places

- Water and the landscape
- Settlement
- Environmental issues

- Local area
- Contrasting UK locality
- Developing world locality
- Locational knowledge

Figure 6 | *Geographical studies involve combining skills, places and themes.*

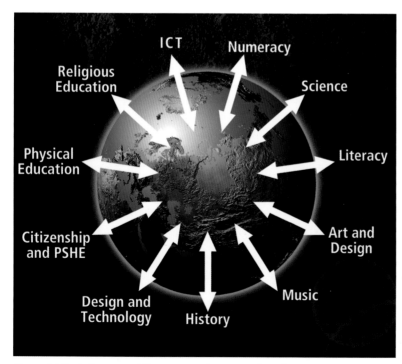

Figure 7 | *Exploring many different curriculum areas through geography. Source: Geographical Association, 2003.*

it was good or better in about 35% of schools. While these are not by any means strong figures, they nonetheless indicate that considerable strides have been made since the late 1980s when geography was only satisfactory or better in just 25% of schools (DES/WO, 1989).

If we look at the nature of the good and better teaching of geography a number of characteristics emerge (Catling, 2003a). One is that pupils should be able to relate their own experience to the geography they are studying, which they find stimulating and which engages them. Pupils seem to respond well to teachers who convey confidence in their geographical understanding and who can support pupils' learning with well-focused guidance. Such teachers tend to:

- be clear about the geographical intentions of units of work and lessons;
- use a variety of creative and engaging teaching methods and activities;
- differentiate the tasks while ensuring suitable challenge;
- make use of appropriate vocabulary;
- employ demanding questioning techniques;
- involve the pupils in debate;
- support work with effective resources; and
- use formative assessment proactively to respond to pupils' needs in lessons.

Rather than being isolated, effective geography lessons link to long- and medium-term planning for geography in the school. There is a clear structure and progression to the units of work from the Foundation Stage to the end of key stage 2, with teachers building on the work of each other, enabling progression in experience and learning to occur. In such schools, a fair and reasonable amount of time (in relation to other subjects) is given to geography and pupils can undertake independent research and follow up personal interests and lines of enquiry. Geography lessons are also, in part, linked to topical occurrences and issues.

Good practice involves encouraging pupils to relate and synthesise information, to analyse and explain their findings, and to evaluate and indicate possible solutions to problems in places and the environment. They also need to look for similarities and contrasts in what they study. This occurs in a school which values geography, as it does other subjects in the curriculum; looks for the appropriate links between subjects and makes good use of them; and ensures that suitable resources are available for pupils to use. It also occurs in schools where the senior management supports the geography subject leader in monitoring and developing geography across the staff. Such schools are not dependent on a single good teacher of geography; the quality and consistency of pupils' geographical learning is the result of a whole-school commitment to enabling pupils to do well in their learning.

It is a matter of some concern that only a minority of nursery and primary schools meet these standards at present. Yet there is the foundation in current geography teaching on which to build. Almost all primary schools have got a geography scheme of work in place, and very many nursery schools have included exploring the world about the pupils in their range of activities and development for the youngest pupils in schooling. Much use is made of the local environment. While it is not always easy to organise, fieldwork is reasonably evident, if not frequently organised. When it happens, it is clear that pupils are strongly stimulated in their geographical learning. Much use has been of locality packs to support the study of more distant places in other parts of the world, and the study of environmental sustainability can be done well through enquiries into the school grounds and local issues. However, much of geography teaching in primary schools lacks challenge and stimulus (Ofsted, 2003), with an emphasis on descriptive activities rather than an engagement in analysis and explanation. There seems to be an inappropriate use of commercial worksheets to show that geography has been taught, rather than a real involvement of the pupils in looking at real places, environments and events, drawing on the news and topical concerns. There are two main needs. First, to raise the expectations that teachers hold of pupils in their geographical learning. Second, to build in teachers the confidence to provide stimulating, imaginative and challenging activities in geography topics. At the core lies the need to take forward more effective curriculum planning and classroom teaching, underpinned by focused and appropriate formative and summative assessment. We need to widen the approaches to the geography curriculum and draw on a more contentious range of topics to move beyond descriptive reports to analysis, explanation and evaluation. There are links here with the development of active geographical citizens.

Going beyond description: developing analysis, explanation and evaluation

Observing and describing places accurately provides a vital foundation for the development of geographical learning and understanding. However, pupils need to move beyond the initial gathering of information in order to appreciate different viewpoints, make judgements and put

Thinking Skills	Examples with a geographical focus
Information-processing skills	Observing in the environment the feature/scene in a photograph Identifying features on maps using pictogram and abstract symbols Reading the landscape/urbanscape of an area through a synthesis of the symbols on a map Reading information from graphs and charts showing, e.g. traffic data, rainfall, temperature Reading information from thematic maps showing, e.g. building and land use, the distribution of shops, etc. Reading cross-sections showing, e.g. river-bed depth
Reasoning skills	Drawing sketch maps to show, e.g. the key features of an area, the distribution of specific features such as houses Drawing diagrams to show, e.g. the sequence and relationship in river and coastal processes Drawing a series of pictures to show, e.g. changes in an area over time Using a set of photographs, e.g. of different sorts of scenes, sort them into urban/rural, view/detail, particular features
Enquiry skills	Examining maps, photographs, sketches, diagrams and charts to generate questions, e.g. about an environmental issue Selecting appropriate graphical resources from which to gather information, e.g. for a locality study, including maps and photographs Making drawings and sketches to show the equipment and processes used in an investigation, e.g. of river processes Drawing maps, pictures or diagrams to show, e.g. a prediction of the possible outcome of an investigation into wind direction and sheltered areas in the school grounds
Creative thinking skills	Creating models and drawings to show, e.g. how a new development might look in the local area on a piece of derelict land Given various resources, creating, e.g. a particular type of home or shop in the home corner Making a poster to show, e.g. ways to tackle an environmental problem, such as litter disposal Create a story-board sequence to show, e.g. the day in the life of a child in a particular community Using a map as a basis, create a wall painting, e.g. of how an area might have looked in an earlier period
Evaluation skills	From a range of published visual sources create a collage of helpful and unhelpful visual images, e.g. of the way places are shown in various travel brochures Use graphics to show what is felt about, e.g. a particular environmental concern, such as 'smiley' or 'dismayed' faces Draw cartoons to show, e.g. views about a decision in a role play related to change in a community Create a visual set of criteria, e.g. to use to judge the 'pollution', cleanliness and care for a street or park

Figure 8 | Some examples of geographical approaches for developing thinking skills. Source: Catling, 2002, p. 7.

forward proposals for improvements. The best learning occurs when pupils are encouraged and challenged to offer their own explanations for phenomena they have observed or identified. In these situations they begin to develop the key concepts which stand at the heart of geography – location, pattern and process – and apply some of the principles and ideas which are unique to geography. If we wish to promote higher order skills of this kind it needs to be featured in our planning. Figure 8 illustrates some ways in which the skills being promoted across the national curriculum can be developed in geographical studies.

Widening awareness of the world

It is crucial that pupils of all ages explore both their known world and the world beyond their experience. The local area is a vital part of geographical study as it grounds pupils in their direct experience. However, geography also involves developing an understanding of more distant people and places and forming an informed appreciation of the planet on which we live. This means that in their work on geographical themes and localities pupils need to encounter places at a range of scales. This needs to be planned for, using carefully selected examples as well as incidental and topical events as they occur in the news and even from the pupils' and your own direct experience.

Widening pupils' sense of the world also involves moving beyond the comfortable to examining some of the problems and difficulties that affect people, places and environments. This may involve local change and developments about which people feel strongly, or problems that people face in other, less-well-off parts of the world. On a local level, a study of the school playground might lead to proposals to make it a more varied, attractive and stimulating place. On a global level, finding out about the causes and impact of poverty on

Figure 9 | *Some possible topics for geography units of study in the early years and primary geography curriculum. Adapted from Catling, 2003b.*

Geography topic	Focus of study
Exploring the school environment	To consider who the school is for, and how and why it is used in this way; school as friendly or fearful; inclusion/exclusion, dark corners, open spaces, crowds and quiet areas, signs and symbols; designing the school space as place; school as a community. Examine the features, spatial layout, use of school building and grounds; look at activities undertaken in school, why where; consider benefits and problems in way schools are laid out, structured, regulated; access, impact of features, layout; examine issues in school and how they relate to the nature of the place, such as peer and adult relationships, access to space, environmental quality, change; ideal school layouts, functions, organisation; proposals for improvement.
Playground and play area space (in and/or out of school)	To examine the use of playground space, included and excluded children, changing use of space; who uses the playground, multiple uses and users of the playground; places in the playground in the mind; creating playground landscapes; fair uses of space, timing; rules for space, activity and relationships; playground traditions; changing playgrounds.
Limits to mobility (in and/or out of school)	To understand the environment experienced by those with mobility difficulties, young, old, disabled, parents of toddlers; feelings of exclusion; inhibitors to movement and access; approaches to improvement; quality review of the environment.
My locality	To explore the local area through children's eyes and experience; what they value; places used and avoided, reasons why; variations between children; what to improve or set aside for children, how to leave it; use of local facilities and services; concerns and hopes for the future; what other and younger children might want of the area; proposals for the future.
Taking care in the local area	To work with a local group to care for, improve and manage an area; contact with local environment groups, why conservation and environment groups do what they do, people's attitudes to and concerns for the environment, why act; examining how children and adults are involved in environmental management and sustainability in other places nationally and globally; work on a particular environmental activity; examining the outcomes and consequences of actions.
Water for people	To investigate water as a fundamental human need; ways to access water, water use, water control, excessive and limited water access/use; people's experience of water access, variations in access, attitudes to water plenty and scarcity; water as a source for agriculture, industry, energy, leisure, environmental quality, household activities, sanitation; water in the community; sources of water; control of water; water aid, community development and aid, personal responses to water need and use; water in the future.
The impact of poverty: far and near	To explore the nature and impact of poverty in different contexts around the world; local poverty issues, poverty in different parts of the world; the effects and impact of poverty; causes of poverty, how poverty is maintained and exploited; views of poverty; attitudes to and feelings about people in poverty; poverty and its impact on the environment; ways to tackle poverty, aid and poverty; who benefits from poverty; creating and overcoming poverty.
Getting into place	To study particular localities in non-UK areas of the world; children's lives in communities and neighbourhoods; the questions to ask, the resources to use; how do different people use places, why these variations; what are people's views of, interests in, attitudes to, concerns for, hopes about the future of their locality; local features, facilities and amenities; the impact of change; local features.
A fairer world	To investigate the issues of fair trade and work; study of a particular product from source to home, e.g. bananas, trainers; employment, work practices; who is involved, their contribution, benefits, losses; trade control, who wins/loses; consumer choices, place of purchase; being a consumer, making choices; commitment to fair trade; attitudes to and use of income, expenditure, personal practices, willingness to change.
Future world	To consider and propose how the local area and other places and environments might become; how we would prefer them to be; personal and group ideas for the future for ourselves, the local community, for others and globally; examining the role of values, attitudes and preferences in thinking about the future; possible impacts of ways the world locally and globally might change and be; alternative scenarios, positive and of concern; the role of and difference between prediction and proposal; how we might change one particular aspect of our home area and of another place.

Photo | Paula Richardson.

people's lives in another part of the world may cause pupils to reconsider, and perhaps modify, the way they live their own lives. Figure 9 illustrates a number of such possible topics.

Active learning for geographical understanding

Geography studies the world as it is today. It is based on what happens in the world directly around us and further afield. We can keep abreast of what is happening locally through our use of personal information, through what we hear from friends, from the local press and radio and television and through fieldwork. We are aware of what is happening in other parts of the world through similar media reports. To these we can add the increasing use of the internet as a source of local and distant knowledge. Geography teaching not only needs to focus on the world about us as it is now, but it needs to recognise that the study of the day-to-day world requires an active approach rather than the passive study of pre-digested information. It also involves critical examination, not unquestioning acceptance.

Pupils are enthused by tasks that provide challenge, use their experience, draw on their understanding, have practical dimensions and involve them actively. Such activities should reinforce and extend their understanding, through new contexts and examples as much as through new ideas and content. There should be variety in the tasks used and they should challenge pupils' thinking, however incrementally.

Geography develops positive appreciation of places and the environment. As such it links strongly to aspects of citizenship education. While the citizenship guidelines are not statutory for key stages 1 and 2, geography, as it always has, clearly contributes to the development of citizenship. There are a number of overlapping elements in the citizenship recommendations which clearly contribute to the development of geographical awareness, understanding and appreciation. These are described in greater detail in chapter 21.

Resourcing geography teaching

Geography is a resource-rich and resource-needy subject. Focusing as it does on the world today, it is essential that the materials we use are up-to-date. Commercial publications provide information about other places but they date very quickly, indeed they begin to date the moment they appear. This makes more immediate resources, such as the news reports on radio and television, newspapers and the internet, all the more important. Pupils need to recognise the limitations of such current, almost ephemeral, material. It may be inaccurate, selective and even contain bias or prejudice, depending on the source. These same concerns need to be kept in mind when selecting published materials for schools, whether CD-Roms, textbooks, information books, photo packs, or video programmes.

Although pupils can create some resources themselves, and many can be obtained very cheaply, or for free, some resources must be purchased. Investment is important. It is essential, therefore, that proper financial budgeting takes place in school. Ordnance Survey maps, atlases, reference books and CD-Roms can be expensive. The budget, therefore, needs to take account of the need to update geography resources annually.

Connecting with key stage 3 geography

Early years and primary geography provides pupils not just with an introduction to geography but with a variety of opportunities to develop their geographical awareness, knowledge, understanding, skills and values. Geography is part of their curriculum for between six and eight years. As pupils move into key stage 3, they will work with teachers whose specialism, or one of whose teaching subjects, is geography. There are overlaps between the key stage 2 and key stage 3 programmes of study; for instance, the four key aspects of primary geography, as well as enquiry and skills (see Figure 2), are all developed further. More specifically, examples of geographical development include increasing the scale at which pupils are studying places and regions, moving from the national to the global, extending studies of water and land/coastscape into geomorphological processes, deepening investigations into settlement and environmental issues, looking more extensively at sustainable development, and making greater use of Ordnance Survey maps (DfEE/QCA, 1999c).

There are a number of ways in which primary schools can support pupils as they move into key stage 3. These include termly meetings between geography subject leaders to share ideas about topics and teaching methods. This will help secondary geography teachers to recognise the importance of the geographical work undertaken in primary schools, and encourage primary teachers to acknowledge the opportunities for pupils to develop their geographical learning in the early years of secondary education.

Conclusion

We are curious about our world, however far we travel in and over it. We are fascinated by its variety, its physical environment, the people who inhabit it, what they do, their cultures and what we make of each other. Geography explores what the world is like, what happens in it, the impact we have upon it, how we can limit the damage we do to it and, indeed, how we can help to make the world a better place. It is a moral imperative that we pass the world to the next generation in as good or better condition than we inherited it and that we equip them to undertake the task of creating a world as fit or better to live in, with less damage and potential harm to resolve than we inherited. In summary, we can say that geography in the nursery and primary school should:

- use and develop pupils' natural curiosity about the world around them, near and far;
- explore and develop pupils' geographical ideas and skills so that they can explore and investigate places and the environment more effectively;
- extend pupils' effective use of language to describe, explain, evaluate and make proposals about places and environments;
- examine and clarify pupils' existing experience and awareness of places and environments;
- develop pupils' existing knowledge and understanding of their own and other places and environments;
- develop pupils' spatial awareness from a local towards a global scale;
- help pupils recognise their interdependence with the rest of the world;
- build positive attitudes towards other peoples;
- help pupils build a global perspective from their local perspective and feed a global appreciation into their sense of the local;

- enable pupils to value diversity in peoples, cultures, places and environments;
- combat ignorance and bias to avoid pupils developing stereotypes and prejudice, and raise awareness of the partiality of our understanding of people and places, such that we may always intend to overcome it;
- foster in pupils the desire to act responsibly themselves and in working with others in order to minimise damage and enhance improvement to wherever they are on the planet, for the good of that area, and for the good of the wider world.

The vital role of geography in the early years and primary curriculum is to support and enable the development of informed, concerned and responsible members of the local and global community, whose sense of wonder, interest in and fascination with the world about them leads to active engagement in sustaining and improving people's lives in their own places, other environments and across the wider world.

References

Catling, S. (1998) 'Geography in the National Curriculum and beyond' in Carter, R. (ed) *Handbook of Primary Geography*. Sheffield: Geographical Association, pp. 29-41.

Catling, S. (2002) 'Thinking Geographically', *Primary Geographer*, 47, pp. 7-9.

Catling (2003a) *The State of Primary Geography*, paper given at the Geographical Association Annual Conference, University of Derby.

Catling, S. (2003b) 'Curriculum contested: Primary geography and social justice', *Geography*, 88, 3, pp.164-210.

DES (1986) *Geography from 5 to 16: Curriculum Matters 7*. London: HMSO.

DES/WO (1989) *The Teaching and Learning of History and Geography*. London: HMSO.

DES/WO (1990) *Geography for Ages 5 to 16*. London: DES/WO.

DfEE/QCA (1999a) *The National Curriculum: Handbook for primary teachers in England*. London: DfEE/QCA.

DfEE/QCA (1999b) *Early Learning Goals*. London: DfEE/QCA

DfEE/QCA (1999c) *The National Curriculum: Handbook for secondary teachers in England*. London: DfEE/QCA

DfEE/QCA (2000a) *Curriculum Guidance for the Foundation Stage*. London: DfEE/QCA.

DfEE/QCA (2000b) *Geography: A scheme of work for key stages 1 and 2. Update*. London: DfEE/QCA.

Geographical Association (2003) *Finding Time for Things that Matter: Geography in primary schools*. Sheffield: Geographical Association.

Ofsted (2003) *Geography in Primary Schools: Ofsted subject reports 2001/02*. London: Ofsted

QCA (1998) *Managing Breadth and Balance at key stages 1 and 2*. London: QCA.

Section 2

Geographical skills

Chapter

What is London like? Busy!

Geographical enquiries and investigations

Enquiry is the process of finding out answers to questions. At its simplest, it involves encouraging children to ask questions and search for answers, based on what they might already know from data sources. As their skills develop, children can move to a more rigorous form of enquiry involving the development and testing of hypotheses' (NCC, 1993).

Learning and enquiry

Would you be surprised if, during an interview, you were asked how you think children learn? It is not an easy question. The most direct response I ever received was 'Somewhere between when I tell them and when they want to!'. What response would you give? And how does your notion of learning relate to your school's geography policy? This may sound like an uncomfortable grilling but these are very important questions.

Over the years teachers have come to recognise that the enquiry approach is a powerful strategy for promoting pupils' learning. Research suggests that meaningful learning occurs when investigations are directed by challenging questions and lead to solutions and/or reasoned viewpoints. Key questions help to guide this process, serving to signpost the paths that lead to understanding. Such an approach can help pupils to manage their own learning, handle issues objectively and use skills with a clear purpose.

The geography national curriculum makes specific reference to the enquiry approach. The programme of study clearly states that at both key stages pupils should be taught to ask geographical questions, observe and record information and express views about their surroundings. The emphasis on enquiry is maintained in the QCA schemes of work where enquiry questions are used to structure individual lessons. Clearly, therefore, those responsible for the curriculum's details were confident that geographical enquiry is a potent strategy for both teaching and learning; so let us investigate it further.

Theoretical background

The enquiry approach has a strong theoretical base. As Roberts (2003) points out, many educationalists follow Vygotsky in arguing that knowledge is not transmitted directly from teacher to pupil; rather pupils learn about the world by actively making sense of it for themselves. Vygotsky's work focused especially on the difference between what children can do on their own and what they can do with support from others. He called this the 'zone of proximal development'.

Vygotsky's work and the notion of the zone of proximal development has implications for geographical enquiries. It challenges the notion of totally independent work, reminds us that pupils will need teacher support when they are working with new ideas, and suggests that enquiries should not be beyond what they can achieve at the time (their zones of proximal

The main characteristics of key questions are that they:

- are an immensely powerful way of planning a curriculum with structure;
- unlock the door to learning about key ideas and processes;
- give direction and structure to a topic, emphasising what pupils are investigating and why;
- are the driving force behind enquiry learning;
- demand enquiry, require investigation and beg more questions;
- may be awareness raising, descriptive, explanatory, speculative, moral/value based or reflective;
- should be shared with pupils;
- provide professional development for teachers generating them;
- stimulate skills development to enable them to be investigated.

Figure 1 | *Key questions.*

development). In practice this means that pupils will need assistance from the teacher both when they are planning their enquiries and when they are conducting and analysing their investigations.

Asking questions

Educationalists have classified the questions posed in the classroom into many different types. As far as geography is concerned, the effectiveness of an enquiry will depend upon the teacher's understanding of key questions and general questioning skills. You might pause to ask yourself the following questions:

- How often are your questions posed clearly and succinctly?
- What proportion of your questions are open and closed?
- Do you deliberately arrange your questions in a logical sequence?
- How are you encouraging pupils to ask their own geographical questions?

It is well worth having a serious word with oneself, and with colleagues, about the use of questions in the classroom and their associated contribution to pupils' learning.

Remember too that asking questions is not restricted to cognitive matters only. Questions that are carefully and sensitively posed can also be effective for exploring the affective domain. Attitudes, beliefs and values are a vital ingredient of geographical studies and the subject offers many opportunities for considering the viewpoints of pupils and of other groups and individuals.

Key questions

Effective geographical enquiry depends initially upon the quality of the key questions posed for the learning experience. There have been many attempts to devise a definitive set of questions for geography. Storm (1989) suggested five basic questions which could be applied in any circumstances:

- What is this place like?
- Why is this place as it is?
- How is this place connected to other places?
- How is this place changing?
- What would it feel like to live in this place?

Notice that, above all, these questions beg more questions and require further investigation before an answer can be attempted.

The characteristics of key, or leading, questions are listed in Figure 1. The origin of key questions must obviously lie in the intended learning outcomes (objectives) of your curriculum and, for most teachers, these will be related to the programme of study for each key stage. Key questions unlock the door to learning about the concepts implicit in the programme of study statements for places and themes. Key questions also stimulate the development of the skills needed for the enquiry.

Stage 1a	Awareness raising
Stage 1b	Generating enabling questions
Stage 2	Collecting and recording information
Stage 3	Processing the gathered information
Stage 4	Drawing conclusions from the processed data
Stage 5	Sharing the learning and affective outcomes
Stage 6	Evaluation by all concerned

Figure 2 | *A framework for managing a focused geographical enquiry*

The best key questions will involve pupils in conducting a sustained enquiry and synthesising information. For example, you might ask them what life is like for the people who live in a specific village as opposed to those who live elsewhere. This question has real scope, unlike one asking how many people live in that village. The latter question can be tackled relatively easily using reference skills and leads to a closed and definite answer. In fact it is subsumed by the first question. To consolidate your thinking, compare the investigative potential of these two questions:

1. When did the Romans occupy England?
2. How do we know the Romans occupied England?

Generally, questions which begin with 'how' and 'why' are more conducive to geographical enquiry than those that begin with 'what' and 'when'.

Progression in enquiry

The precise wording of a key question can be significant. For instance, 'Why are there differences in land use in our local area?' encourages a higher level of geographical thinking and analysis than the question 'How is land used in our area?'. You will need to be aware of progression in geographical enquiry. It has been suggested (CWW, 1993) that teachers should consider the following parameters:

■ the enquiry focus
■ the degree of pupil involvement in planning
■ the range of viewpoints considered in issues
■ the nature of the skills needed for the investigation
■ the range and type of resources

Add the progression in key questions and you have an excellent tool for monitoring this component of the curriculum through the key stage(s).

Photo | *John Halocha.*

Focused geographical enquiry

There is an important distinction between a focused enquiry route (Figure 2) and general enquiry which can permeate the curriculum. In Figure 2, notice how Stage 1a is an orientation phase when pupils relate their previous experiences and existing knowledge to the chosen enquiry. This is followed by a focusing phase (Stage 1b) when initiative-taking begins in earnest and a range of enabling questions is raised as the enquiry process gets started.

Stages 2, 3 and 4 mark the information phase and should be the time of intense and purposeful geographical activity. Factual statements and generalisations emerge during Stage 4 as pupils attempt to answer the key question. Interpersonal skills are practised during the important Stage 5, the outcome of which can influence the responses during Stage 6 when feelings are revealed, including self-evaluation. These last two stages represent the reflective phase. Such a structured and substantial piece of work could be part of topic work or make up the bulk of the study unit.

A well-organised, well-resourced and sharply focused geographical enquiry makes serious learning great fun for your pupils whether you are working in the classroom, school grounds

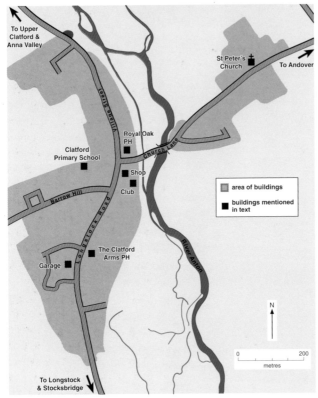

Figure 3 | *The local area of Clatford CE Primary School, Goodworth Clatford.*

or off-site. Procedural skills are practised as the enquiry process moves into action with individuals and learning teams engaging in activity-based tasks. You know it makes sense!

General enquiry

Unlike focused enquiry, a strategy of general enquiry should permeate the geography curriculum in your school. Posing challenging and open questions encourages the investigative and higher modes of thinking in the classroom. However there is a caveat: beware of over-dependence upon this approach; it can become a predictable recipe and restrict the variety of teaching and learning strategies used by the teacher. As one pupil reportedly advised a newcomer to the class: 'Don't look at anything in the room or she'll ask you a searching question about it!'. Many pupils pick up a great deal through direct teacher transmission; how you relate your knowledge and experience can be decisive.

An example of geographical enquiry: traffic problems in a school locality (key stage 2)

Background

Local issues lend themselves admirably to the enquiry approach as they are real and immediate. With this in mind, Clatford CE Primary School, near Andover in Hampshire, decided to investigate with year 6 the traffic problems created in the school's vicinity (Figure 3) by the arrival and departure of pupils. This issue related to the pupils' own school and was one in which they were likely to be personally involved – promising ingredients for a learning experience. Of course, this issue is not unique to the school and its village, Goodworth Clatford.

Approximately 12 hours of curriculum time were allocated to the topic and links with mathematics, ICT and English were recognised and established. Opportunities to revise, practise and learn geographical skills were identified and teacher enthusiasm was channelled into the enquiry process.

Four key questions emerged from discussion of the learning objectives and formed the framework for the enquiry.

Question 1: Why are journeys made?

Question 2: Where do our pupils come from and how do they get to school?

Question 3: What effect does our coming to and leaving school have upon the village?

Question 4: What can we do to ease or solve any problems that may be caused?

RESULTS OF QUESTIONNAIRE; GETTING TO SCHOOL

A total of 117 questionnaires was returned, representing 73% of the current number on roll. There were responses from all six teaching groups and the relatively large sample can be regarded as representative.
The following data was obtained

ORIGINS OF PUPILS COMING TO SCHOOL

Anna Valley	33%	Andover	26%
Upper Clatford	20%	Local village	19%
Chilbolton	1.5%	Longstock	0.5%

MODE OF TRAVEL TO ATTEND SCHOOL

Car 94%   Walk 5%   Bicycle 1%

DIRECTION OF ARRIVAL BY CAR

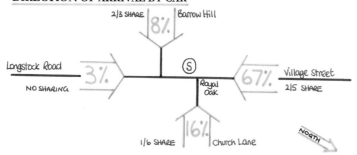

REPORTED TIME OF USUAL ARRIVAL IN THE MORNING
(by car)

0825	2%
0830	12%
0835	14%
0840	24%)
0845	28%) 72% arrive within 10 minutes
0850	20%)

Figure 4 | *Results of questionnaire on 'getting to school'.*

Why are journeys made?

The topic opened with a discussion about journeys, including movements on the Earth's surface, spanning journeys within and beyond the classroom and within and beyond the home: helicopters passing over the school from nearby Middle Wallop; moles tunnelling under the school's playing field; clouds moving across the sky making shadows over the landscape; the River Anton flowing through the village; earwigs moving when a log is disturbed; getting to school and back home again. A matrix logging scale of journeys against, on, under and above the Earth's surface stimulates pupils to think geographically.

Reasons for such movements were discussed through questioning and many pupils began to grasp the following generalisations about journeys.

They:

- are made for a reason and have a purpose;
- can be of different lengths;
- take place at different times;
- use varying modes;
- can have peaks or rush-hours;
- can occur on, above or below the Earth's surface.

It was decided not to include the key idea that routes, for various reasons, often deviate from the most direct path, with main routes tending to be more direct than minor ones.

Where do our pupils come from and how do they get to school?

Following the exploratory session, pupils mapped their journeys from home to obtain groceries, clothing and medical care, one map being produced for each service. The journey to school was then the focus of investigation and the second key question stimulated the crucial enabling questions of 'What do we need to know?' and 'How do we find out?'.

Figure 5 | *Logging the traffic outside the school.*
Photo: Tina Horler.

A simple questionnaire was proposed to obtain the raw data: origin of journey and route followed; mode of transport; possible sharing of transport; and approximate time of arrival at school in the morning. Pupils soon realised the need for a simple format and agreed to observe anonymity by just recording the year group on the returns. In addition to ticking boxes, one section of the questionnaire invited parents/guardians to state their own feelings about taking and collecting youngsters from school (it also informed parents of the work being carried out by year six). The questionnaires were distributed to all year groups with great anticipation and returns were grabbed avidly for processing.

The total number of returns represented almost three-quarters of the pupils and from such a representative sample it was possible to produce a summary which involved mathematical as well as drawing and design skills (Figure 4). One important finding was that seven out of ten pupils usually arrived at school during the ten minutes before the bell.

With at least nine out of ten arriving by car and two out of every three coming along Village Street from the north, this led naturally to a closer look at things in the vicinity of the school gate.

Now came the time to organise a traffic count with a clear purpose in mind and to establish the real scale of the scene. The pupils measured the widths of the roads near the school and calculated the amount of land available for vehicles in the parking area opposite the entrance at the Royal Oak (the landlord offers this facility at dropping and collecting times). The average width and length of a car was also calculated. In addition pupils devised their own recording sheet for logging the arrival and departure of cars and organised coverage of the study area on four consecutive mornings in early March from 0815 to 0850 hours (Figure 5). It was raining the first morning!

Figure 6 presents data collected by pupils on the four mornings. They also reported

Weather	rain	dry	dry/cold	dry
Day and date	Tu 4/3	We 5/3	Th 6/3	Fr 7/3
Total no. of vehicles logged	259	225	339	223
No. coming from Anna Valley direction	114	85	92	101
No. coming from Longstock direction	100	65	195	70
No. coming down Church Lane	45	75	52	52
No. using Royal Oak car park	22	20	48	30
No. parked in road outside school	64	65	111	84
No. parked in Church Lane	0	7	0	1
% of cars arriving after 0835 hours	83	94	84	70
% of cars parked more than five minutes in road	3	11	2	2

Figure 6 | *Results of fieldwork investigation into traffic near Clatford CE Primary School.*

Cars parked across driveways of private houses, leaving the occupants unable to leave their property while the car was there. Many of these cars were left unattended for several minutes, some up to 10 minutes.

71 Longstock Road has had its garden spoilt with wheel marks where people have pulled in to the side to pass other cars.

People do U-turns at the T-junction.

People park on the zig zag lines.

Cars are parked outside the old people's bungalows for up to 10 minutes when parents take their children on to the playground and then stand and watch them play. This leads to other cars having to park even further up the village, thus increasing congestion.

During the period from 0830 to 0855, the road is reduced to single line traffic. When the buses arrive to take children to Stockbridge, there is a virtual standstill, as the road is totally blocked.

During this period mentioned, there can be as many as 19 cars per minute passing or parking.

One car was left unattended with its engine running.

One car was left unattended on the road with two small children inside.

Parents chat while their cars are parked on the highway which is obviously congested.

At no time was the Royal Oak car park full.

Figure 7 | *Pupils' observations of traffic problems at Clatford CE Primary School*

Movements Compared

On class 6 maps showing journeys from home to buy groceries, clothes and for medical care, people travel in different directions, although they are moving for the same purpose. They have different destinations.

Although we travel for the same purpose to Clatford School, we don't go to different destinations. Instead we all converge on one place, although we may come from different directions.

This pattern of movement will lead to traffic congestion.

Looking at traffic movement outside Clatford School.

Traffic coming from Longstock has been found to be greater than that of Anna Valley / Upper Clatford in our traffic survey. However we have found out that most people coming to school come from Anna Valley / Upper Clatford, so we can only assume that the majority of traffic coming from Longstock passes school for a different reason than to drop off children there. This reason might be that they are coming along the Longstock Road and Barrow Hill, turning right and driving into Andover. However this does make the traffic congestion outside Clatford School even though they do not park there. The busiest time is from 8:35 to 8:55.

The busiest day is a Thursday, when 14 cars pass the school per minute. We think that Thursday is the busiest day because it is market day in Andover. On Thursday 339 cars passed or stopped outside the school. Unfortunately most people park outside the school rather than in the Royal Oak car park. Because of laziness people do not park in the Royal Oak. They could park on Church Lane, but not on the corner because drivers cannot see round the corner of Longstock Road.

As I have said, I think that it would be a good idea to put a policeman on duty between 8:35 and 8:55 at the T-junction outside school. We also could have a no-stopping-for-more-than-3-minutes rule on the road outside the school. The policeman would ensure this. It would be helpful to have 2 mirrors at the junction were Church Lane meets the Longstock Road and the road from Anna Valley / Upper Clatford, so that drivers could see round the corner.

Figure 8 | Traffic survey by Laura May, year 6, Clatford CE Primary School near Andover.

events and human behaviour during the survey mornings (Figure 7). Pupils witnessed several examples of dangerous driving, illegal parking, lack of thought for local residents, lack of thought for other parents, a small example of gridlock, a relatively high density of traffic and the peaking of traffic.

What effect does our coming to and leaving school have upon the village?

The key issue had now emerged and most of the pupils were part of it! One youngster deduced that the problem must be worse at the end of the school day when the infants depart at 15.20 hours and Y3-6 come out ten minutes later. 'At least other people are not coming home from work at the same time' said another. Most pupils grasped the fact that all the traffic arriving and departing in such a short time had a clear impact on the school's immediate locality. It was decided that next year's class could build upon these investigations and take a close look at the traffic situation in the afternoon, as well as checking if the current situation had improved or worsened. The latter was a perfect prompt for the final key question.

What can we do to ease or solve any problems?

The last key question contributed to the reflective phase of the enquiry. Several residents had complained about traffic chaos in the mornings, how their drives were blocked and their own journey to work hindered. The Parish Council had debated the matter and contacted the County Highway Authority who stated that while there had been no reported accidents at or near the T-junction, they would agree to mark white lines around the junction. Suggestions for alleviating the traffic problem included imposing a 20mph speed limit near the school, encouraging more pupils to cycle to school, raising awareness of the problem regularly in school newsletters and improving the provision of bus transport.

It was a salutary experience for the pupils to realise that problems of this kind cannot be solved in a trice and that some changes in drivers' behaviour would go far towards mitigating the effects upon the village. The last word comes from Laura May (Figure 8).

Photo | Paula Richardson.

An example of geographical enquiry: Parking problems in our local area (key stage 1)

The previous example was devised by primary school staff with help from a local geography consultant, before the availability of study units from QCA. For this example, the QCA scheme of work Unit 2: How can we make our local area safer? is particularly relevant (Figure 9).

Four key questions drive the study unit which is aimed at key stage 1 and provide a structured pathway through it: 'Is our school on a busy road?', 'Is parking a problem?', 'How is parking controlled?' and 'How could the area be made safer for children?'. Match these against the characteristics of key questions listed in Figure 1. Consider the possible responses from an average Y6 pupil to the first three questions and compare the questions with those generated for the Clatford pupils. Can you see how key questions need to be carefully created for different ages and abilities and how they can require different depths of investigation and involve greater skill development? Look again at the previously listed parameters for considering progression in enquiries.

The QCA documentation clearly relates the key questions to the learning objectives (what children should learn) and offers possible teaching activities leading to (possible) learning outcomes. The expectation that most pupils will be able to carry out a small local survey, draw some specific conclusions from the evidence they find and discuss ways of tackling the parking issue in their locality could apply equally well to the Y6 investigation described. In theory, the general progression in geographical experiences and learning over the intervening years must lead to greater sophistication of responses, outcomes and skills practised from the Y6 pupils. The latter may even have begun to twig the overlap with other subjects!

Teachers who are unaware of the QCA units at key stages 1 and 2 should waste no time in acquiring them. Some professional development will come from studying and using them intelligently and the natural step is to use the enquiry model provided to create, over time, your own units. One always gets a kick from creating quality material and you learn so much in the process.

Postscript

This chapter has attempted to analyse a fundamental element of the geography national curriculum in a nutshell and to amplify it through exemplars. Like effective INSET, has it provoked thought, challenged you professionally and sown the seeds of change in pedagogical attitudes and behaviour?

Clearly, geographical enquiry is a teaching methodology, part of the teacher's toolbox, and is a professional skill that needs to be developed in tandem with the crucial ability to think

Learning objectives	Possible teaching activities
Children should learn	
Is our school on a busy road?	
■ about the character of a place ■ to ask geographical questions ■ geographical terms	■ Discuss with the children what makes a busy or quiet road. ■ Arrange for the children to complete a simple traffic survey on the road outside the school. ■ With the children's help, label a wall display of photographs of the road outside the school to show aspects related to traffic, e.g. road signs, road markings. ■ Ask the children to think about their own road at home and decide whether it is quieter or noisier than the school road. ■ Encourage the children to think up their own questions about traffic around the school.
Is parking a problem?	
■ to carry out a small-scale investigation about parking in the local area	■ Discuss with the children what makes a 'fair' test in a survey, e.g. times, frequency, place. ■ With the children's help, design and carry out a survey of the numbers of cars parked in the street. Ask the children to present the results as a graph, using simple graphing software, and analyse them. ■ Ask the children to consider questions like: Are the parked cars there all day? Where do people go when they park their cars?
How is parking controlled?	
■ to observe, recognise and describe the main ways in which parking is controlled ■ to undertake simple mapping tasks	■ Discuss with the children the ways in which parking is controlled, e.g. yellow lines, pedestrian crossing, 'lollipop' men or ladies, clearways. ■ Visit the road outside the school and ask the children either: to record on a map the various ways used to control traffic; or to observe the parking controls and then on return to the classroom draw a map from memory to show the observations. ■ Discuss with the class the accuracy of the two methods.
How could the area be made safer for pupils?	
■ to express views about making an area safer ■ to recognise ways of changing the environment	■ Ask the children to identify methods of making an area safer, e.g. cycleways, pavements, fencing, no parking zones, road signs, pedestrian crossings, and to think about how the school grounds and other streets they know are made safe. ■ Ask the children to make use of all the evidence they have collected (photographs and survey results) to write a letter to the transport department at the local council to ask about the possibility of a safety feature, e.g. a pedestrian crossing, being constructed.

Figure 9 | *Geography scheme of work Unit 2: How can we make the local area safer? Source: QCA, 2000. Reproduced with kind permission of Qualifications and Curriculum Authority.*

geographically. As one of the Ofsted annual reports declares: 'Enquiry skills are best developed when addressing a real problem… The best enquiries build cumulatively on pupils' previous learning and explore the issues in ever increasing depth' (Ofsted, 2002, p. 7).

References

Curriculum Council for Wales (1993) *An Enquiry Approach to Learning Geography at Key Stages 2 and 3*. Cardiff: CCW.

Garner, W. (2002) 'Questioning enquiry', *Primary Geographer*, 48, p 35.

National Curriculum Council (1993) *An Introduction to Teaching Geography at Key Stages 1 and 2*. York: NCC.

Ofsted (2003) *Geography in primary schools: Ofsted subject reports series 2001/02*. London: Ofsted.

DES/WO (1990) *Geography for Ages 5-16*. London/Cardiff: HMSO/WO.

DfEE/QCA (2000) *Geography: A scheme of work for key stages 1 and 2. Update*. London: DfEE/QCA.

Roberts, M. (2003) *Learning through Enquiry*. Sheffield: Geographical Association.

Storm, M. (1989) 'The five basic questions for Primary Geography', *Primary Geographer*, 2, pp. 4-5.

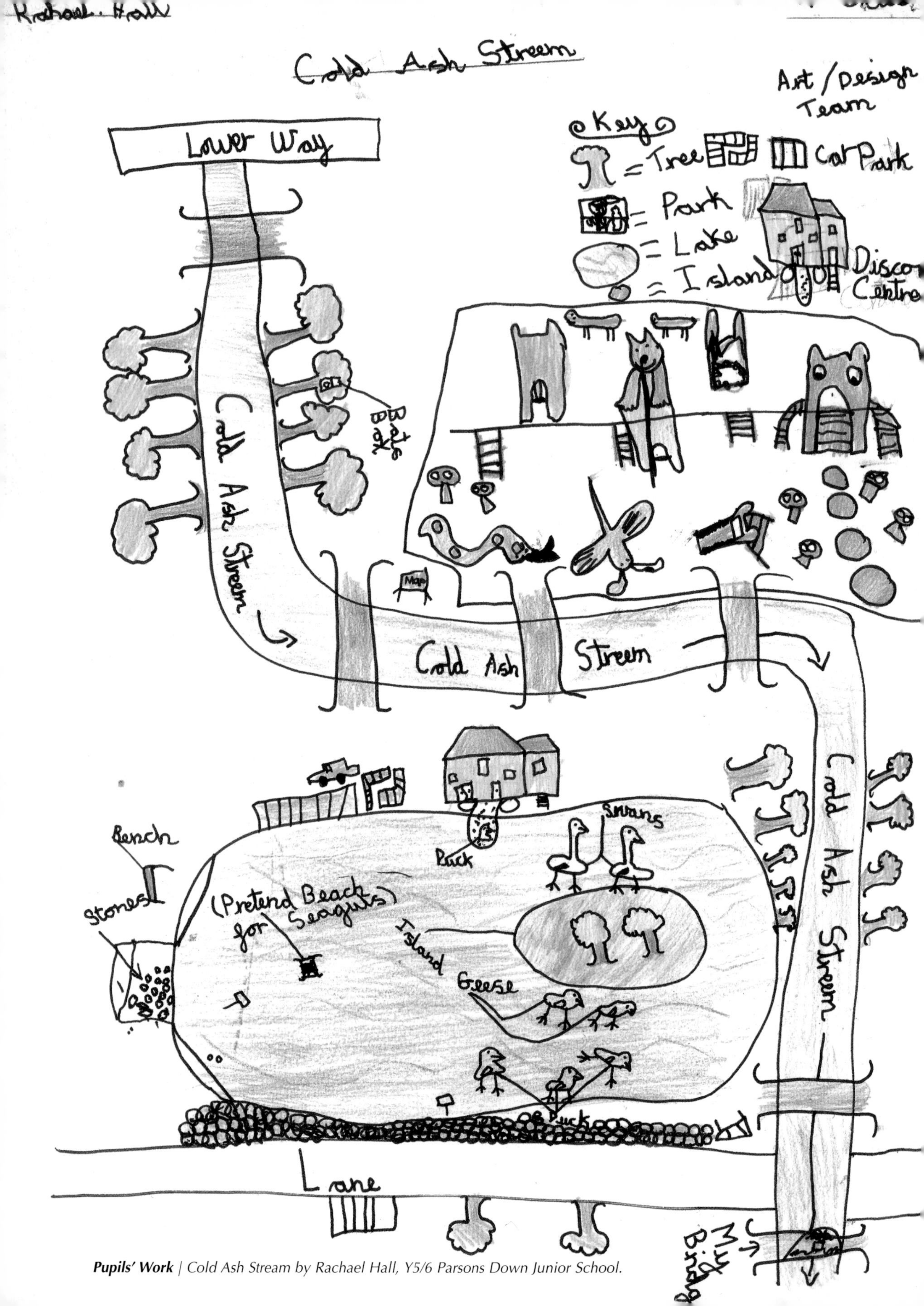

Rachael Hall

Cold Ash Stream

Lower Way

Art / Design Team

Key
= Tree
= Car Park
= Park
= Lake
= Island

Discov Centre

Cold Ash Stream

Cold Ash Stream

Map

Cold Ash Stream

Bench

Stones

(Pretend Beach for Seagulls)

Island

Duck

Swans

Geese

Ducks

Cold Ash Stream

Lane

My Bridge

Pupils' Work | *Cold Ash Stream by Rachael Hall, Y5/6 Parsons Down Junior School.*

Mapwork skills

Photo | *Paula Richardson*

The need to record, revisit and pass on information to others is of fundamental importance to all our lives, and one of the most effective ways of achieving all these things is by making and using maps. Maps and plans are simple, clear and decisive means of communication. A classroom plan provides a view of something that children can see in its entirety. It also shows how the space in a room is used and organised. A plan of the whole school uses the same techniques but involves abstract thought because it is not possible to see the complete internal arrangement of the building except by mapping or map-based modelling.

Why are mapwork skills so important?

Once a plan is created then distributions, locations, associations of areas, links and communications, routes and resources are revealed. It is easy to see why mapping is held to be the most powerful mechanism of geographical enquiry. It has improved our lives in immeasurable ways but it can also be seen to have encouraged activities which radically affect and exploit the environment. Whichever way one looks at it, map skills are key skills for the modern world.

How do I start?

The suggestions in this chapter recognise that while there is tremendous scope for the mapping enthusiast, many teachers will find themselves constrained by lack of time and curriculum pressures. The ideas and activities suggested here have been designed with this in mind. In order to make your own selection you will need an overview of what you hope to achieve.

Our overall objectives are that we want our pupils to learn how to describe where they are and record information in their local environment, to be able to plan routes and journeys, to extend this ability from the locality to the rest of the UK, and to make sense of the daily bombardment of media messages by acquiring a basic familiarity with Europe and the world map. Mapwork is made much more exciting nowadays through the use of aerial photos, satellite images, internet searches and an expanding range of interactive computer programmes. Most adults will use modern technology as part of their daily lives when they buy travel tickets, plan routes, make weather checks and so on. What we are attempting to do is to encourage children to do the same.

Some teachers will feel they now have enough general guidance to plan a programme of mapwork experiences for themselves. What follows are more detailed suggestions and practical activities for those who would like a little more.

What map resources should I have?

The following list gives an idea of the resources you might want to consider. As a general rule, acquire the resource in each category that is easiest to get hold of and is most likely to be used regularly because it's actually useful!

1. Basic plan of the classroom: let pupils of all ages make their own plans of the classroom. The older ones should do this to scale with some accuracy. However, there are times when a basic, accurate, copiable blank outline, which you have prepared yourself, will be useful.

2. A school plan: if any building work has been done recently then your school may well have an accurate scale plan. If not you may have to make your own.

3. Local street plans.

4. Ordnance Survey maps: the Explorer maps at a scale of 1:25,000 now cover the whole UK mainland and contain a wealth of detail including open access land and leisure activities. A set of six will be useful for class group activities.

5. UK maps: a UK wallmap should be displayed somewhere in the classroom for instant, regular reference. Road atlases at 3 or 4 miles to the inch (a lesson in itself here) can also be sourced very cheaply.

6. Europe base map: these can be obtained from book of photocopiable geography activity sheets.

7. World base map: these are easy to obtain as above.

8. Atlases: pupils should have access to a set of up-to-date atlases. Remember that you will need a balance of physical maps, political and thematic maps.

9. Globes: A traditional globe on a stand is always useful. Make sure that the printing and colours are of good quality. There is a range of inflatable globes up to sizes as big as the pupils which can really excite their imaginations.

10. Satellite images: are now widely available and can inspire a variety of lessons. For instance, images showing the UK at night can stimulate discussion on how light pollution is now blotting out the stars in urban locations.

11. Informal local maps: Collect all the local leisure maps you can from Tourist Information Centres, libraries and leisure outlets, or make your own (Figure 1). See if you can find a postcard with a picture map of the attractions in your area. Town trails and leaflets describing walks are especially useful as they are often supplemented with pictures and explanations. These maps tend to be 'informal' in that they emphasise key features at the expense of accurate distance and scale. There are excellent discussions to be had about this.

12. Transport route maps: Bus and train timetables contain route diagrams showing communication networks, e.g. there is a map of the entire network in the all-UK railway timetable.

13. Historical maps: The local main library may have an archive room, or try the nearest museum. Previous editions of OS maps are also useful.

14. Aerial photographs: local newspapers may well have a collection, or there are CD-Roms for whole counties available commercially.

15. Photopacks: sets of photographs that portray themes such as 'coasts and mountains' or world locations are widely available.

16. Electronic maps: virtually all the resources suggested so far are also available electronically through the internet, CD-Roms and interactive computer programmes. It is valuable for pupils to experience these facilities. The Ordnance Survey and QCA websites are good starting points.

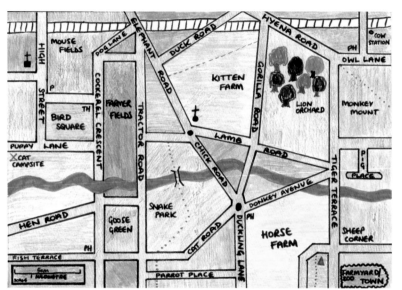

Figure 1 | *'creative co-ordinate' game*

What do I need to know to read a map?

Most households have an atlas of one kind or another. Even so, many adults find using a map a demanding and a complex task. It could be that they were not taught map skills in an efficient and progressive way. It could be that they have not had enough practice in using maps. If children are to develop confidence with maps they need constant practice in using and making maps and plans in real situations. Equally, time on task, the age of the learner and the complexity of the material are all crucial factors in learning.

Maps, like all methods of communication, have their own rules. These specific ways of imparting information can be a barrier to understanding. In teaching map conventions it is best to start with maps of the immediate locality. Pupils will be familiar with this area and they will be able to make constant comparisons between what they know and what is on the printed sheet. However, in order to read a map the user needs to be aware of the various conventions. All the items in the following list will, at some stage, be new and possibly confusing to pupils. Try to present each idea carefully using a mixture of direct teaching, practical work and discussion:

1. Location grids: Grids were first pioneered in Roman times by Ptolemy to enable places and features to be located on maps. The simplest is the alpha-numeric grid which uses letters to describe vertical columns and numbers for the horizontal rows. This gives a simple letter/number description for each square on the grid. This sort of grid is commonly used for street plans. It leaves the user to do their own search within the square for the place they want. Ordnance Survey maps use a more sophisticated system of four- or six-figure co-ordinates to locate features within a square. Set realistic goals for pupils before worrying about complex co-ordinates. They should be able to locate themselves on school and local plans and be able to use alpha-numeric grids with confidence. Use real situations like the many 'You are here' visitor boards in most town centres to develop the work.

2. Plan view: All formal maps show features in plan view (as though seen from directly above). Plan views are much less of a problem to children nowadays as images of the Earth from space on television are common. However, in making their own plans younger children especially need help with viewing 3-D objects from above and translating them into plan view. A set of photographs can help them appreciate the differences. If the school has a digital camera you could make your own set.

3. Colour, words and lines: Maps are a highly efficient way of using colour, words and lines to create an almost instantaneous mental picture of an area. Colours are immensely useful and pupils should explore the different ways in which they are used to show countries, height above sea-level or the different types of road. There is no standard system and the only thing that cartographers seem to agree on is to use blue for the sea.

The way that words are written is also significant. The size and boldness of the type usually indicates the size and importance of the settlement. The Ordnance Survey uses Gothic script for historic sites. Other words such as 'Pennines' may be overlayed because they refer to an area rather than one specific point.

Lines provide the linkages across maps and describe the routes of roads, railways, rivers and boundaries. Larger scale maps may also show tracks, footpaths, and lines of power cables. A system of grid lines is often added for reference, sometimes drawn in blue. Pupils need to appreciate that some lines depict features that exist on the ground while others exist only on the map.

4. Symbols and numbers: Symbols allow information to be displayed in a concise way and reduce the need for labels which may be confused with place names. The majority of formal published maps have a key explaining the meaning of the symbols that they use. On thematic maps, symbols may have statistical significance and be crucial for interpreting data.

Numbers on maps may provide information about roads, heights of hills and distances between places. Here too colour is used for slightly different types of information. Most pupils enjoy working with symbols as they are often tiny pictures or stylised outline shapes (Figure 2). You need to encourage them to make keys of their own symbols, make collections of symbols and learn how to consult and interpret a map key.

5. Scale: Maps 'shrink' the real world to make it fit onto a piece of paper. Maps which show large areas have to shrink the world much more than maps which show small areas. The amount by which the map has shrunk reality (the scale) is indicated by the representative fraction. For example, a representative fraction of 1:1000 means that one centimetre on the map represents 1000 centimetres on the ground. Maps of this scale will show the outline shapes of individual houses and gardens. By contrast, a map which is drawn to 1:120,000,000 (the scale used in many world maps) will show very little detail indeed. Children need time to appreciate the difference between large- and small-scale maps. You can practice by drawing plans of familiar objects, starting with a scale of 1:1 and then reducing the scale for larger objects. When it comes to measuring distances on maps encourage children to use a ruler and the scale bar which usually appears with the key. When it comes to calculating distances that involve twists and bends, such as the distance by road to different places in the local area, a piece of string will be helpful to get around the corners.

6. Compass direction: Since the sixteenth century it has been the convention to orientate maps to the north. Children need to be very secure with their ideas of 'left' and 'right' before embarking on more complicated compass directions. North and south take precedence over east and west in describing intermediate points such as NE and SW. Giving instructions using compass points is not easy especially if you are facing south and therefore moving 'down' a map. If the four cardinal points of the compass are on display

Spaces in school

1. Draw the correct symbols in each area of the school plan.

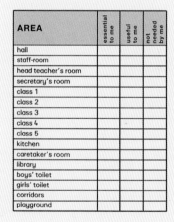

learning resources

cleaning

study area

activity area

offices

cooking

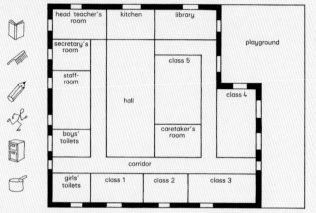

2. Imagine you are in Class 3 of this school. Look at each part of the building in turn and decide how useful it is to you. Colour the chart below using the following key:

```
              essential to me ...............red
KEY    useful to me ..............yellow
              not needed by me .........blue
```

AREA	essential to me	useful to me	not needed by me
hall			
staff-room			
head teacher's room			
secretary's room			
class 1			
class 2			
class 3			
class 4			
class 5			
kitchen			
caretaker's room			
library			
boys' toilet			
girls' toilet			
corridors			
playground			

3. Show your results on a pie chart. Put in the red first, then the yellow, then the blue.

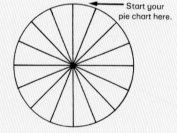

Start your pie chart here.

4. Colour the school plan using the same code.

5. On a separate piece of paper redesign the school leaving out all the areas that are not essential to you.

Figure 2 | *Investigating spaces in school. Source: Scoffham et al., 1995.*

in the classroom then maps can be orientated with the true direction. Pupils of all ages need to discuss direction to put their thoughts into words and explanations. It is important with younger pupils to establish a directional and relational vocabulary such as 'beneath', 'near', 'distant', 'steep', 'high', 'over there', 'above', 'straight on', 'up', 'down', 'back', and confidence with 'go left or right'. Pupils of all ages should be able to describe route instructions in words. Small-scale group activity in the immediate area contrasted with the routes taken by explorers or by characters in a story can develop this work.

What is a sensible sequence of teaching and learning?

Hopefully you are already beginning to establish your own set of priorities for developing and promoting map skills. However, it's useful to have an overview of the way learning generally develops in this area. Table 1 outlines some of the key stages. It must be stressed that this is a developmental framework rather than a year-by-year syllabus. Each row indicates a desirable development of understanding and accomplishment. If you are working with pupils who are proficient in the skills in the middle of the sequences, then you should introduce them to more complicated skills. However the early parts of the sequence should also be revisited to assess their underlying competence, fill in gaps and to see if any misconceptions have arisen.

When teaching map skills you need to set the pupils challenges which involve the use of maps and plans. In this way they will come to appreciate that maps are a natural and absorbing part of everyday life. Here are some simple enquiry questions which will help to prompt discussion:

Table 1 | *A sequence of mapwork skills and activities.*

	MAKING A START			MAKING PROGRESS		A SENSE OF ACHIEVEMENT	
1 KNOWING WHERE YOU ARE (LOCATION)	Where am I in the room? Location of objects	What parts of the room do I use? Positions on a class plan	Where can these things be? Location on a school plan (with grid)	How can I describe where I live? Using a local street plan (with grid)	How can I describe my local area? Using an OS map ~ (discuss grid)	How can I describe where in the UK I live? Using a UK map (use grid)	How can I describe where the UK is in the world? Using a world map (with grid)
2 MAKING AND USING MAPS AND PLANS (PER-SPECTIVE)	What can we tell from a drawing? Drawing objects	What is a plan? Drawing a plan view	How can I show places on a plan? Plans from models and layouts	How do I get to School? Mental maps	Are there reasons for the way places are arranged? Recording information on a school plan	What can I tell from an aerial photograph? Comparing local plans, OS maps and aerial photos and historic maps	How can maps help people understand my views? Using world and country maps and satellite images to locate and convey information
3 SYMBOLS, LINES AND COLOUR (CONVENTIONS)	How do things group together? Giving general classifications to sets of objects	What can you tell from a symbol? Using symbols	Can I make effective symbols? Creating symbols	Where are the busiest places in school? Adding colour, lines and symbols to a school plan	What landmarks might help me remember a school trip? Using an OS map to plan a school trip	How might the weather affect my day? Weather maps	How does my way of life affect other parts of the world? Atlas thematic maps
4 DISTANCE (SCALE)	How far is it? Using Distance vocabulary	Where are they? Using positional vocabulary	What journeys can I make? Route instructions	What are the effects of different scales? Drawing to scale	How does scale affect a plan? Using a scaled school plan	How are different maps of the same area useful? Maps at different scales	What's special about the UK? Scale of continents and oceans
5 ROUTES AND JOURNEYS (DIRECTION)	What is a route? Defining a route	How can I use right and left? Directional vocabulary	What journeys can I make? Routes with direction and distance	How useful are compass-point directions? Using compass points	From which direction do most children travel to school? Routes on OS and UK maps	What is the best way to explore the local area? Writing a local trail	How can I find my way? Orientation
6 USING ATLASES AND GLOBES	What are globes and maps? Describing maps and globes	How do maps show land and sea? Distinguishing land and sea	Why are maps and pictures useful? Relating maps and pictures	Where are these events happening? Locating major world features	How good am I at using an atlas? Using index and contents	Do I live in a region? UK atlas maps	Are European countries about the same? Use world atlas maps
7 THINKING SKILLS	Naming locations	Classifying areas	Identifying associations	Describing resources	Comparing systems	Analysing change	Generalising interactions

■ 'How can I help a visitor to find my classroom?'

■ 'Which is the quickest way from my house to school?'

■ 'Which is the safest way?'

■ 'Why might the two routes not be the same?'

■ 'Would I enjoy a holiday on the Isle of Arran or the Isle of Wight?'

■ 'What would be similar and what different about the two experiences?'

Remember, the idea is not to be ever pressing on into new and unknown territory. Map skills need to be used constantly and practised. Remember too that children – and adults – will need to apply the whole sequence of skills if they are transposed to a new situation. When staying at a field study centre, for example, everybody (both pupils and adults) will need to go back to the beginning to establish their sense of location, direction and place. The point of teaching map skills is to make this an efficient and intuitive reaction; a crucial ability in a society where travel and mobility are at a premium.

Finally, keep in mind that children do not always progress in lockstep; within the sequences pupils will be at different stages of understanding. Telling right from left is a good example. There will be pupils in top juniors who are unsure while others happily made and used the distinction years before. To sum up, don't take anything for granted, don't limit the opportunities you offer pupils, and do use all the skills in real-life, practical, enquiry situations as much as possible.

Here are some suggestions for these practical activities. Each one appears in Table 1 listed alongside the relevant idea in the skills sequence.

Practical activities

Knowing where you are (Location)

1a. *Where am I in the room?* Help pupils to develop a vocabulary which enables them to describe where they are by reference to other people and fixed objects. Collect useful words and display them for reference during a group discussion.

1b. *What parts of the room do I use?* Discuss the different areas in the classroom such as a reading area or an activity area. Colour a class plan to show these areas. What space is left? Is it possible to give it a description too? Could there be a better arrangement?

1c. *Where can these things be?* Make a 'Treasure Hunt' around the school. Give pairs of pupils a school plan with numbered locations. On a separate sheet or set of cards are puzzle descriptions of an object or feature at each location for pupils to identify ('What is made of wood and can be sat on?'). After the hunt add a simple grid to the plan and work at identifying the puzzle locations by using grid references.

1d. *How can I describe where I live?* Draw concentric circles on a local street plan with the school at the centre. Extract information from each ring. Are any particular buildings indicated? Are there open spaces, footpaths or railway lines? Where do the children live? Collect the different terms for road such as street, avenue or close. Are there reasons for using these different words? Overlay a grid to locate features.

1e. *How can I describe my local area?* Use the local OS Explorer map to decide what might be the extent of your local area. Then work in groups to extract all the information possible about what is in the area. It should range from churches and telephone boxes to

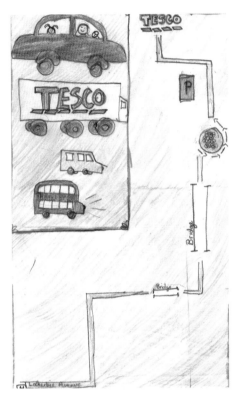

Pupils' work | Y4, Abbey Primary School, Leicester.

car parks and nature reserves. Discuss the OS grid at a level appropriate to the group.

1f. *How can I describe where I live in the UK?* Use both physical and political maps to describe the location of the school in relation to the rest of the UK using specific features from both sorts of maps. If the maps have grids, are they the same? How useful is a grid reference in writing a detailed location description?

1g. *Where is the UK in the world?* Write a description of where the UK is on a globe or world map using at least five different location features and a grid reference.

Making and using maps and plans (Perspective)

2a. *What can we tell from a drawing?* Pupils draw familiar objects and also draw round items such as their hands. Discuss the ways in which drawings differ from the real thing. Why is it useful to have drawings? When you draw around objects why are some results more recognisable (a hand) than others (a matchbox)?

2b. *What is a plan?* Help pupils to make a model of a real or imaginary room in a box. Draw around the model furniture, cut out the outlines and make plans of the room.

2c. *How can I show places on a plan?* Make picture maps of a story, or the school grounds or an imaginary village or island. The pictures should be linked by roads, rivers or paths. Use a coloured background to give an idea of town, country or water.

2d. *How do I get to school?* Draw a route map of the journey to school showing landmarks and features including turnings and crossroads. With so many pupils now reaching school by car, try to encourage them to make notes to describe their journey as they are actually travelling.

2e. *Are there reasons for the way places are arranged?* Colour-code rooms and areas on a school plan to show places used for learning, places for administration and places for health and welfare (including play). Are there reasons for the way the school is planned? Could it be arranged in a better way?

2f. *What can I tell from an aerial photograph?* Which local plan or map best compares with a particular aerial photograph? Why are maps necessary at all?

2g. *How can maps help people understand my ideas?* Write a short report, including maps, about a world issue, such as the great distances much of our food is transported.

Symbols, lines and colour (Conventions)

3a. *How do things group together?* What is a forest, river, cliff or mountain? Assemble photographs which children can sort into sets.

3b. *What can you tell from a symbol?* Make symbols for the different areas of the classroom or the different rooms in the school. What sort of things do you have to think about to make a simple symbol for a place?

Photo | *John Halocha.*

3c. *Can I make effective symbols?* Create and add symbols to a school plan to indicate key places and activities around school. Select the most effective ones to create a useful school plan for visitors.

3d. *Where are the busiest places in school?* Put the pupils in pairs, give them a school plan and coloured markers and ask them to plot all the journeys they make to different parts of the school during a day. Where are the busiest places? If pupils from a class at the other end of the school make similar journeys, do the busy places remain the same? Can the congestion at busy places be relieved?

3e. *What landmarks might help me remember a school trip?* Working from an OS map, encourage children to create a landmark-spotting trail of the route you intend to take on a class trip. Were all the features spotted or did some turn out to be much more prominent than others?

3f. *How might the weather affect my day?* Discuss how weather maps provide snapshots of constantly changing weather conditions. How can you read the signs and symbols to gain clues about how the weather pattern might develop?

3g. *How does my way of life affect other parts of the world?* Use thematic atlas maps to identify global links.

Distance (Scale)

4a. *How far is it?* Make up a game, perhaps in teams. 'Fetch something from the most distant part of the room', 'Which is the nearest pencil to you?', and 'Put a shoe in the middle of the room'. Children might make up their own in groups or pairs.

4b. *Where are they?* Draw something in the middle of the room, something as far away as possible, something in between. Try the same exercise outside.

4c. *What journeys can I make?* How many different journeys from the room can be made going through only one door? How many if you go through two doors? How many through three? Which is the longest and shortest journey for each of these? How do they compare between one another? If possible plot the routes on a school plan.

4d. *What are the effects of different scales?* Draw a building or a face on squared paper of different sizes. Provide 1cm squares, 2cm squares and 4cm squares. If necessary the pupils can draw their own grids. Which was the most useful size on which to draw the object? Why?

Photo | John Halocha.

4e. *How does scale affect a plan?* Provide three sizes of squared paper with squares of 1cm, 2cm and 4cm. If 1cm equals a metre draw a plan of the classroom. Do the same on paper with 2cm equalling a metre and then 4cm equalling a metre. Which is the best scale for the class plan and why? How would you express the scale in numbers?

4f. *How are different maps of the same area useful?* Compare an OS map, a street plan and a tourist or other specialist map of your local area. Create a map of information for your pupils.

4g. *What's special about the UK?* Compare the UK with a country in each of the continents outside Europe.

Routes and journeys (Direction)

5a. *What is a route?* Ask pupils to walk the most direct way from their seat to the door. Can they describe it? What would the longest way be? How could you record these routes?

5b. *How can I use right and left?* Ask pupils to list the things they can do just with their left and then right hands. Then try to describe routes around the room using left and right direction instructions. Some pupils may find it useful to hold a 'l' and 'r' card in each hand as they go.

5c. *How many different journeys from the room can be made going left through one classroom door?* How many going right? Compare and plot on a school plan.

5d. *How useful are compass-point directions?* Pupils use and make up and follow routes around school using N, S, E, W.

5e. *From which direction do most children travel to school?* Using a local map or plan plot the addresses of class members. From which general direction do the majority of children come?

5f. *What's the best way to explore the local area?* Devise a local area trail.

5g. *How can I find my way?* Find the nearest open space or patch of countryside. Work in groups each with an adult. On a map showing just the perimeter of the area plot the paths to be found and useful landmarks.

Using atlases and globes

6a. *What are globes and maps?* As a whole group look at a globe and a large wall map and make a class list of all the words that pupils associate with each. Try to agree on a definition of a globe and a map perhaps as a puzzle to take home to try on an adult.

6b. *How do maps show land and sea?* Discuss land and water on various maps. Help pupils to make simple imaginary maps of their own, including some water.

6c. *Why are maps and pictures useful?* Use a large physical map of the world and a pack of pictures of world landscapes. Use clues to decide where the picture might have been taken.

6d. *Where are these events happening?* Make a class list of world places and events known to the pupils and then locate them on globes and atlas maps.

Photo | *Tina Horler.*

6e. *How good am I at using an atlas?* Select one word for every letter of the alphabet from an atlas index. Is each a country, a place to live, or a feature? In which continent is it located?

6f. *What region do I live in?* Look at regional maps of the UK. How has the extent of your local region been decided upon? Is your local area typical of the regional character?

6g. *Are European countries all the same?* Use physical and climate maps of Europe to explore the range of landscapes and weather.

What are the wider benefits of developing mapwork skills?

When developing mapwork skills it is helpful to recognise that maps may be used in three basic ways. First there are personal applications such as sketching a route to show friends how to reach your house. Second there are the practical skills of reading a map to find your way around an unfamiliar environment. Third there is the enormous potential of maps to give information about places that are impossible to visit. As the popular cliché has it, one picture is worth a thousand words. The visual presentation techniques used in maps convey information quickly and in a context.

1. Using maps for personal information

The first steps to map making are often when a pupil tries to show the inside of a room or house, when they try to draw a plan of their classroom, indicate how they get to school or simply make up imaginary places. It is important to provide many opportunities to encourage and develop this personal mapping which is one of the unnoticed skills of everyday adult life. Having an overview of where we live and being able to record that information in various simple ways encourages awareness and attachment to place. These are qualities crucial to modern living where individuals can sometimes feel confused, lonely and marginalised. A major aim of teaching is to sharpen and augment our mental map of where we live and to develop a fine-grained sense of place and the ability to operate effectively and fully within it.

2. Using maps for practical information

People quickly reveal whether or not they are regular map users. Many people have little idea

Geographical concepts which can be developed through mapwork include:

■ direction (e.g. which way to go to reach places)
■ distance (e.g. how far it is from one place to another)
■ distribution (e.g. how furniture is distributed in the classroom)
■ access (e.g. where post boxes are located)
■ patterns (e.g. variations in weather across a site or region)
■ resources (e.g. the location of minerals and other natural materials)
■ change (e.g. new housing developments in a town or village)
■ system (e.g. evidence of transport, energy or water infrastructure)
■ interaction (e.g. the impact of people on their surroundings and vice versa)

Figure 3 | *Maps and plans can be used to develop a range of key geographical concepts.*

of the main features in their local area, the best way to reach places, where they might walk safely or how to reach the countryside. On the other hand, a map reader will know how their neighbourhood relates to other local settlements, the main features and leisure facilities, the range of alternative routes for vehicles or walking, and have a sense of the character and context of local issues. As far as teaching is concerned, practical activity is the basis for mapwork understanding and environmental action. It promotes the ability to explore and consider possibilities, to make realistic journeys and find places, and to experience the sense of control and achievement that comes from venturing forth with certainty.

3. Using maps to develop thinking skills

Maps have the power to turn the abstract ideas, which we form in our heads, into visual reality. Only a handful of people will ever actually see the UK complete from space. Even fewer can expect to see the whole world in one go except as a map. Maps and plans have a potential for radically extending our understanding by portraying the layout and organisation of the school, revealing the network of roads in a town or region or showing the distribution of natural vegetation such as forests and grasslands. Maps contextualise information within defined spatial boundaries, allowing us to make comparisons, formulate plans and develop generalisations. The identification and analysis of patterns, processes and relationships stands at the heart of geography. The way that mapwork can help to promote attitudes and values is equally important. Pupils need to develop both local and global perspectives, and appreciate the sequence of change that links the past to the future. As their understanding deepens, so their values will mature (Figure 3).

Table 2 puts in sequence these three aspects of mapping – the personal, the practical and the potential – describing a variation in emphasis at certain ages. It is built up of simple assessment statements derived from and referenced to the ideas and activities in Table 1. In Table 2 the activities are re-worded to give a profile of what an individual pupil might have achieved. This can be used for self-assessment or as a handy list for a teacher to take stock of progress. The two tables together show how a mapwork skill can be described, taught as a geographical enquiry and, finally, assessed in terms of how effective it has become as part of a child's life skills.

The personal reaction to place is about direct experience, visual reaction and attachment. The child is at the centre of their world and feels that others see things exactly as they do. It is an exciting time when children's sensitivity to their surroundings is combined with an emerging appreciation that places can be described and organised.

Table 2 | *How children relate to maps.*

PERSONAL INFORMATION	PRACTICAL APPLICATIONS	SEEING POTENTIAL
1a. I can describe where I am in the room.	1c. I can use a school plan.	1g. I can describe where the UK is in the world.
1b. I can describe the different areas of the room.	1d. I can describe where I live.	2f. I can compare maps and photos.
2a. I can talk about what a drawing shows.	1e. I can use a map of my local area.	2g. I can use maps to communicate my ideas.
2b. I can explain what a plan is.	1f. I can use a map of the UK.	3f. I can understand a weather map.
2c. I can record information on a plan.	2e. I can interpret information on a school plan.	3g. I can use maps about themes and issues.
2d. I can describe my journey to school.	3c. I can make symbols for my own sketch maps.	4f. I can select maps at different scales.
3a. I can group things into sets.	3d. I can add colour and lines to maps.	4g. I appreciate the scale of continents and oceans.
3b. I can describe what a symbol might mean.	3e. I can extract information from an OS map.	5f. I can help others to explore the local area.
4a. I can describe how far away things are.	4d. I can describe how scale changes maps.	5g. I can use clues, maps and equipment to locate where I am.
4b. I can describe where things are.	4e. I can use a scaled school plan.	6f. I can describe the region I live in.
4c/5c. I can describe a short journey.	5d. I can use compass points for directions.	6g. I can describe some similarities and differences between European countries.
5a. I can plan a short route.	5e. I can plot routes on maps and plans.	
5b. I can use directions like 'left' and 'right'.	6b. I can explain the special uses of colours on maps.	
6a. I can describe what a globe or a map is.	6c. I can compare maps and pictures.	
	6d. I can use an atlas to locate places.	
	6e. I can use the index and contents of an atlas.	

The practical application of mapwork skills is about understanding and explaining the environment. It relates to the growing sense of confidence in children that they have a measure of power and control over where they go and what they do. It is about awareness and action and the development of ideas and strategies.

The potential use of maps links to assessing and explaining the patterns, processes and relationships in the world; it is about values and attitudes, the local and the global, the sequence of changes from past to present and into the future. It is about thinking globally and acting locally.

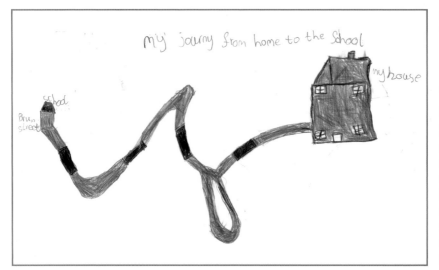

Pupils' work | *Abbey Primary School, Leicester*

What does research tell us?

Developing mapping skills with young children has both attractions and significant hurdles. Research reveals that children may not appreciate aspects of scale, distance and direction until the later years of primary school. However the visual nature of the medium makes maps and plans accessible for recording and exploring ideas from the earliest years onwards.

Palmer (1994) describes how children around the age of five tend to draw pictures in which known places are linked in some way (topological maps) but which show few other map characteristics. By contrast, by the time they are ten years old children are likely to produce much more accurate and detailed maps which involve abstract ideas of scale, direction and symbolic representation (Euclidean maps). Citing findings from Blades and Spencer (1986), Palmer concludes that no child in school is too young to be introduced to the appreciation and use of maps. Interestingly, the ability to orientate a map seems to be a key step to competence in mapwork and should be worked at from the earliest years. The years between six and eight are particularly crucial in moving children from the impressionistic use of maps to more abstract understanding and most children will happily work with the idea of routes and with symbols even before then.

Ultimately Palmer accepts that there is no strong agreement on the best way of introducing maps to young children other than that it should be done within the familiar context of the children's daily lives. Knight (1993) makes the interesting point that children should be encouraged to develop some kind of active overview of how to read maps by describing what they show. This can be done from the earliest years with the children's own picture maps and is a vital part of the development of thinking by describing, comparing, explaining and generalising.

Catling (1998) reports how teaching children mapwork in the context of the local area can radically improve their knowledge of local features and develop their confidence to the point where they are willing to recognise the extent and limitations of their knowledge. Whether there is a difference in mapping abilities between boys and girls remains open to question. In the past researchers found that because boys were freer to range across the locality they were more proficient at mapping (Hart, 1979; Matthews, 1992). It may well be that because children are often ferried around by car nowadays both girls and boys have a less subtle and more restricted view of where they live. There is potential here for your own action research in the classroom. Keep examples of the mapwork you do with your classes across the years. It will develop into a valuable resource bank of ideas as well as an archive of research material for future use.

Conclusion

Give pupils as much exposure as possible to maps of all kinds. Set little problems of orientation in real situations and constantly encourage pupils to draw picture maps of their own location, landmarks, journeys and of those in stories. It is vital for pupils constantly to work on the school site and in the local area so that they can locate, relate and record the places, features and resources familiar to them.

Remember also not to take any aspect of the interpretation of maps, plans, globes or atlases for granted. Seek instead to involve pupils in discussion and encourage them to put their thoughts into words and sentences. Listen carefully for partial understanding and misconceptions. It may take children considerable time, for instance, fully to grasp the hierarchy that underpins the components of an address. The relationship between villages, towns, cities, regions, countries and continents is not immediately obvious. Furthermore, as Wiegand (1998) has found, the ability to sketch a broadly accurate representation of the continents is present in only ten per cent of upper juniors. So don't expect too much and try to provide contexts that link mapwork to other ideas such as environmental issues.

Finally, you need to help pupils to establish a clear vision about the usefulness of map skills as key skills for life. Today's children – tomorrow's adults – need the practical ability to locate and orientate themselves. They should be able to convey spatial information to other people in a succinct and accurate way. They also need to know how to extract information from maps so that they can describe and interpret world processes and events and the context in which they occur.

References

Blades, M. and Spencer, C. (1986) 'Map use in the environment and educating children to use maps', *Environmental Education and Information*, 5, pp. 187-204.

Catling, S. (1998) 'Children as mapmakers' in Scoffham, S. (ed) *Primary Sources: Research findings in primary geography*. Sheffield: Geographical Association, pp. 10-11.

Hart, R. (1979) *Children's Experience of Place*. New York, NY: Irvington.

Knight, P. (1993) *Primary Geography Primary History*. London: David Fulton.

Matthews, M. (1992) *Making Sense of Place*. Harvester: Wheatsheaf.

Ordnance Survey website: http://www.ordnancesurvey.co.uk

Palmer, J.A. (1994) *Geography in the Early Years*. London: Routledge.

QCA website: http://www.qca.org.uk

Scoffham, S. (1998) 'Places, attachment and identity' in Scoffham, S. (ed) *Primary Sources: Research findings in primary geography*. Sheffield: Geographical Association, pp. 26-7.

Scoffham, S., Bridge, C. and Jewson, T. (1995) *Schoolbase Geography: Copymaster book 2*. Huddersfield: Schofield and Sims Ltd.

Wiegand, P. (1998) 'Understanding the world map' in Scoffham, S. (ed) *Primary Sources: Research findings in primary geography*. Sheffield: Geographical Association, pp. 50-1.

Aerial Photographs

Geography

Land Use
Blu-tac a centimetre grid transparency over an aerial photo of the Brenchley and Paddock Wood area.

Use the ordnance survey map of the same area and write down the co-ordinates of these locations.

Transport (orange on Ordnance survey maps).
Gedges Hill, _T18,S18_
Pixot Hill, _H7 & H8_
Pearson's Green Road, _M10, M11_
Castle Hill, _Q12,Q13,Q14_

Industry & farming (red on Ordnance Survey maps).
Widmore Farm, _P9,P10_
Gedges Farm, _P12,P14_
Goshen Farm, _P8,P8_
Moatlands Farm, _S10, S11_

Recreation and open space (green on Ordnance Survey maps).
Market Heath, _B16,C16_
Brenchley Village Bowling Green, _E16,E15_
Great Wood, _T6_

Residential (grey on Oranance Survey maps).
....Owen's.......house, _D9_
The Police house, _F13_

Service Areas (black on Ordnance Survey maps).
Brenchley Church, _C8_

Pupils' work | *Courtesy of Stephen Scoffham.*
Photo: © Getmapping plc (www.getmapping.com).

IN THIS CHAPTER YOU WILL FIND KEY IDEAS ON
GRAPHICACY • MAPWORK • MISCONCEPTIONS • PHOTOGRAPHS • PROGRESSION

Images in geography: using photographs, sketches and diagrams

Chapter 9: Images in geography: using photographs, sketches and diagrams

Visualise this! A young child is constructing a complex model from boxes, tubes and copious amounts of glue. When a visitor asks 'What are you making?' he replies matter-of-factly, 'I don't know, I haven't seen it yet!'. This gives delightful insight into a child's thinking. The boy needs to see what he is making. He isn't able to visualise the outcome.

There is a lesson here for geography teaching. We spend a lot of our time trying to describe settlements, landscapes, patterns, processes and other aspects of the world but do we actually succeed? Part of the problem is that geographical language uses 'ordinary' words such as river, town, mountain or storm. When we use these terms we tend to assume that children can create, in their own mind, images similar to ours; that they share our visualisation or understanding. But are we sure this is so? What does a child 'see' when we talk, for example, of the 'currents in the river'?

Using visual resources, either first-hand through fieldwork, or through pictorial and photographic images, can help a child really see the geography.

What is graphicacy?

Children increasingly make sense of their world through visual images which, for young children, provide more information than text. The skill of interpreting pictorial forms of spatial information is known as graphicacy. It is an essential form of communication, a life skill which has been described as the 'fourth ace in the pack' alongside, and arguably as important as literacy, numeracy and oracy (Balchin and Coleman, 1965). Graphicacy is a form of visual literacy which can be compared with print literacy in that both employ similar processes: identifying, decoding and interpreting symbols. Furthermore, both print and images depend on prediction, observation, supposition and narrative skills.

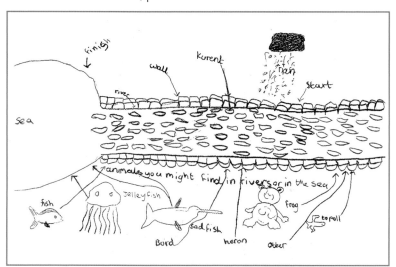

Pupils' work *| But teacher said there were currants in the river!*

Photo | John Halocha.

In developing their graphicacy skills pupils will need to use signs, symbols, diagrams and photographs (including aerial photographs). They will also use maps, atlases and globes and make maps of their own (see Chapter 8). They will refer to these resources separately or in combination, when they use photographs to draw diagrams and field sketches, compare an atlas with a globe or use maps to locate photographs.

Understanding pictures and photographs

As adults we consciously or subconsciously interpret spatial information when we look at pictures in, for example, holiday brochures, catalogues, newspapers and estate agents' windows. We are gathering information about shape or form, scale, size and distance, as well as aesthetics. This is why illustrations are a powerful tool in geography teaching, and why they are essential for learning about localities and processes.

We tend to assume that pictures are easy to understand, requiring little skill, and we often take it for granted that children see what they are asked to look at, and that they see what we see in pictures. But this is not so. The skill of looking at, understanding and interpreting pictures has to be taught through planned directed study. Research tells us that young children do not see a picture as a whole but as a series of apparently unconnected details selected at random (Long, 1953; Mackintosh, 1998). They also notice foreground and large background objects and tend to ignore the middle ground. Children need guidance in selection of detail, training in appreciation of size and in perception of the whole rather than the parts.

We know that, generally, responses to photographs occur in an identifiable sequence. Reception pupils focus first on big things, things they know, recognise and can name, then size and colour become important. By year 2 pupils focus on lots of detail, especially foreground detail, including things relatively unimportant to the understanding of the photograph, and not necessarily the main essence of the picture. By year 4 they tend to concentrate on the main essence and associated objects. Detail is less important to them than getting basic understanding of the picture, whereas by year 6 pupils get a grasp of the whole picture and are able to generalise.

Using photographs

There are a number of strategies and techniques which will help younger pupils to 'see' the content of a picture or photograph. To begin with you can encourage them to:

■ develop the specific vocabulary for identifying different features by describing, discussing and labelling photographs;

■ give the picture a title, to help generalise the 'whole';

■ see all of the picture: look at the foreground, the middle ground and the background, the right and the left, using appropriate positional language, e.g. near to, next to, far from;

■ place labels with specific vocabulary (which you have provided) around pictures as appropriate;

Photo | Kate Russell.

■ label or annotate a black and white photocopy, referring to the coloured picture;

■ label a 'field sketch' or line drawing of the photograph that you have provided (later the children will make their own).

Once you have helped pupils to 'see' photographs, they need to be helped to gather information from them, to 'read' them. This requires them to look closely, to focus on relevant detail, to categorise and compare. Encourage and help pupils to:

■ sort geographical pictures into sets: e.g. river/not river, local/not local, like/don't like, and explain their sorting. Postcards often provide a good range of images for this activity;

■ sequence collections of photographs: e.g. sequence pictures from a story, put a series of photos of local shops in order, sequence the landmarks seen on a recent walk, order photos of a river from source to mouth;

■ join or overlap successive photographs (of townscape or landscape) into a panorama: e.g. a set of photographs taken from the top of a tall building or hill (to do this, take a series of photos with about 25% overlap, as you turn round on the spot);

■ match pictures: e.g. same building taken from different perspectives or elevations, same view taken in different weather/seasons, and explain. (This can sometimes be resourced with postcards.);

■ appreciate size, scale and distance, e.g. talk about how big, near or far the elements are;

■ handle increasingly complex or unfamiliar photographs, encouraging both comparison and generalisation;

■ make a 'field sketch' or line drawing of a photograph, identifying key elements, buildings, landscape features: e.g. by tracing (use a plastic pocket) or freehand;

■ label, annotate, colour elements of (photocopy of) a field sketch to indicate reading of photograph: e.g. identify different types of land use such as agricultural, industrial, residential;

■ take their own geographical photographs, to use for practical activities.

Interpreting photographs

The information pupils gain from photographs is influenced by their past experience, geographical vocabulary, their imagination, preconceived ideas and stereotypes. These often need challenging. You can help pupils to interpret photographs by organising questions into categories. Here are some suggestions which you could apply to the photo in Figure 1:

■ Observation and description (Makes children interrogate the picture)
What can you see? What is it like there?

■ Interpretation, explanation and classification (Requires children to locate learning in their own conceptual framework)
What are they doing? Why are they doing it outside?

■ Comparison (Requires application of learning to a new situation)
Do people do this in your local area?

■ Evaluation (Requires children to make judgements, express opinions and evaluate information)
Would you like to live there?

Figure 1 | *Women harvesting rice near Mandinari, The Gambia.*
Photo | *Margaret Mackintosh.*

■ Extrapolation (Requires children to extend their understanding to determine implications or consequences of natural processes or human actions)
What would happen if they had machinery?

Chambers (1996) gives a slightly different, but equally useful, sequence of open questions providing progression:

■ Concrete   What can you see?
■ Descriptive   What are they doing?
■ Speculative   Why are they working in a group?
■ Reasoning   Why are they doing it by hand instead of using machinery?
■ Evaluative   Is this the 'best' way for these people to do this work?
■ Problem solving   How could their work be made easier or more efficient?

Be creative in your use of photographs, use tried and tested activities and develop your own. The teacher's notes in most resource packs provide many ideas for appropriate activities. These include: labelling, titling, making friends, describing, sequencing, 'good' and 'bad' adjectives, speech or thought bubbles (what is s/he saying/thinking?), questioning, cropping/masking, 'outside the picture', freeze frame, matching sets, drawing photographs, alternative views, comparisons. Whichever you choose, think about the purpose of the activity. Is it helping pupils to see, to read or to interpret the photographs? What graphicacy skills and geography are they learning?

Ultimately, pupils need to appreciate that because geography is based on people's attitudes and values some questions simply do not have a 'right' answer. Indeed, the photographer and the people being photographed will probably have different interpretations of what is happening even though they are both there at the time. Pupils will also discover that because you may not know the answer you will join in the enquiry or discussion with them. Encouraging pupils to work in pairs or small groups to explore and compare their own ideas (and use the vocabulary) can be productive since each pupils' 'reading' or understanding of a picture is different, influenced by their own attitudes, perceptions and previous experience.

What do you need to do?

1. Collect large numbers of geographical photographs for use throughout the school. Your collection could be grouped according to themes such as: general views (urban and rural

views, landscapes, townscapes, human and physical features); settlement (homes, occupations, buildings, journeys, transport); environment (aesthetics, issues, changes); weather (types and effects); water in the environment (rivers, sea, lakes, reservoirs, canals) National and local newspapers and old calendars can be a good source. So too are key stage 1 big books and photopacks focusing on UK, European and other localities. Try to include aerial as well as ground-level views and images showing different weather conditions and seasons. Pictures of issues and changes are also likely to be particularly valuable. Geography is not just about describing places, it also involves finding out about the forces that are influencing and affecting them at a deeper level.

2. Train children to use – to see, to read, to interpret and to question and challenge – pictorial resources.

3. Develop strategies for progression and differentiation in the use of illustrations. One possibility is to use enquiry and issue-based questions with photopacks. Another option is to devise a sequence of questions aimed at developing pupils' conceptual understanding (see Chapter 7).

Pictorial Maps

Pictorial maps, using the term 'maps' loosely, often provide a delightful and useful transition between photographs and conventional maps of places. They help children make and understand the link between the horizontal and vertical view of the environment around them.

There is considerable variety of pictorial map. Some are artistic interpretations of oblique aerial photographs. Some are simplifications that include 3-D drawings of key buildings and landmarks but codify surrounding land-use with colour. One note of caution - Wiegand (2000) discussed the use of picture atlases and, alongside showing their value, emphasised how important it is to check what children understand the pictures to show. He illustrated this by reporting how children interpreted a picture of a red car, which they all recognised as a car, near Birmingham on a map of Great Britain. 'That's where there are a lot of cars, it's a traffic jam', 'It's a car park', 'That's where you buy cars, it's a garage' or even 'They just like red cars at that place'. Very few identified the map-maker's intention of 'It's a place where cars are made'.

Others are conventional vertical views, with roads networks embellished with 3-D drawings of buildings. This format starts with playmats (with distorted scale), but it is also available as attractive commercial maps.

Many places now display pictorial maps in prominent positions on 'You are here' notice boards and there are abundant tourist brochures with attractive pictorial maps and diagrams. Experience suggests that children enjoy using these. They help their understanding so that negative attitudes to maps don't develop.

This progression of pictorial resources – eye-level, oblique and vertical photographs and postcards, pictorial maps of all sorts and conventional maps, atlases and globes – should not be used in isolation to teach graphicacy skills or mapwork. Their use should be integrated into geographical work as a tool, in a similar way to how a dictionary is used in literacy – as a tool that communicates meaning about processes, patterns and places.

Aerial photographs

As adults we do not often use aerial photographs, although there has been renewed interest in them with the publication of several books of oblique aerial views of landscapes, towns and coastlines. Commercial postcards and calendars are useful sources and satellite images are now widely available.

Oblique and vertical aerial photographs have different characteristics. In oblique photographs features are easier to identify, but there is distortion because the foreground detail is larger than the background. In vertical photographs shapes, sizes and patterns are easier to identify, and since there is no distortion the outlines can be traced and compared directly with a map.

Children have a ground-level view of the world. Understanding maps – the geographer's main tool – requires an understanding of a vertical view of the world. This is a big conceptual leap which pupils need help to make. Aerial photographs provide a link between the real world which children see and the abstract world of the cartographer. The photograph has the merit of showing what is actually there, not just the things the map maker has selected to include. The progression from ground level to low angle, high angle and vertical perspective helps children make the transition. You can illustrate this by starting with young children's experiences of toys and models. You can then progress to familiar environments such as the school and local streets before looking at unfamiliar places.

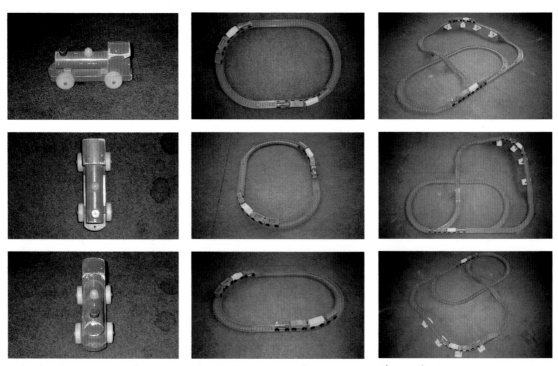

Can the pupils sort the photographs into sets of different perspctives and then work out which way the photographer was facing?

Figure 2 | *Progression in perspective using toys and models.*
Photos: Margaret Mackintosh.

Toys and models

Take photographs from different perspectives and have multiple copies printed for pupils to sort and match (see Figure 2).

Take oblique and vertical aerial photographs of different arrangements (e.g. of train set, Lego models) for pupils to reconstruct, using the photograph as a pictorial layout diagram or map.

Real environments: school, school grounds, area around school

Identify human and physical features including buildings (landmarks) and routeways; recognise plan shapes, colour, texture, tone and pattern.

- Label an outline drawing of a photograph.
- Add detail to an incomplete fieldsketch.
- Make a sketch map of part of a photograph.
- Recognise different land uses.
- Trace a vertical photograph to make a map.
- Select and add detail to the map.
- Match a photograph to a real map, orientating appropriately.
- Identify relative and actual position, compass direction, scale and distance.
- Identify the time of day or year when the photograph was taken.

Plester *et al.* (2003) have shown that the youngest pupils can benefit from using aerial photographs. In the school environment they can recognise the correspondence between a photograph and the landscape. Using a photograph rather like a map they can identify features, locate themselves and locate places. Using a photograph like a treasure map they can locate treasure!

Unfamiliar environments

Undertake similar activities to above, extending learning to new situations, especially settlements, rivers, environmental changes and issues. Map patterns made by features (e.g. town buildings, road networks, rivers, fields), and patterns of land use; Compare with own area, identifying similarities and differences; Compare vertical and oblique views.

Satellite images

Identify features and patterns e.g. settlement, transport, rivers, coastline, land use; Link with vertical, oblique and ground level photographs; Describe routes; Link with maps and globes.

What do you need to do?

1. Build up a collection of vertical and oblique aerial photographs, e.g. of classroom equipment (furniture, toys and models); the school (some taken from nearby tall buildings, if possible); the school grounds (possibly taken from the school roof – close-up oblique); the local area; other localities.

2. Collect postcards, calendars and books showing oblique aerial views – there are some excellent ones available.

3. Compile a range of satellite images. These are often produced as posters and postcards (Barnett *et al.*, 1994). You can also make video recordings of national weather forecasts and other TV programmes that involve remote sensing techniques.

Signs, symbols and logos

In adult life we gather spatial information from road signs, locational and directional signs at airports and other public places and spaces, many of which include internationally accepted symbols. We also discover the location of significant buildings, including specific shops, railway stations, National Trust properties and so forth, by identifying logos, both in the real world and on maps. And, of course, we gather information from weather forecast maps by interpreting symbols.

There are many practical ways in which pupils can learn about signs in your school. To begin with they can look for examples of signs, symbols and logos that are already in use. For example, there will be directions to the Head's office, and signs on the toilet doors. You might also encourage pupils to:

Photo | *The Land's End & John O'Groats Company.*

■ Erect signposts or symbol-posts around school, the pupils inventing the symbols, indicating direction and distance to, for example, the library;

■ Erect a playground signpost or symbol-post, indicating direction and distance to nearby landmarks (e.g. the town centre, the secondary school), to nearby towns and villages, and to more distant localities (e.g. London, Edinburgh, New York), providing opportunities to incorporate compass and mapwork;

■ Devise symbols to indicate the location of equipment in the nursery or reception class – link these with symbols for pupils to place on a model or plan of the classroom;

■ Put appropriate symbols on various spaces and places in and around school;

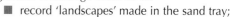

- Place 'traffic' signs at key points to encourage safe movement in and around school: older pupils, identifying a movement problem in school, might like to study and 'solve' it with the use of signs and symbols;
- Create their own weather forecast symbols and charts, collect newspaper weather charts, and watch television weather forecasts.

Diagrams

In adult life we interpret diagrams showing spatial information when we use a car maintenance manual or a dress-making pattern, assemble flat-pack furniture, wire up electrical equipment, plan a new kitchen or study diagrams in textbooks or newspapers. These are all two-dimensional representations of three-dimensional structures, requiring the ability to interpret perspective.

Children need to be taught to 'see' and 'read' diagrams in much the same way as they need to be taught to 'see' and 'read' photographs. Developing the skill to understand two-dimensional representation of three-dimensional structures starts with toys. Among other things you could encourage pupils to make toys from assembly diagrams (e.g. Lego, Meccano and Airfix), and create models from diagrams (e.g. with boxes or wooden bricks). As well as developing geographical skills this work will provide many cross-curricular opportunities in technology, maths and science. You might also encourage children to:

- record 'landscapes' made in the sand tray;
- make drawings/diagrams to record the layout of play situations (e.g. the home corner set out as a café, a travel agents, a shop);
- represent familiar objects such as building blocks, perhaps constructed to represent local buildings, in both side and plan view;
- add labels to postcards and photographs to identify and highlight key features.

One type of diagram which has a unique role in geography is the field sketch. Field sketches attempt to summarise aspects of an urban or rural landscape using hand-drawn outlines and supporting notes, e.g. different areas in a school playground or the course of a river in a mountain environment. Making a field sketch depends not so much on artistic ability as ability to generalise and identify key features in a real world setting. You can practise making field sketches in the classroom working from photographs before taking pupils outside (Figure 3).

Photo | Paula Richardson.

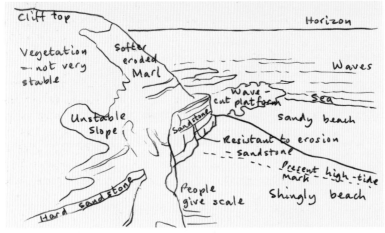

Cliff top   Horizon

Vegetation — not very stable

Softer eroded Marl

Waves

Wave-cut platform   Sea

Sandstone

sandy beach

Unstable Slope

Resistant to erosion — sandstone

Present high-tide mark

Hard Sandstone

People give scale

Shingly beach

Figure 3 | *Practise identifying key features on field sketches from photographs.*
Photo: Margaret Mackintosh.

Developing a sense of place

Everyone sees and experiences a place differently. Is one person 'right' and another 'wrong'? Of course not. We all have different angles and viewpoints. People are increasingly aware of this nowadays and recognise the validity of multiple perspectives. In the same way we need to remember that each pupil will interpret photographs and perceive places differently, depending on their own experiences, opportunities and interests, and those of their family. They will compare the view with places they know, and make judgements. It follows that in building up your own photograph collection you should try to assemble as wide a range of images as possible. Include the backstreets as well as the town centre. Photograph bad points as well as good things, not just key features. There will be lots of possible responses to these images – a different one from each pupil – and they are all right. Value each pupils' perceptions and ideas.

You also need to encourage pupils to think about places as social spaces or areas where people go about their daily lives, getting to know different districts, streets and buildings as they pursue their various interests and opportunities. Everyone has a different set of social interactions which contributes to their individual sense of place. These interactions will influence how children interpret photographic images of places as they use their geographical imaginations. Ask pupils 'What is this place like for you, your family, other people?' and expect a wide range of valid responses.

Remember too that when they look at photographs children interpret visual images as text. They will be using the images to tell themselves a story with all the distortion and bias this involves. To help your pupils to appreciate that photographs are taken with a purpose or point of view ask questions like 'Who has chosen these views?', 'How do they make you feel?', 'What message is given?', and 'Do the photographs tell the whole story about the place or event?

It is important to help pupils to appreciate how hard it is to communicate what a place is really like. If they appreciate this in their own locality and in culturally familiar 'western'

Photos | *Margaret Mackintosh.*

environments they will be better placed to understand the limitations of the images in photopacks of culturally unfamiliar distant localities. Ask pupils to make a photopack for their own local area. How will they decide which photographs to include, for example which type of housing, which shops? How will other people interpret them? Discuss whether a photopack can ever tell the whole story about a place.

You can illustrate the way that visual images select and distort information in this simple classroom activity which also works well in INSET sessions. Ask pupils to bring photographs to school from local or national newspapers of a major geographical event such as an earthquake, volcanic eruption, flooding, wind farm proposal, dwindling fish stocks or deforestation. Add photographs which you have collected yourself of other major news items which have happened in the same time span. Mount each one on A3 paper and arrange them on desks around the room. Use the photographs as 'text' to tell a story. When pupils have looked at all the pictures ask questions like:

- What is it about?
- Are all the photographs about it? (to eliminate non-exemplars)
- What is the place like?
- Who is involved?
- What would you like to ask them?
- What do the people think about it?
- What did it feel like for them?
- What is the story? (Sequence the photographs – e.g. before, during and after rescue from flooding)
- Why did it happen?
- Could it happen again?
- What is the future?
- What do you feel about it?

These particular questions were used with photographs of flooding in Mozambique.

Photo | *Per-Anders Pettersson/UNEP/Still Pictures*

Photo | *Paula Richardson.*

Thinking about the future

Change is a key concept in geography. As well as changes which happen on a daily, seasonal or longer time span, change also involves considering the future and anticipating different possibilities. Young children can be introduced to the idea of change by using Jeannie Baker's delightful picture book *Window* (Baker, 1991). This shows how the view from a boy's bedroom window changed as he grew up and eventually moved away – maybe because of the changes.

Discuss with pupils the changes, likes, dislikes and alternatives from their own bedroom or school window. Consider what might happen in the future – next year, in five years', ten years', twenty five years', a hundred years' time. Not only will this help them to understand how change happens, it will also help them to see how they can influence change to some extent. This work links very well with children's strong concerns for the environment. Experience suggests that they like this because, as with the work on geographical imaginations, there isn't just one right answer. The pupils' different opinions can be valued and discussed. There is particularly useful illustrated material, including classroom activities, in *Educating for the Future* by David Hicks (1994). Children see their world as it is now. Helping them to visualise the future is vital if they are to appreciate how important it is that intelligent decisions are made in the present.

Understanding geography depends on visual images. Children can see geography through fieldwork, film and video, photographs, diagrams and maps of all sorts. Children usually see their world at ground level, in close-up – the small picture. Each image is like a piece in a jigsaw. They also need to use oblique and vertical aerial photographs to see their world from above and from a distance – the big picture. This helps them to make sense of their immediate surroundings, and localities further afield, to see how each individual jigsaw piece contributes and fits in to the whole picture.

Useful resources

■ Newspaper photographs

■ Television news clips

■ Postcards, including oblique and aerial images

■ Calendar photographs

■ Photographs of your local area, from ground and elevated levels – keep old photographs to identify change

References and further reading

Baker, J. (1991) *Window*. London: Red Fox.

Balchin, W. and Coleman, A. (1965) 'Graphicacy should be the fourth ace in the pack', *Times Educational Supplement*, 5 November.

Barnett, M., Kent, A. and Milton, M. (eds) (1994) *Images of Earth: A teachers' guide to remote sensing in geography at KS2*. Sheffield: Geographical Association.

Chambers, B. (1996) 'Step by step', *Junior Focus*, April, pp. 2-4.

Hicks, D. (1994) *Educating for the Future*. Godalming: WWF UK.

Long, M. (1953) 'Children's reactions to geographical pictures', *Geography*, 38, pp. 100-07.

Mackintosh, M. (1998) 'Learning from photographs' in Scoffham, S. (ed) *Primary Sources: Research findings in primary geography*. Sheffield: Geographical Association.

Plester, B., Richards, J., Shevelan, C., Blades, M. and Spencer, C. (2003) 'Hunt from above', *Primary Geographer*, 51, pp. 20-1.

Wiegand, P (2000) 'cARTography', *Primary Geographer*, 42, pp. 16-17.

Fieldwork

Fieldwork or outdoor activities can take place in and around every school. Working outdoors provides pupils with a new way of looking at their environment and allows them to develop a curiosity about where they live as well as teaching them to become more observant and thoughtful.

This chapter outlines ways of developing fieldwork and gives practical suggestions about how to organise visits. There is also important safety advice for geography co-ordinators and recommendations about where to go for further information relating to out-of-school activities.

Fieldwork is using an outdoor experience to provide an environment to test out ideas and hypotheses, and to help pupils to extend their understanding of the world around them. However, as teachers, we know it is more than that. It is allowing our pupils to see for themselves and experience things as they are in the real world: to hear traffic noise, see the rainbow and feel the wind is all part of the magic of 'going out'. Exposure to television, advertising and other media has extended children's awareness of other places and environments, but they may nevertheless remain uncritical, unobservant and unchallenged. Even well-travelled pupils often make little use of either observation skills or the experiences they have of their own locality.

Fieldwork or out-of-classroom experiences offer you the opportunity, through a structured pathway, to help pupils to become observant, to develop the skills of recording, analysis and deduction and, it is hoped, to develop enquiring minds. Over a hundred years ago the great geographical educationalist Sir Archibald Geikie made the important link between research enquiry using observation and recording in the field, and good practice in the primary classroom. In the 1960s the Plowden Report (Ministry of Education, 1967) reaffirmed that 'exploration of the school locality as a teaching laboratory' is an essential part of developing pupils' geographical knowledge. Recent Ofsted reports continue to stress this message. *The Curriculum in Successful Primary Schools* (Ofsted, 2002), for example, was firm in its declaration that in the exemplar schools selected for detailed study 'the curriculum was enriched by first-hand experiences, including visits locally and further afield'.

As adults we know that it is 'the doing' which can help us to reach a fuller understanding of things; for children this need 'to do' is even greater. Fieldwork gives them real practical opportunities to see, hear, feel and understand for themselves. This practical experience can use a range of environments such as the interior and exterior of the school itself, local streets and buildings or more distant localities including places visited during residential field visits.

How can I plan and organise fieldwork experiences?

Out-of-classroom experiences can provide stimulating support for any topic, serving either as an introduction to a topic, to enhance work in progress or to pull threads together at the end. Sometimes it is really important for you to take the pupils outside without doing too much preparation and to ask them to look with fresh eyes at a place or an activity. Questions such as 'What is happening?', 'What is this place like?', 'What do you think about this?' will often

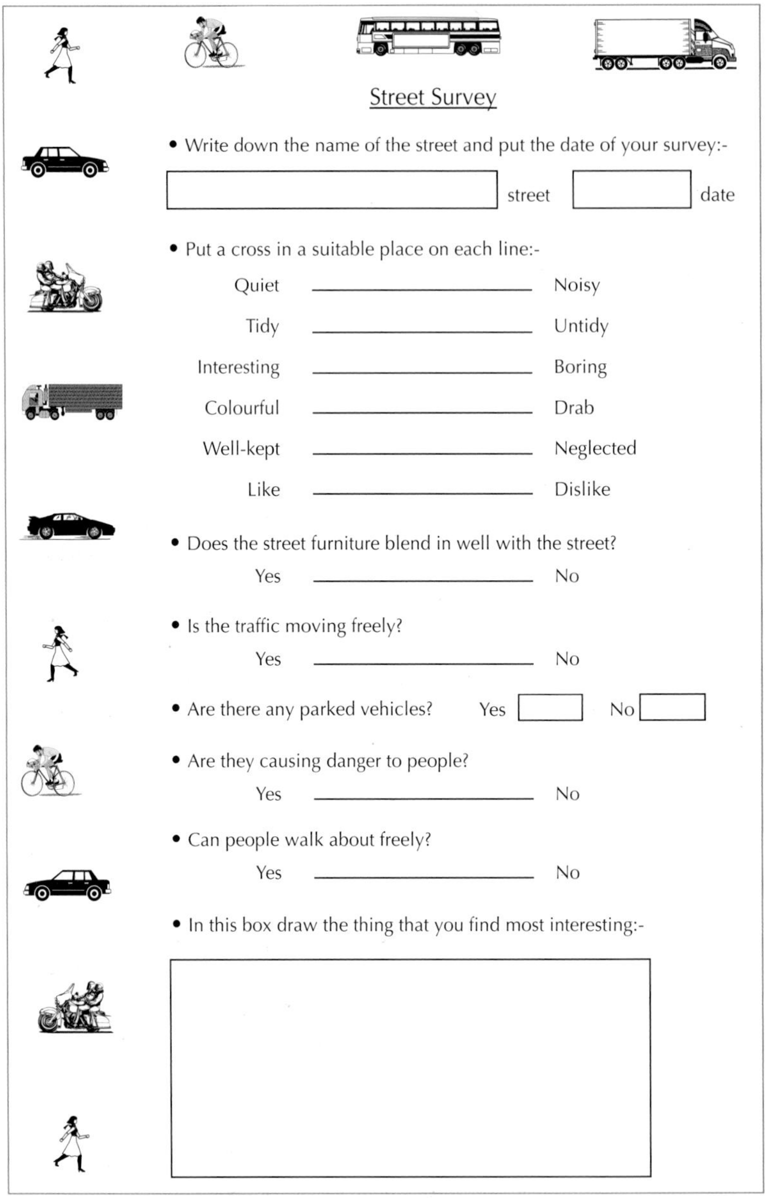

Figure 1 | *A sample questionnaire for use in an urban area.*

Street Survey

- Write down the name of the street and put the date of your survey:-

  [   ] street   [   ] date

- Put a cross in a suitable place on each line:-

Quiet	———————	Noisy
Tidy	———————	Untidy
Interesting	———————	Boring
Colourful	———————	Drab
Well-kept	———————	Neglected
Like	———————	Dislike

- Does the street furniture blend in well with the street?

  Yes ——————— No

- Is the traffic moving freely?

  Yes ——————— No

- Are there any parked vehicles?   Yes [   ]   No [   ]

- Are they causing danger to people?

  Yes ——————— No

- Can people walk about freely?

  Yes ——————— No

- In this box draw the thing that you find most interesting:-

  [   ]

elicit spontaneous and interesting responses. These kinds of conversations and discussions will help to give pupils moments of awe and wonder and excite their curiosity and thinking. At other times it will be important to ensure that pupils have an activity to complete or to practise a specific skill such as sketching or map reading. The most important thing, however, is that you consider a variety of approaches for your work outdoors. At the Foundation Stage pupils will need many experiences of 'seeing it' for themselves, and the local area is excellent for looking at places, people, jobs and services. Even at this young age pupils should also be taken further afield on occasion, for example to the seaside, the countryside/a big town so that they can experience the unknown and the unfamiliar.

Introducing pupils to a variety of places will help to challenge their understanding and their perceptions and open their eyes to the nature of different places. Remember that older pupils may repeatedly visit the same locality in their fieldwork over time so it is important to ensure that there is progression in terms of activities, analysis and intended learning outcomes to keep the study of the same area fresh and interesting. When you organise geography fieldwork you also need to strike a balance with other curriculum areas. History, art, maths, English and other subject areas will all overlap with what you are doing. It makes sense to develop these links but you also need to have clear geographical objectives otherwise the focus of what you are doing will become blurred and the geographical work may become too diluted to be useful.

When they undertake fieldwork pupils can work in a variety of groupings, with a partner, in a small group, as a whole class, or even with pupils of other ages and abilities from the rest of the school. All too often when pupils are taken outdoors they end up as slaves filling in worksheets. Try to avoid this and encourage them to observe, sketch and think about what they are seeing. Pupils also need to reflect on what they observe and appreciate that when it comes to the environment there are very few situations where it is possible to be definite about what is 'right' and 'wrong'.

Digital cameras are an excellent way of recording what the pupils see. One of the advantages of these more modern cameras is that pupils can select the pictures they want to keep and review the places they need to revisit and photograph again. Pictures of a place at different times or seasons will provide pupils with a visual record of the concept of change. The essence of a good geographical experience is the extent to which the pupils will be challenged in terms of analysing the reasons for these changes and arriving at conclusions about what they have seen.

Fieldwork provides many opportunities for pupils to investigate things for themselves as a homework activity, and bring their findings into school for discussion and analysis. Data collection is another valuable activity for pupils to do at home. The questionnaire in Figure 1 is an excellent example which can be completed from outside their houses. Alternatively, each pupil asked two people to complete the questionnaire as a homework exercise the class would quickly assemble a significant data set. The pupils could then analyse the results and draw conclusions using a prepared database on the computer. Not only does work of this kind represent good practice, but you can collect first hand data without even having to organise a visit!

Which places can I use for fieldwork?

One way of structuring fieldwork is to start with the pupils' immediate surroundings and gradually broaden into less familiar territory as they become more confident with each area and better at conducting their enquiries and investigations. However, it is important that pupils have the opportunity to experience different locations throughout their school years so that very young children, for example, are not simply confined to the school grounds but are also given experiences in places further away. A simple 'nested hierarchy' of fieldwork areas is shown in Figure 2, and the different potential of each area is examined in detail in the paragraphs which follow.

School grounds and the school buildings

The school buildings and surrounding grounds provide an initial stimulus for observing, recording and enjoying work outside the classroom. They also provide a useful context in which to practise sketching, estimating and evaluating aspects of the environment. This means that when pupils are further afield time is not wasted in learning how to perform these skills; they can be put to use directly, and related to the work in hand.

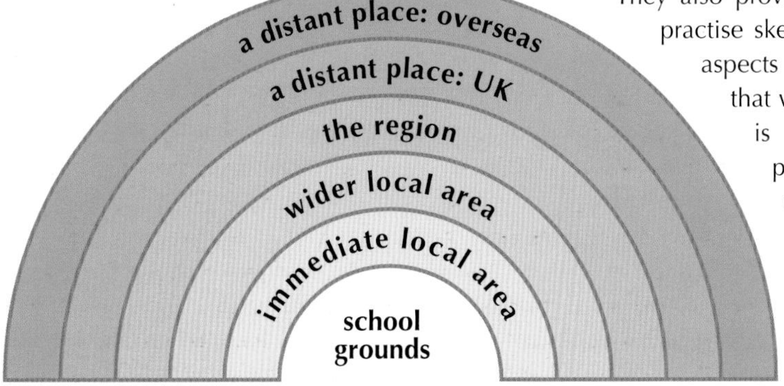

a distant place: overseas
a distant place: UK
the region
wider local area
immediate local area
school grounds

Figure 2 | Broadening fieldwork horizons.

Urban

Investigation of local shops/shopping parades/corner shops.

Comparison of buildings/designs/materials.

Assessment of litter problems, recycling issues.

Environmental walks to assess quality of the environment using an index.

Investigation of the impact of traffic in the area – How is parking/speeding controlled? Is it effective?

Visit to a building site – monitor progress.

Identify and collect logos/signs – do patterns emerge in the provision of services?

Map the land use of the area using a key.

Map routes to school/pupils' houses.

Design/locate a leisure centre.

Rural

Complete a seasonal walk (each term to record changes).

Investigate the village/local services provided.

What is the effect of any closures, e.g. local shops?

Map the land use of the area using a key.

Identify/visit areas of change, e.g. a new building under construction, a road/bridge development.

Examine housing types/styles/ages.

Map where pupils live – routes to school or town.

Interview residents – What is good about living in the area? What are the drawbacks? What changes would they like to see?

Walk to assess environmental quality of the area using an index.

Investigate traffic problems.

Figure 3 | *Lines of enquiry for urban and rural fieldwork.*

Activities in the school grounds could include:

■ finding different kinds of building materials;

■ 'dating' the area on a map – Which parts are older/newer?

■ using map keys;

■ finding pleasant/boring/warmest/coldest spots around the grounds;

■ making up directions/routes round the site;

■ sketching the views from the site;

■ doing an environmental audit of the site;

■ improving/changing the site;

■ identifying the impact of seasonal/time changes on the site.

Immediate local area

The potential of the local area will vary enormously from school to school, but some of the following themes will be suitable for investigation:

Photo | *Tina Horler.*

■ Streets, roads, buildings

■ Local shops

■ Recreation, park areas

■ Woodlands, fields

■ Water features

■ Transport links – car parks, traffic

■ Local viewpoints

■ Industrial estates

■ Pupils' home areas

■ New developments, e.g. building sites

■ Service provision in the area

The local area has the benefit of being accessible and can therefore be visited frequently, cheaply and for a variety of reasons (Figure 3). Individuals and groups can revisit to check evidence and observe changes over time. However it is important that out-of-classroom geography activities are not seen to be restricted to the study of the local area on foot, while other curriculum areas, such as history, involve more exciting visits by coach to castles, famous houses and open-air sites!

Visits to more distant places in the region, the rest of the UK and overseas

Travel to places away from the immediate vicinity of the school provides a number of benefits and experiences for pupils. They begin to see their own area in a wider context and can identify the inter-relationships which exist between them. They may see some contrasting environments and be able to compare how things change from place to place. The journey itself will also be part of the experience. It is quite possible that some of the pupils will never have been on a train journey before. Finance and school organisation have a major part to play in determining how often visits of this type can take place. However, it is very important to allow pupils to have these experiences to broaden their horizons in both the curriculum and the social sense.

One way of setting up visits further afield is to make links with another school and offer a series of return visits. The potential educational value is huge and the project can be very attractive financially. For example, if each school hosts its visitors for the day, the only outgoings would be transport costs to the school. Many primary schools are also taking longer residential visits to a centre either in Britain or overseas, often in France. These can be very exciting experiences for pupils and the personal and social benefits for those who go away from home are immense, even on a short stay. It is also possible to have just a one-night stay which could be suitable for younger pupils and which can still be very rewarding and exciting but without the greater financial commitment of a longer visit.

What guidelines should I follow?

In order to plan effectively for fieldwork and other out-of-school visits it is important that a policy or set of guidelines is established and used by all staff when planning off-site visits. By establishing links to the curriculum this will ensure that fieldwork or visits become part of the learning process rather than an optional extra which happens at the end of term. These guidelines will need to incorporate LEA/ Governor guidelines which already exist and have been approved by the governing body of the individual school.

Figure 4 offers some ideas that might be included in the policy. Each might only be a sentence or so, thus ensuring that the document as a whole is concise and helpful to those who use it. The *Fieldwork File* resources (GA/FSC, 2005) provide guidance on all aspects of fieldwork safety.

A policy document might consider:
1. The importance of working outdoors
2. The aims and objectives of fieldwork
3. How fieldwork supports the curriculum, including cross-curricular links
4. Suggestions for locations and possible activities/ideas for progression
5. Assessment opportunities
6. List of fieldwork equipment/where equipment is stored in school

A list of resources such as:
1. LEA guidelines/country code/health and safety guidelines
2. Templates of letters to parents/adult helpers/booking guidance
3. Chart for completing a site risk analysis
4. Medical forms
5. Mini-bus information

Figure 4 | *Towards a school policy. After May et al., 1993.*

How can I plan a visit?

Visits beyond the school grounds must always be planned with great care and with close attention to health and safety issues. The following checklist highlights the essential elements but detailed guidance will be available from the LEA which will help you though the process of organising a visit. An excellent pack entitled *The SAFE pack* is available from Norfolk County Council and includes a CD with downloadable sheets for use in planning (NCC, 2001).

The following is a possible checklist for planning fieldwork visits:

1   Permission from headteacher ☐
2   Choice of location – aims, objectives, dates ☐
3   Refer to school/LEA documents, insurance, costings, etc. ☐
4   Book visit/transport, check timings ☐
5   Fill in required forms/letters to parents, medical forms ☐
6   Do site visit and risk assessment ☐
7   Prepare pupils – activities, clothing, equipment ☐
8   Brief staff and adult helpers ☐
9   Keep good accounts of monies collected ☐
10 On the day: leave lists of participants at school, know emergency procedures, brief staff, organise equipment and who is responsible for bringing it back ☐
11 Follow up: letters of thanks, payments, classwork, evaluate visit and note any changes to be made ☐

Risk assessment and child protection issues

One of the most important aspects of pre-fieldwork preparation is visiting the chosen site to assess risks, to identify safety issues and to gain up-to-date information. For example, you will need to find out about parking, toilets and wet weather alternatives. Figure 5 provides guidelines which you might like to follow. Once you have completed the sheets you can file them as a resource when you return to school and update them as needed.

There are a number of stages to carrying out a risk assessment which ideally should be done not too long before the visit takes place.

A risk assessment should be undertaken for each new site that you use and updated for places you have visited previously (Figure 6). This is a valuable exercise that helps to identify issues which might create problems and helps you to decide if the area is relatively safe to visit. As well as recording possible hazards you can evaluate them on a scale from one to five, with five providing an unacceptable risk.

A risk assessment is not merely a paperwork exercise. It also needs to be shared with the pupils and helpers so that they too are aware of potential dangers. One way of doing this is to ask the pupils to identify what they believe are the main risks before they go on a visit. For example, one class that was about to visit the Pennines decided that the chief dangers they faced were (a) getting lost (b) falling over (c) falling into water. Their recommendations included staying in a group with a grown up, wearing walking boots with a good grip and taking special care near the edges of rivers, lakes and streams.

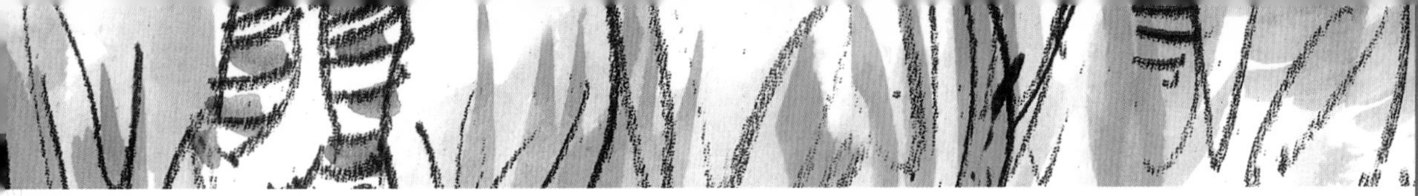

Figure 5 | *Fieldwork visit record sheet.*

Location: _____

Address: _____

_____

Tel: _____

Date visited: _____   Year group: _____

Number of pupils: _____   Topic links: _____

Number of adults: _____

Education facilities

Education officer

Educational resources/worksheets

Teacher's notes/guidebooks

Handling materials

Loans

Video/tape commentaries

Talks/activities

Practical details

Cost involved

Journey time

Parking facilities

Lunch room

Education room

Toilets

Shop

Aims of visit

On-site activities and organisation
(Attach any useful material)

Risk Assessment completed   Yes/No
(Attach assessment)

Overall risk comment

Follow-up work

Evaluation/comments

RISK ASSESSMENT QUESTIONS

Place:
Date of assessment:
Name of person making the
assessment:

1. What are the hazards which could
   occur here?
   (Narrow pavements, many people,
   steep banks, deep water, slippery
   grass, hidden bends, etc.)

2. What could happen as a result of
   a particular hazard?
   (Pupils pushed off pavement, difficult to cross the road, slip and fall,
   not be able to see the view.)

3. What is your evaluation of the situation? Is the place safe to use?
   (Cross the road further down, use
   the pavements which are wider,
   use the pedestrian crossing at X,
   give clear instructions at point Y to
   remind pupils to take care, put staff
   in correct positions, use X Street as
   an example rather than Y Street.)

4. Date and sign the assessment
   details.

Add any further information
(Where will the first aid kit be?
Emergency telephone numbers. If
there is an accident who will deal
with it?)

5. Review
   When will a review of the location
   be needed?

*Pupils' work | Shelley
Primary School, Horsham.
Photos: Paula Richardson.*

*Figure 6 | Stages to consider
in a risk assessment exercise.*

Falling over wearing walking boots
with grip, keep to the path and walk
sensibly, don't mess about near the edge

falling in don't go near the edge,
walk carefuly, take care, dontmess about,
don't run off,

Any helpers who have significant access to pupils during the field trip will need to give their consent to a Criminal Records Bureau check. This is now a standard procedure and your LEA or governing body will have a system in place. Be sensitive in the way you explain what needs to be done. After all, it is for the benefit of the helpers themselves and will give peace of mind to parents whose children may be in their care. You also need to see that any adults who accompany the pupils are aware of the code of behaviour expected of them in relation to the pupils in their care. And most important of all, make quite sure they know what the visit is about and what you want the pupils to learn from it. A short briefing paper for helpers would be useful.

How can I make fieldwork progressive?

The local area is relatively accessible and cheap to use for fieldwork and pupils may find that they visit it several times during their primary school years. This can result in the area, or certain features of it, being overused. For example, a pupil might be asked to study the local shopping area several times over, often using similar activities. For this reason liaison between staff who teach different year groups and key stages is crucial. Figure 7 gives an example of how a study of a local shopping centre can progress from a simple analysis of the area (for lower primary) to a more in-depth exercise (for upper primary). This approach takes account of pupils' increasing maturity and understanding in various ways, such as:

■ A move from closed to more open questions

■ An increase in problem-solving or decision-making activities

■ A movement from one to several variables

■ A move towards collaborative work and more independent learning

One of the striking aspects of outdoor work is that it can be tailored to meet the varying needs of individual

A shopping parade – lower primary age

1   On your base map, write in the types of shop, or colour them in and make a key in the margin.

2   What is the use of the floor above most of the shops?

3   Which types of shops are missing from here?

4   Is parking easy?

5   How many parking spaces are there?

6   How long do cars tend to stay here?

7   Why is this?

8   Where do you think the shoppers come from?

9   What can you do here besides shop? (telephone, recycling banks, postbox, etc.)

A shopping parade – upper primary age

One of the shops is empty. You have decided to rent it and open a ...

1   You need to decide what service or shop is needed here. How will you do this? Question shoppers? How many? Look at other shopping possibilities in the area? What is the market for your service/shop?

2   You have to decide on a window display, signs, layout, uniforms, opening hours.

3   Plan for your staffing – write an advert for staff (look at examples in local newspapers). Think about costs and rates of pay.

4   Plan your opening date. How and where will you advertise it? Local paper? Local radio? Free gifts?

5   How will you make sure your shop/service survives?

Figure 7 | *Progression fieldwork tasks in a similar location.*

pupils within a group. Teachers often find that some pupils are able to investigate the environment and analyse evidence to quite a high level even though their skills of reading and writing are relatively poor. It is important to offer these pupils a range of ways in which to record their findings so that their geographical attainment is not masked by their ability to read or write. You may find that asking them to record evidence using a tape recorder, digital camera or by choosing a scribe within the group can help the geographical understanding to shine through. It is also important to see that you challenge pupils who are more able than average. One way of doing this it to pose further questions, speculate on different 'what if' scenarios, and invite independent research back in the library or using the internet. Fieldwork provides the evidence for what exists in the environment at that precise moment and so it is important for all the participants to analyse the findings, discuss the reasons for the answers and think about how things might look in the future. All pupils can then participate by making a contribution to the final presentation of their work.

Are activity sheets useful?

While it is important to allow pupils to observe and use all their senses without too much distraction from worksheets, there will always be times when a structured approach will be helpful. It may well be that some pointers to stimulate questions/identify features/ observations may be more appropriate at certain times on a sheet which then can be recycled and used again, rather than a sheet for writing on. The following are some simple rules for making a worksheet as useful as possible. The best test for a successful worksheet is to ask yourself if you would enjoy/be able to do it yourself!

1. The sheet should always be clear and well laid out.
2. The text should not be too crowded or contain too many tasks or directions on the same page.
3. There should be a variety of activities, including an open-ended one if possible.
4. Questions which begin with How? Why? Which? Where? tend produce better answers than those which begin Do? Does?
5. Answers should require a variety of styles of response: draw, estimate, decide, underline, count, and so on.
6. Vocabulary should be straightforward.
7. Pupils should be encouraged to use their senses whenever appropriate and safe: hear, see, smell, feel, taste.
8. Pupils should be asked for their views and give reasons for the things they like/dislike.
9. Cover survey sheets with plastic and get pupils to mark their answers with drywipe pens so the sheets can be re-used.

Resources for fieldwork

It is important to build up resources for local fieldwork, not least to help and inspire other colleagues. The resources should be centrally located and it is helpful to have several copies of the most useful items. It is also essential to keep adding to the bank of items with, for example, photographs of changes in the area, and newspaper cuttings.

Photo | Paula Richardson.

Resources are divided into three types:
1. Visual, written and other basic resources – for local fieldwork
2. Equipment – for use in the field
3. Information regarding places to visit

Visual, written and other basic resources

Maps	at scales of 1:50,000, 1:25,000, 1:10,000, 1:2500; local street maps; theme maps
Booklets	local directories, e.g. Kelly's; guidebooks; leaflets about area
Photographs	ground-based of selected places/items/festivals; sequence photographs of, e.g. building sites; aerial photographs, both professional and amateur; postcards; newspaper pictures
Statistics	about weather (local information), traffic; environmental issues, conservation in local area. The last three, plus local Census data, are often available through local Borough/Council departments
Audio tapes	local people talking about changes/living in the area
Newspapers	cuttings about a variety of local issues
Previous work	data collected on previous fieldwork visits as evidence of change – and for making comparisons
Artefacts	relating to local geology/industry

Equipment

Key stage 1

- clipboards
- tape measure
- magnifying glasses
- metre rulers
- non-standard measures, e.g. foot
- stop-watch/clock
- compass
- rock hardness testing equipment
- thermometer
- large-scale base maps
- plans, e.g. school, shopping area
- simple soil testing kit
- simple hand-held anemometer
- wind sock
- rain gauge
- soil profiles in jars

Key stage 2 – as for key stage 1 plus:

- recognition cards for trees, etc.
- quadrats
- soil augers
- water pollution testing kit
- maps (1:50,000 and 1:25,000)
- noise meter
- wind gauge
- more sophisticated soil testing kit (to include chemicals)
- cloud identification charts
- clinometer

Photo | John Halocha.

Activities

Out-of-classroom activities should take advantage of the chance pupils have to see, measure, identify and reason out things for themselves. Fieldwork is an opportunity to focus attention on a small area and develop the skills of recall, thinking and prediction as well as techniques of sketching, map reading, measuring, and so on. It is also important to help pupils to look at the detail as well as the big picture, for example, looking at how the brickwork of a house is constructed and then working out how to show this on a sketch. You can help pupils to achieve excellent field sketches by asking them to label features which are hard to draw, thus reminding them that artistic merit is not the main function of a field sketch!

The two activities below suggest some ideas for using the local area in a creative way with pupils.

Picture activities

1. Take pictures of specific places in the immediate area, e.g. church, set of traffic lights, shops, memorials. On a large, possibly hand-drawn, map of the area displayed in the classroom, ask the pupils to locate the photographs correctly. Ask them to check these on their way to and from school and discuss their findings the next day.
2. Take a series of photographs in the area or in school, and cut them in half. Ask pupils to sketch in what is missing or complete a simpler activity such as matching them up. Again ask them to check when they are outside. The checking is important as it involves pupils in decision-making aspects of the activity.

What is our place like?

Tell the pupils they are responsible for planning a short walk for some visitors round their local area. They will have to choose the most interesting places to show them in a given amount of time. The pupils will need to discuss what makes their area interesting/ special/memorable. If you decide that the visitors come from abroad it will provide an opportunity for pupils to decide what characterises their area and makes if different from elsewhere.

Walk the pupils round with a view to deciding which places/views/buildings will be best to include. The outcome might be a guided tour, brochure or perhaps a self-guided tour put onto tape.

bar

Photo | *Paula Richardson.*

If you can organise some visitors to undertake the tour this will provide the pupils with a chance to develop their social skills as well as their geographic ones! The evaluation of the success of the project will be important too. What went well? What didn't go so well? What will we change if we do it again?

Conclusion

Fieldwork and outdoor activities offer the opportunity for interesting and innovative teaching and learning: we need to make the most of it to bring a real and practical dimension to our pupils' geographical experience. The argument for fieldwork has also been dramatically strengthened in the past few years by neurological studies of the brain. We learn best, it seems, from rich multi-sensory environments that provide a range of messages and meanings. Immediate feedback is also highly desirable (Scoffham, 2002). Fieldwork provides both of these things. It is thus a vital part of the arsenal of teaching strategies available to teachers and thus a key technique for promoting geographical learning.

Fieldwork allows pupils to see what is really happening rather than what is supposed to happen. A quiet street may become a busy thoroughfare for traffic in the rush hour as drivers try to by-pass a bottleneck of traffic further back in their journey. It is the personal observations and investigations which can help pupils to understand how and why their environment is a constantly changing place. We should remember that for many pupils a fieldwork visit may be one of the most exciting and memorable events of their lives (Chambers and Donert, 1996). When did we last give our pupils a fieldwork experience to remember?

References

Chambers, B. and Donert, K. (1996) *Teaching Geography at Key Stage 2.* Cambridge: Chris Kington Publishing.

DfEE (1998) *Health and Safety of Pupils on Educational Visits.* London: DfEE.

GA/FSC (2005) *Fieldwork File.* Sheffield: Geographical Association.

Ministry of Education (1967) *Children and their Primary Schools: A report of the Central Advisory Council for Education (England) (Plowden Report).* London: HMSO.

Norfolk Area Child Protection Committee (2001) *The SAFE pack.* Norwich: Norfolk ACPC.

Ofsted (2002) *The Curriculum in Successful Primary Schools.* London: Ofsted.

QCA (1998) *Geographical Enquiry at Key Stages 1-3.* London: QCA.

Scoffham, S. (2002) 'Neuro-geography', *Primary Geographer,* 47, pp. 4-6.

Smart, J. and Wilton, G. (1995) *Educational Visits: The practical management.* Leamington Spa: Campion Communications Ltd.

Websites

Adventure Activity Licensing Authority (www.ada.org/guidance) provides helpful material about visits, booklists and information about licensing a range of Activity and Field Centres

Norfolk ACPC www.acpc.norfolk.gov.uk

Shells lay motionless on the beach like jewels on a golden carpet.

Eels swim in the vast cold sea.

Algae clings to pieces of drift wood.

Seaweed tangles around your feet as you step into the water.

Herring gulls perch on the grassy cliffs spying on the local fishermen.

Oceans roar as they crash into the shore.

Rockpools, full of life waiting to be explored.

Every day the tide delivers a parcel of surprises for visitors to investigate.

IN THIS CHAPTER YOU WILL FIND KEY IDEAS ON
CONCEPTS • DISPLAYS • ENQUIRIES • FIELDWORK • INSERVICE TRAINING •
MAPWORK • NATIONAL CURRICULUM • PROGRESSION • VOCABULARY

Geography and language development

Once upon a time, a master and one of his most useful and well-regarded servants began to have a deep discussion about their importance to one another. Each felt they were more important to the other. Fortunately, they were sensible enough to see that this kind of argument could have gone on for ever without either being deemed to have won it. So, being basically faithful allies in spite of their difference in status, they simply had to agree that they were indispensable to one another.

The last few years have seen similar discussions take place between teachers with curriculum responsibility for English and for geography as they have tried to make sense of the relationship between their respective subjects in planning for the primary phase. Ever since the Dearing Report of 1993 (Dearing, 1993) a strong re-emphasis on the core subjects, and particularly on the development of basic English language skills, has been a priority in primary schools. This has been significantly reinforced by the subsequent introduction of the National Literacy Strategy and by schools' own desire to develop PSHE and now, potentially, by primary MFL initiatives. However, many schools, in the wake of recent Ofsted inspections, have also found themselves reprimanded for the narrowness of their curriculum. We are entering a new phase which demands some return toward curricular 'breadth and balance' and, it is to be hoped, one which also stresses creativity in both teaching and learning.

This chapter begins from the premise that, however much we have been encouraged to see literacy as the supreme priority, English and geography should not be viewed as master and servant, but as mutually supportive elements in a holistic approach to children's learning. It aims to show how geography lessons, while still pursuing specific geographical objectives, can also be part of the solution in raising literacy standards rather than yet another demand on hard-pressed teachers. Geographers investigate our world, local and global, by asking questions, discussing issues and solutions and presenting findings. The ability to listen to others' views and to communicate in speaking and writing is essential to good geographical work. At the same time, by providing varied, stimulating and relevant learning experiences for pupils, geography offers some of the richest and most relevant contexts through which these essential literacy skills can take root and flourish.

Where is geography on the curriculum map?

Geographers are well known for their anxiety about defining the position of their subject in the curriculum and justifying its continued existence to others who may ask them some taxing questions. Is it an arts subject or a science subject? Is it social science or economics? Why does local geography fieldwork sometimes become enmeshed with history? Are we actually studying maths when we undertake a litter or traffic survey? Why is some of the information we need in studying the landscape really geological, biological or technological?

Geography is about **places** and the relationship we **humans**, as the Earth's most influential inhabitants, have with them. At the heart of geography is the simple idea that all places are 'distinctive' or different from one another. The reasons for this distinctiveness can be traced to the fact that every place has a unique set of **physical** (natural) and **human** (manufactured) **features**, which are not found in exactly the same combination in any other place. Description and comparison of **places** and their characteristic features, using the **skills** of geographical enquiry, are the starting point for most geographical studies whether at a high academic level or in an infant classroom. Explanation of the similarities and differences between places demands an understanding of **processes** within the physical and human domains and leads us towards recognition of **patterns**, which help us to generalise our conceptions of the planet. Geography also offers flexibility. It can start from the study of a particular place, or set out to study a particular process, or begin from a theme or issue in the environment.

Figure 1 | *A working definition of geography.*

Eminent geographers have made sense of these questions by visualising geography as a bridge which links the arts subjects and the sciences, or even as the bridge of a ship from which the whole curriculum can be surveyed and navigated! They often describe geography as an 'integrative discipline', one which uses knowledge from many subjects and brings ideas together to develop a multifaceted understanding of our world. Figure 1 provides a working definition of geography. The terms highlighted in bold are critical to interpreting the national curriculum requirements.

What can the content of geography contribute to literacy development?

Understanding that geography is a flexible subject that occupies a focal position on the curriculum map helps us to see why it can offer such an breadth of opportunity for literacy and language development. When studying geography, we need not only to use technical and scientific language to explain places, features and processes, but also to employ descriptive, evaluative, aesthetic and emotive language in expressing personal responses to places or issues. Units of work which set out to explore issues about the quality of an environment, for example, demand that we combine scientific explanation about the causes and consequences of environmental damage with evaluative and possibly even emotional responses regarding its impact on people. Such work also promotes information-seeking skills from non-fiction texts or the internet and may encourage the use of stories, myths and fables and poems. It inspires writing in all these genres too. Most importantly, geographical studies provoke us to employ these different aspects of literacy work together, and in doing so bring the distinctive elements of the different genres into sharper focus.

When the national curriculum was introduced, school geography and literacy work benefited hugely from a surge of activity in the production of new pupils' information books, geographical stories as well as poems, photopacks, audio, video, and CD-Rom resources. The increasing use of the internet and e-mail contacts between schools around the world has opened up further exciting arenas for using and exploring language in geography lessons and geography through literacy. The impending MFL initiatives may be seen as a further opportunity, already being trialled in a proposed QCA unit which combines the study of geography with early French language experience in the study of a francophone locality.

What geographical work can do for pupils' progress in literacy, then, goes far beyond the very important, though rather limited, aims expressed in the general teaching requirements of the national curriculum. There we are reminded that 'Pupils should be taught in all subjects to express themselves correctly and appropriately and to read accurately and with understanding' (DfEE/QCA, 1999, p. 36). As good teachers I'm sure we would want geographical responses to aspire to the best possible standards of grammatical and syntactical accuracy, good quality vocabulary and creative flair!

How does the process of geographical enquiry link to literacy development?

While understanding the relevance of what geographers study is important to considering geography's relationship to language work, so too is the knowledge of how geographers work. The national curriculum documents tell us that even in setting up simple geographical activities with the youngest children we should teach through geographical questions and investigation. This 'enquiry approach' is central to all geographical work and is well exemplified in the range of QCA schemes of work for geography which adopt a medium-term planning framework based on 'key questions'. The enquiring process begins when pupils are motivated to ask pertinent questions about their surroundings. Finding the answers to these questions involves setting up an investigation within which they begin to comprehend, process and present information in the form of description, explanation, hypothesis and evaluation. So not only will they speak, listen, read and write about geography, but the very processes of thinking, of developing new concepts, of refining ideas and of testing them out on others to confirm understanding are heavily language dependent. In essence the approach which is advocated mimics the professional geographical research process at a level accessible to young children. It also ensures that good primary school geography relies on first-hand experience of the outdoor environment (see Chapter 10).

This strong emphasis on fieldwork experience has transformed the way that geography is taught in many primary schools. Also, outdoor work has always been regarded as a powerful stimulus to self expression. The unsolicited response in Figure 2 was written at the computer late one evening at a County Durham residential centre by a pupil with limited family opportunities for travel or literary support. She had been taken, during a geography field visit, to see the spectacular waterfalls at High Force and Low Force for the first time. It is a committed, emotional response to the landscape, partly descriptive but with some embryonic explanation.

Yet it also contains elements of scientific comprehension, some newly-acquired specialist vocabulary, an element of story and a personal communication with her teacher. Within it there is ample evidence for those tracking her development

High Force

At first I felt frightened and worried then it started to grow on me. It was really fun I loved it but the power of the water is amazing. How a 11 year old boy survived is unbelieveable it is a miracle that he is alive today. The white spray is all over the place particularly over the plunge pool. What more can I say just it's fantastic. The height of the thing is high let me tell you. I'm only glad sir had a tight hold of me because I felt dizzy when Sarah took a picture. The force of the water is incredible the speed of the thing is fast it just zooms down. Thankyou sir for keeping a tight hold of me.

Low Force
Wynch bridge

Dirty water flowing down, the foam sticking to the rocks, the rocks at the side look like steps. The sound, sounds like when you run the tap but much louder. It was BRILLIANT
But some people find it scarey and are frightened but not me
p.s I also felt surprised because the last time I came it was much cleaner some people do not care what they throw into the water.

Figure 2 | *A personal response to two waterfalls by Paula Cook, Consett, County Durham.*

Key question	Concept	Language skills
What is it?	Naming and identifying	Developing new vocabulary as pupils encounter an increasing variety of physical, human and environmental features and processes. Stories, poems, information books and dictionaries are important sources of words and ideas for all age groups.
Where is it?	Locating places and features	Understanding and using prepositions of place (in, beside, behind, under, etc.) out of doors and on pictures. The description of relative locations leads to the use of specialised vocabulary of absolute location (compass points, grid references, latitude, longitude) on globes and maps.
What is it like?	Describing and comparing	Developing descriptive language from visual and other sensory stimuli, progressing through increasing clarity, improved vocabulary and attention to detail, and through the ability to describe and compare at an increasing range of scales. Acquiring and practising comparative and superlative adjectives and adverbs and noun hierarchies.
How did it come to to be like this?	Explaining	Sequencing an explanation or hypothesis. Weighing and ordering ideas to enhance meaning. Offering alternative explanations using the language of tentative reasoning. Drawing on an increasingly wide range of geographical factors in framing ideas using drafting and editing skills.
How is it changing and what might happen next?	Hypothesising and predicting	Using tentative language to predict, speculate and suggest hypotheses about places and the processes of change. This may include expressions of uncertainty, hope, anxiety, delight, anger, relief and persuasive language in relation to environmental issues. Oracy is highlighted in debate, drama and role play.
How do I feel about it?	Evaluating, expressing opinion and caring	Employing expressive language to convey opinions of, or evaluative, aesthetic emotional responses to places (e.g. landscape appreciation, conservation). Could involve prose, poetry, opinionated/journalistic writing or oral responses such as speech making.

Figure 3 | *How key questions in geography link to basic literacy skills.*

of parallel progress in language and geography, as well as some indication of her immediate literacy needs. Best of all, being directly inspired by nature, it is written with spirit, a quality so very difficult to achieve in a book-based lesson.

Developing language and literacy through the geographical enquiry process

A recurring theme in this handbook is the planning of geographical work through a sequence of geographical questions usually referred to as the 'key' questions. The question sequence shown in Figure 3 is recommended in the national curriculum, used as a structural framework by the QCA schemes of work and has been widely adopted by teachers as a powerful planning tool. It has theoretical roots in Bloom's taxonomy of educational objectives (Bloom, 1972) and helps to clarify the purposes of geographical study and to support the tracking of progression in learning.

As pupils move through this enquiry sequence learning develops from simple recognition and description towards explanation and hypothesis, being continually revisited in different projects and making increasing levels of demand on their thinking processes. It is worth noting that the final question in the sequence does not necessarily demand a higher level of thinking, for it demands an evaluative response, and this can be attempted even by the very youngest learner.

We can also envisage how progression occurs within the context of each question. For example, when young pupils who are looking at a river are asked a 'What is it like?' question, they will answer using simple known words such as 'the water bends' whereas more experienced pupils could offer 'the river's channel is meandering'. Similarly a 'Where is it?' question may at first elicit only simple place references such as 'over there' or even self-referenced ones like 'behind me'. But these responses might be expected to progress with teaching, from relative to absolute locations, as described in four- or six-figure grid references. The overall implication of these two ways of tracking progress is that pupils who are operating at differing national curriculum levels may be encouraged to take part in the same enquiry at the same time but, through the judicious intervention of a teacher who is aware of their capabilities, can progress their understanding at a pace which is appropriate for them. It is worth taking time, perhaps as an INSET activity, to trace in detail the relationship between the skills demanded in a geographical question sequence and those defined as indicative of achievement in the national curriculum level descriptions.

Engaging in enquiry with pupils does not, however, involve confronting them with a mechanical set of structured questions. It involves helping them to consider these types of questions, and framing them in ways which they find understandable. The general aim is to induct young pupils into the excitement of an enquiry which has relevance to their own lives, such as 'How can we make our local area safer?', so that they begin to ask their own questions and to research answers for themselves. By tracking the nature of their questions the teacher will again be able to trace progression.

Planning geography and literacy work

It is important when planning for geography and English in tandem to be clear about time allocation for the two subjects. A day's fieldwork and follow-up may consume a complete half-term's geography time, yet there may be several different aspects of geography which need to be addressed within that half term. So there may be a temptation to dismiss the fieldwork as being too time-consuming. However, if we can be more confident of saying precisely what language development is accruing from our work in the field, then opportunities are opened up for deeper, more rewarding, multi-purpose outdoor work.

An excellent way to begin any new geography topic is to elicit what the pupils already know and what they want or need to know. As well as defining starting points for the teacher, this approach involves pupils in framing statements and questions and, with the teacher's support, in refining and grouping them. As part of this process the teacher or more able pupils will model different ways of composing and writing (e.g. Y1,T1, SL4) and there will probably be opportunities to discuss the structure and punctuation of

Planning for Progression

From:	To:
writing in geography	
■ single words and phrases;	■ longer sustained pieces of writing;
■ writing with a simple single focus (e.g. describing a street scene);	■ writing with several sub-sections (e.g. describing stages in a survey);
■ writing for a single audience (e.g. teacher).	■ writing for many different audiences (e.g. other pupils, local newspaper).
reading in geography	
■ text with low readability age;	■ text with high readability age;
■ text selected by the teacher;	■ text found and selected by pupils (e.g. on CD-Rom);
■ reading a text, all of which is relevant;	■ reading a text from which relevant information has to be extracted (e.g. entry in atlas);
■ texts from a single source (e.g. KS1/2 book).	■ texts from many different sources, possibly conflicting (e.g. brochures).
speaking and listening in geography	
■ short answers to closed questions (one response expected);	■ answers to open questions for which pupils explain their thinking;
■ discussion in class or small groups;	■ speaking aloud in front of a class in discussion or role-play;
■ listening to or giving a narrative account (e.g. story of a journey);	■ listening to or giving an analytical account (e.g. how the water cycle works);
■ listening to and using simple vocabulary.	■ listening to and using specialised geographical vocabulary.

Figure 4 | *Summary of the use of geography in language development.*
Source: SCAA, 1997.

Materials that are natural
Sand   sea weed   stones   grass

Materials that are made by people
Plastic bucket   rope   tins   Plastic box ✗   bottles

Pupils' work |
 Courtesy of Greg
 Walker.

statements and questions viewed together (e.g. Y1 T3 SL7). This work might well be a 'literacy hour activity' and can certainly be justified as such even though it may have a geographical purpose.

Figure 4, compiled as a result of a joint Geographical Association and School Curriculum and Assessment Authority (SCAA) initiative provides a neat way of outlining the progression of language skills through reading, writing and speaking in geography. It is evident from this that not only can progression be planned for both subjects together, but geography can play a huge role in literacy development. Teachers may consider using this framework as a tracking device to ensure that geographical work is indeed providing appropriate support for each main element of the literacy strategy.

Geographical vocabulary

Geography is frequently criticised for having an obscure, specialised vocabulary which is inaccessible to non-specialists. Primary teachers attending INSET courses are often particularly anxious about specific terms used in physical geography. So it is important to ask how much specialist vocabulary is appropriate to primary pupils' needs and how we should go about introducing it. There are many pupils who might become quickly excluded from understanding by just one or two unfamiliar words, and in this respect a non-specialist may be more sensitive to their plight than a 'geographer' who is liable to slip technical words into a conversation unwittingly. However, there are other pupils who revel in terminology, if only to impress, and it would be a pity to stifle this enthusiasm. There are only occasional examples of technical words in the national curriculum, and there is little clear research to guide us through this issue, although a useful series of vocabulary lists was produced in *Primary Geographer* by Bridge (1993), Scoffham and Jewson (1993, 1994), and Scoffham (1998), to guide practice. Agreeing with colleagues a set of words to be developed within each topic, possibly with pre-defined levels of difficulty, is a profitable aspect of whole-school planning. The implications for some purposeful dictionary work are evident (on-going word level work, e.g. Y3, T1, WL13 and T2, WL19).

We have little knowledge of what understanding pupils come to school with, although all teachers have come across examples of pupils knowing the same words but attaching different meanings to them. Wiegand (1993) notes that urban pupils often think the term 'hill' (which is three dimensional) has the same meaning as 'slope' (which is two dimensional), perhaps because they rarely see entire hills not obscured by buildings. Devlin (1994) identified homonyms as a particular problem in geography. Most children attached the word 'peak' to a cap and might therefore visualise a mountain peak as a protrusion from the side rather than the summit. We can easily forestall such misunderstanding by mentioning a more usual meaning whenever we introduce a new term, e.g. 'The top of a mountain is called the "peak". Can you think of any other things that are called a peak?'

In introducing river terminology to pupils, teachers are surprised to find that new terms like 'tributary' are more easily absorbed that familiar words such as 'source', 'channel' and 'mouth' which have a double meaning. A 'channel' is where you find a favourite programme, a 'mouth' is very useful for eating and talking, and 'source' conjures up a very familiar red bottle! Similarly, in the north east of England a 'geyser' is a slang term for an old man so the vision of hot geysers spouting out of the ground conjures an interesting spectacle! We therefore need to think about how we can deliberately dismantle the known meaning of homographs and homonyms before building the new one.

All this will not only offset geographical misconceptions but offer a contextualised opportunity to teach about homonyms, homophones and homographs (e.g. Y3, T3, WL14) and to introduce some of the ways in which words are used colloquially in different geographical regions. For example, words such as 'stream', 'burn', 'beck', 'piddle' all refer to the same landscape feature but enrich and enliven our vocabulary (e.g. Y5, T3, WL9). Pupils might also explore the cultural origins of regional dialect as part of their vocabulary acquisition work (Lewis, 2001).

It may also help if we are more specific about how we use particular terms. Using 'countryside' to refer to rural areas avoids confusion with 'country' in the sense of a state until pupils are mature enough to have the dual meaning of 'country' explained. Finally, giving simple derivations of terms is possible for older juniors – 'cumulus' is like 'accumulate', and 'tributary' is from the same root as 'contribute' (e.g.Y5, T1, WL8).

A good CPD activity relating to the acquisition of geographical vocabulary is for year group teachers to agree suitable key vocabulary lists for the different topics and within these lists to highlight homophones, homonyms, homographs and dialect words and those which have a variety of meanings. A shared approach will benefit both the geographical learning and the whole process of vocabulary acquisition. A fundamental need is for a more detailed understanding of pupils' geographical vocabulary development, something that could be achieved for your school by keeping a shared notebook of piecemeal evidence of pupils' understanding to inform your shared planning and practice.

The language of location

Expressing position is a fundamental geographical skill which begins with the gradual acquisition of place language and progresses towards sophisticated knowledge of mapping conventions and the ability to define absolute location.

Map reading itself has much to offer literacy skills. Skimming and scanning are necessary approaches to reading a map, offering a contrasting technique to sequential left-to-right reading, and so helping to develop awareness that in reading for information different approaches to text are needed (e.g. Y2, T3, TL16/17). Also, young children need to understand that places and features such as mountains and even farmers' fields have names just as people do. Maps are therefore a tremendous resource for consolidating teaching about proper nouns and capital letters (e.g. Y1, T2, SL7).

The skills involved in describing geographical location can also be developed through stories. For example, the progress of events in a story is sequenced in place as well as time and, if we encourage pupils to map the story, their comprehension and visualisation of spatial relationships can be assessed. Pupils' own story writing may be enhanced by encouraging them to invent settings and make imaginary maps, and this in turn can support their ability to visualise sequenced events. Resources such as the Red Riding Hood picture map in Folens *Geography 1 Big Book* (Harrison and Havard, 1994) provide an excellent stimulus for this kind of work.

Using language to describe features and places

Retrieving information from pictures is another essential geographical skill with implications for language development. Here, experience shows that children pick out details relevant to their own immediate interest, such as animals, people, vehicles and other foreground details, rather than the surrounding landscape (see also Mackintosh, 1998). We therefore need to develop strategies for directing their attention. One possibility is to cover part of the picture, arouse pupils' curiosity by asking them to guess what is there, and then reveal it only when their attention is fully focused. It is also important to teach the vocabulary that will enable them to say where things are, e.g. near part, far part, distance, foreground, background, left and right.

The ability to describe accurately is a general language skill which teachers constantly monitor as vocabulary grows. Outdoor work within geography can greatly enhance the skill of describing if it deliberately focuses on the full range of sensory stimuli in the environment. You can focus pupils' attention by asking questions such as 'Is it rough or smooth?', 'Does it have a smell?' and so on. Even when we are limited to secondary sources, studies of unfamiliar places, based on multimedia packs, can still invoke descriptive writing which involves empathising with the people who live there. For example, 'How do you think the inhabitants of Chembakolli feel when the monsoon breaks?'. Most teachers are well practised at taking children into a picture by using questions to help them to imagine sensory experience: typical questions include 'What do you think you could hear in this place?', 'Do you think you could feel the hot sun on your face?', 'What would happen here if the wind began to blow hard?' (see Chapter 9).

Activities where imaginary islands and countries with geographies, culture, laws and customs are invented by groups of pupils also give great scope for language development through descriptive invention (e.g. Y4, T2, TL10) from making lists of products to writing 'national poetry'. They are also excellent vehicles for assessing pupils' geographical

understanding of the constituent features of different kinds of environment and offer huge scope for creativity across the curriculum. You might consider some or all of the following; What are the laws of your country? (PSHE); What is your language like? (MFL); Do you use the Euro or some other currency? (maths); How does it work? Have you any famous artists, architects, musicians, engineers, scientists? What did they do that made them famous? The Birmingham DEC advisory materials, e.g. 'Themework' (Birmingham DEC, 1991), offer many further ideas which can extend this work towards the understanding of international relations and global citizenship.

Using story books and poetry

The use of picture story books in Foundation Stage and primary geography is now well documented (TIDE~DEC, 2002). Among the most popular are Mairi Hedderwick's Katie Morag stories, based on the island of Struay, a fictitious

By the Sea

The lashing waves,
Crash against the cliffs
And eat them away,
Never sleeping night or day.
At night the moon glistens on the sea
And makes the fish silvery
Under the waves they dash
In and out
Arrows of swift light
On a summer day
Children play on the beach
Eat rock
And run along way
To reach the cool waves
To paddle, dip toes
Splash
They never think these same waves
Crash on cliffs, and eat rock too-
        Did you?

Figure 5 | *Responding to a geographical experience with creative writing.*

place inspired by the real Hebridean island of Coll. Coll now forms the main locational focus of QCA Unit 3: An Island Home and an excellent new geography picture pack, *Discover Coll* (Wildgoose, 2003; see also Chapter 8) and map (Geographical Association, 2004).

The Geographical Association has also recognised the value of teaching Foundation Stage geography through fictional characters. It has produced a wide range of materials focused on Barnaby Bear, which include geographically-grounded literacy hour activities. These publications include whole-class big books, annotated with respect to both subjects (Jackson, 1999-2003), and differentiated little books with geographical themes for guided reading groups (Lewis, 2000-2003). Another addition is *Barnaby's Local Area Album* (White, 2003) an information text in scrapbook format which acts as a model to support early descriptive and evaluative writing about your own local area (e.g. Y3, T1, TL11).

For older pupils there are many texts which use powerful writing and illustrations to raise interest in geographical issues. Some well-known titles are *Peter's Place* (Grindley and Foreman, 1995), *Window* (Baker, 1991), *Where the Forest Meets the Sea,* (Baker, 1989), *Welcome Back Sun* (Emberley, 1993), *At the Crossroads* (Isadora, 1999), *One World* (Foreman, 1990) *Rainforest* (Cowcher, 1988) and *My Grandpa and the Sea* (Orr, 1990).

Reading extracts from longer stories to introduce ideas about place is yet another way of stimulating curiosity and geographical interest. *The Railway Children* (Nesbitt, 1906), especially the description of the landslide, and *Earthquake* and *Hurricane* (Salkey, 1965; 1968) are both wonderful examples of powerful geographical description. Conversely, pupils

Photo | *Paula Richardson.*

can be encouraged to visualise the stories they write themselves in terms of place as well as time and to illustrate them with maps. If you have the courage to tell a story rather than reading it, the potential to expand the geographical content will be entirely in your own hands! Even re-telling fairy tales and nursery rhymes has great potential for thinking about the world, as demonstrated by Harrison and Havard (1994), and Scoffham and Jewson (1992).

The special, evocative power of poems such as *The Sea* by James Reeves (Farjeon *et al.*, 1970) or the strong messages of almost any of the environmental poem collection in *Earthways Earthwise* (Nicholls, 1993) can also be harnessed to develop understanding of landscapes, processes and issues, and pupils can be encouraged to respond to their own geographical experiences by writing original poetry (Figure 5). If personal writing of this kind is shared through class plenaries, it can also be used to develop pupils' capacity to understand that places evoke strong human emotions which differ from person to person. Creating an anthology of geographical poems at the whole-school or LEA level can prove an enjoyable CPD activity and one which offers the promise of immediate application to classroom practice.

Making comparisons and offering explanations

In the progression from simple geographical description of a single place or feature to more complex comparisons, the need for comparative language arises. Geography gives pupils many opportunities to practise manipulating comparative forms such as 'smaller than', 'more industrialised than', 'not quite so cold as', and so forth. There will also be opportunities to introduce comparative hierarchies or word ladders like 'dwelling, hamlet, village, town, city, metropolis', or 'hill, mountain, range, chain and system'.

As pupils progress towards secondary level it becomes increasingly important for them to acquire the skills of logically sequenced explanation (e.g. Y4, T2, TL24/25). By using varied approaches to classroom practice and providing frequent opportunities for spoken interaction and presentation you can enhance both primary geography and language work. Activities in which pupils propose changes to their environment and adopt the role of a planner or developer are rich examples of contexts in which pupils explain and justify their intentions and present ideas for class discussion. The judicious use of writing frames in geography (Weldon, 1998) can help pupils to move from these spoken explanations towards well-organised written sequences.

Adopting the enquiry approach implies a constructivist stance to learning geography, so we must be prepared for many early 'explanations' to be only partially tenable. It is vital for pupils to know that expressing uncertainty does not imply weakness in their understanding but is rather a recognition of the complexity of some geographical processes. Tentative language is a strong feature of all geographical explanation, even at high academic level, not only because of the range variables that influence places, but also because there are still many gaps in our understanding of the world. So the subject offers great scope for hazarding explanations: 'It might be because the weather …','It could possibly make a difference to …', 'Maybe it will be washed away or …'. Realising that there are often several possible and equally tenable explanations offers great comfort to non-specialist teachers too. In a climate of true enquiry there is no place for an all-knowing expert. The teacher's role is simply to apply the tacit knowledge and sharper logic of adult experience to the exciting task of helping pupils gradually to refine their own geographical reasoning.

Enquiries into the way places are changing and evolving also offer great opportunities for self-expression as they invoke personal viewpoints and engage the emotions as well as the intellect. Pupils' own views may be presented through speech, drama, articles for class newspapers, letters, prose or poetry and will include opinion, persuasion, uncertainty, hope, anxiety, relief, anger and delight. The power of discussion in supporting concept development and refining understanding highlights oral language as an essential element of all environmental geography work.

A good example of how to study a local issue was when pupils from years 5 and 6 at Lumley Primary School, County Durham, working with education students from the University of Sunderland, undertook a land use survey of their village and identified four pieces of open land which might well come under pressure to be developed in the near future. In role as development and construction teams, they set about planning new land uses for these sites following a discussion to establish what their village most needed – a supermarket, light industrial units, a leisure centre and a drive-in McDonald's! Having surveyed the sites and agreed on which environmental features were worth preserving they drew detailed exterior and interior plans, manipulating a great variety of geographical variables from car parking space to noise pollution. A 'public meeting' (role play) followed in which plans were presented orally and debated for their merits or otherwise by pupils acting as politicians and NIMBYs. The heated argument which erupted from some proposals and the eloquence with which controversial issues, such as the removal of half the school field to accommodate the leisure facility, were debated, provided ample evidence of the power of a personally experienced geographical context in promoting oracy. Looking at this from the standpoint of geographical pedagogy, the activity was one in which a holistic experience of geography, incorporating skills (map reading and map construction), places (the immediate locality), themes (settlement, environmental quality), patterns (land use conflicts) and processes (environmental change), was experienced and enjoyed by all the participants.

Developing a library for geography

Of the many kinds of geographical stimulus mentioned in this chapter, books are arguably still the most important resource in most primary schools. In spite of the increasing use of appropriate internet sites, the maintenance of a library to support geography is still a priority for the subject co-ordinator. It should include as wide a collection as possible of children's fiction organised to support individual topics. This is best visualised as a shared task for the staff and a judicious co-ordinator can harness substantial progress for all by setting up a CPD activity in which colleagues are encouraged to identify and evaluate the school's existing materials in the light of long-term planning or a pre- or post-Ofsted action plan.

Most schools will find that their existing stock abounds in picture story books with striking settings which support the acquisition of geographical vocabulary and present concepts in a form which is enjoyable for pupils. It is important to ensure, however, that the pleasure of stories is not compromised by an over-zealous teacher trying to extract the last grain of geography from the pictures! The secret of using stories well rests on enjoying them with the pupils in an atmosphere of relaxed anticipation, with no unwelcome geographical interruptions. After all, good stories bear much re-reading and this in itself is a powerful aid to consolidating the geographical language and ideas they contain. You can always return to particularly useful picture spreads if their content is vital to the lesson, and pupils can be invited to develop geographical ideas in a follow-up discussion which does not interrupt the story itself. These comments confirm the legitimacy of pursuing geographical as well as literary ends in the literacy hour. If the purpose of literacy is to attach meaning to text then it may just as well be geographical meaning as any other.

Many poetry collections which schools already have in stock include poems with geographical potential. In addition to the CPD activity suggested above, a co-operative search (in-service time, perhaps) or a competition for older pupils to amass suitable examples, could be organised. A selection of poems relating to different aspects of geography could be word-processed in large print onto cards (a good job for adult helpers), illustrated by pupils and laminated for classroom use. Poems printed on OHT film can be overlain on enlarged photographs to make attractive shared reading resources and the potential value of the interactive whiteboard in this context has yet to be realised. Pupils' own geographical poems might also be added to these collections as time progresses.

The collection of information texts for geography is more problematic. Ideally, each unit of work requires a selection of books at different reading levels. To optimise their use they need to double as resources for teaching information skills, so they should all contain adequate contents pages and glossaries. Books which are to be used by pupils themselves have to be strongly bound. A good variety of styles of presentation – text and pictures, comic strips, speech bubbles – has more appeal to children, as do books, referred to as 'faction', where characters to whom the pupils can relate act as guides through the factual information. Question and answer formats are particularly helpful for geography since they reflect the enquiry idea. Vetting for outdated or stereotypical images, bias and prejudice needs to be carefully done, too. It is important to be aware that some modern environmental texts contain images, of whaling for example, which many pupils find distressing and what better way to raise this awareness than to ask colleagues to bring one book they would use and one they would not use to a CPD session for discussion?

For obvious commercial reasons, publishers have been unable to embrace the demand sparked off by the national curriculum for information about specific localities in other parts of the world. Instead, non-fiction, place-based series tend to concentrate on whole countries or continents. Such resources are invaluable sources of background material but are not adequate in themselves for detailed locality studies. A good source of support has come in the form of resources from organisations such as ActionAid, Oxfam, the Development Education Centres, and the Geographical Association itself (Figure 6). These have greatly aided the quest for materials detailed enough to foster empathetic understanding of other communities.

As your library expands, making colleagues aware that it can be used in both geography

and language lessons will enhance its use. Where resources for foundation subjects are scarce there is a strong case for some deliberate co-operative planning in the buying of books between the language and geography co-ordinators. Another source of reading material in the classroom is display. Most geographical topics provide excellent inspiration for colourful assemblages of pupils' work and these are at their best as a resource for literacy development, where the pupils themselves have been actively involved in their planning and presentation. Of particular value are interactive displays, which provide things to do as well as to look at. Participation can be invited by the use of bold captions based on geographical questions, changed periodically to refocus attention and renew interest. Finally, one should not forget the value of traditional geographical resources (globes, wall maps, atlases, aerial photographs and weather charts) as a stimulus for the development of geography and language skills.

Some suggested types of lesson

A number of suggestions are offered below, in which the session or topic time is deliberately organised to facilitate related geography and English language teaching. The list is not exhaustive but may spawn many other ideas of your own.

- A geography session leads into a creative writing follow-up such as writing a poem after fieldwork or writing a descriptive postcard from a distant locality.
- A literature experience is used to stimulate a geographical enquiry, e.g. a reading of *Dear Daddy* (Dupasquier, 1986) leads into an exploration of the globe and the plotting on a world map of all the connections between the class and other countries.
- The literacy session is about reading for meaning but the context is part of the current geography topic and uses geography information texts.
- The geography written work is designed for a particular audience, e.g. after researching weather stories older pupils write stories about weather for younger ones or prepare a display for a parents evening.
- A speaking and listening activity compares a geography video of St Lucia with an audio tape on the same theme.

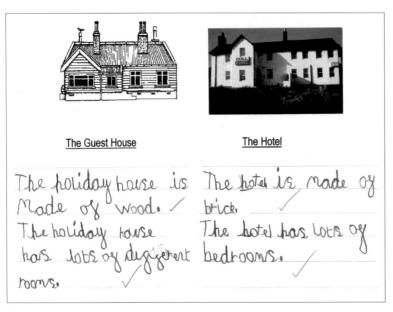

Pupils' work | *Courtesy of Greg Walker.*

■ A geography topic concludes with the making of an information book about it, involving aspects of book production and page design (ICT), and preparing a contents list and glossary in language time.

■ Groups of children are invited to invent a country and make a map of it. They consolidate 'features vocabulary' and learn about proper nouns while plotting imaginary towns, rivers and mountains. They also explore ideas about what a country is by inventing a national flag, currency and language, and by writing laws, inventing folk tales and traditional recipes, and composing imagined biographies of the country's most celebrated inhabitants.

Attractive as it might be for teachers with demanding timetables to make time economies by combining English and geography, this chapter is not about making space in an overcrowded curriculum. Nor is there any intention to represent geography as the sole vehicle for language development. Using language well is a vital life skill. It supports our understanding, brings alive our thoughts and gives expression to our beliefs. English not only shares a central place on the curriculum map with geography but also fundamental educational goals. Both subjects are strongly concerned with communication and seek to nurture our growing connections with fellow humans in other parts of the world. Geography provides a tremendous range of exciting real-world contexts in which pupils can be offered an inspiring range of language experiences. But equally, to give pupils the best possible experience of geography we need to use quality literary resources and offer pupils the fullest possible opportunity for self-expression in speech and writing within our lessons.

References and further reading

Baker, J. (1989) *Where the Forest Meets the Sea.* London: Walker Books.

Baker, J. (1991) *Window.* London: Red Fox.

Birmingham DEC (1991) *Themework.* Birmingham: Birmingham DEC.

Bloom, B. (ed) (1972) *Taxonomy of Educational Objectives: Handbook 1 cognitive domain.* London: Longman.

Bridge, C. (1993) 'Word sets', *Primary Geographer*, 15, p. 22.

Cowcher, H. (1988) *Rainforest.* London: Walker Books.

Dearing, R. (1993) *The National Curriculum and its Assessment*. Final Report Dec 1993. London: SCAA.

Devlin, M. (1994) 'Geographical vocabulary in the junior school', *Issues in Education*, 2, 1, pp. 26-30.

DfEE/QCA (1999) *The National Curriculum: Handbook for primary teachers in England*. London: DfEE/QCA.

Dupasquier, P. (1986) *Dear Daddy*. London: Puffin.

Emberley, M. (1993) *Welcome Back Sun*. London: Little Brown.

Farjeon, E., Reeves, J., Rieu, E.V., Serraillier, I. and Graham, E. (1970) *A Puffin Quartet of Poems*. Harmondsworth: Penguin.

Geographical Association (1998) *Primary Geographer* 32: 'Focus on Language'. Sheffield: Geographical Association.

Geographical Association (2004) *Coll map*. Sheffield: Geographical Association.

Geographical Association (2004a) *Tocuaro map*. Sheffield: Geographical Association.

Graham, J. and Walker, G. (2003) *Discover Coll*. Coalville: Wildgoose.

Grindley, S. and Foreman, M. (1995) *Peter's Place*. London: Andersen Press.

Harrison, P. and Havard, J. (1994) *Folens Geography 1 Big Book*. Dunstable: Folens.

Hedderwick, M. (1994) *Katie Morag and the New Pier*. London: Random House.

Hedderwick, M. (1994) *Katie Morag and the Two Grandmothers*. London: Random House.

Hughes, S. (1988) *Out and About*. London: Walker Books.

Isadora, R. (1993) *At the Crossroads*. London: Red Fox.

Jackson, E. (1999-2003) *Barnaby Bear Big Books*. Sheffield: Geographical Association.

Lewis, L. (2000-03) *Barnaby Bear Little Books*. Sheffield: Geographical Association.

Lewis, L. (2001) 'What's in a name?', *Primary Geographer*, 45, pp. 23-4.

Mackintosh, M. (1998) 'Learning from Photographs' in Scoffham, S. (ed) *Primary Sources: Research findings in primary geography*. Sheffield: Geographical Association, pp. 18-19.

Nesbitt, E. (1906) *The Railway Children*. London: Penguin.

Nicholls, J. (ed) (1993) *Earthways Earthwise*. Oxford: Oxford University Press.

Orr, K. (1990) *My Grandpa and the Sea*. Minneapolis, MI: Carolrhoda Books.

Salkey, A. (1965) *Earthquake*. Oxford: Oxford University Press.

Salkey, A. (1968) *Hurricane*. Oxford: Oxford University Press.

Schools Curriculum and Assessment Authority (SCAA) (1997) *Geography and the Use of Language*. London: SCAA.

Scoffham, S. (ed) (1998) *Primary Sources. Research findings in primary geography*. Sheffield: Geographical Association.

Scoffham, S. and Jewson, T. (1992) 'Geography through Nursery Rhymes', *Primary Geographer*, 11, p. 2.

Scoffham, S. and Jewson, T. (1993) 'A glossary of terms for key stage 1 geography', *Primary Geographer*, 15, p. 2.

Scoffham, S. and Jewson, T. (1994) 'A glossary of terms for key stage 2 geography', *Primary Geographer*, 16, p. 2.

TIDE~DEC (2002) *Start with a Story*. Birmingham: TIDE~DEC.

Weldon, M. (1998) 'Using writing frames in geography', *Primary Geographer*, 32, pp. 12-13.

White, K. (2003) *Barnaby Bear's Local Area Album*. Sheffield: Geographical Association.

Wiegand, P. (1993) *Children and Primary Geography*. London: Cassell.

Welsh Office (1991) *Non-Statutory Guidance for Geography*. Cardiff: WO.

My Design For Pound Lane Park

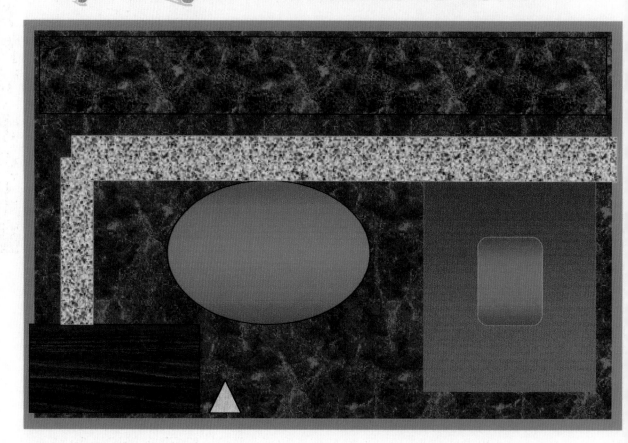

Key

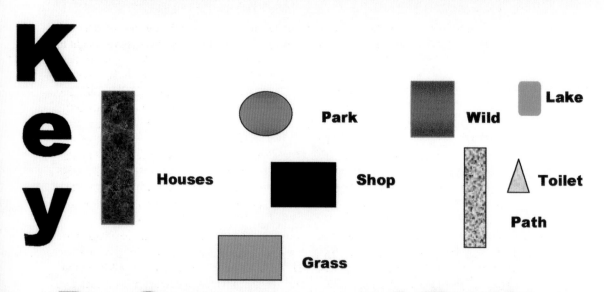

Houses

Park

Shop

Grass

Wild

Lake

Toilet

Path

By Gemma and Bethany

**IN THIS CHAPTER YOU WILL FIND KEY IDEAS ON
ENVIRONMENT • GOOD PRACTICE • ICT • NATIONAL CURRICULUM •
PLANNING • SPECIAL EDUCATIONAL NEEDS**

Geography and ICT

This chapter looks at the opportunities and some of the challenges for using Information and Communication Technology (ICT) in geography in primary classrooms, set in the national context for ICT in education. Opportunities for using ICT in the theme of the mountain environment are explored. A range of websites are suggested; some of these are aimed at pupils, others are to help teachers in their planning and professional development.

The world of education is changing rapidly. Over the last decade many primary schools have undergone huge transformations in terms of pedagogy, staff and physical appearance. Educational changes have been many, rapid and revolutionary; but none have been more dramatic than in the field of ICT.

To begin with, technology itself has advanced considerably. Personal computers are faster, sleeker and more affordable, as are laptops and portable machines, both of which have liberated users from the tie of a central location – a feature which is particularly valuable in the context of fieldwork. Computers are often networked together enabling software and printing to be shared and making it easier to save and retrieve pupils' work. Peripherals such as digital cameras, interactive or Smart boards, scanners and data projectors are also increasingly found in primary schools. Developments in software have brought about huge improvements in the range and quality of computer programs available to pupils, including simplified versions based on 'grown up' software suites. There has also been the immense growth of the internet which, coupled with an increase in educational websites, has led to further innovation.

The government has a strong commitment to raising standards in the use of ICT, and massive, centrally-funded initiatives, often supplemented by local resources, have sought to make these technological advances accessible to all primary pupils. At the same time initiatives such as the National Grid for Learning (NGfL) have led to a huge increase in the range and quality of hardware in schools, including networks, computer suites and other peripherals. As a result the ratio of pupils to computers in primary schools now stands at 8:1 (DfES, 2004). Many teachers have also benefited from 'laptops for teachers' schemes. There has been a concerted drive to improve teachers' ICT skills and their confidence in using ICT across the curriculum. There are now plans to provide additional software to schools through e-learning credits, with schools selecting computer programs or on-line services from a range of approved titles and providers. These initiatives are to be warmly welcomed; they constitute some 'joined-up thinking' as one development cannot be seen in isolation from the others and schools need support in all areas.

Despite these developments Ofsted have found that 'in only one school in five is good use made of ICT in geography' (Ofsted, 2002). We can only surmise the reasons for this. The lack of resources and staff expertise in schools may be a factor. Teachers also note the lack of time available for geography and as a consequence may focus on the basics rather than extending the subject through ICT. Another explanation could be that primary teachers regard ICT predominantly as a subject in its own right and not as a cross-curricular tool.

ICT in the national curriculum

ICT in primary schools dates back to the early 1980s when the first computers were introduced. In 1990 ICT was included in the national curriculum as part of the general teaching requirements. Now ICT features in the present version of the national curriculum both as a subject in its own right and as a cross-curricular tool to aid teaching and enrich learning. The importance of ICT is highlighted as follows:

> *Information and communication technology (ICT) prepares pupils to participate in a rapidly changing world in which work and other activities are increasingly transformed by access to varied and developing technology. Pupils use ICT tools to find, explore, analyse, exchange and present information responsibly, creatively and with discrimination. They learn how to employ ICT to enable rapid access to ideas and experiences from a wide range of people, communities and cultures. Increased capability in the use of ICT promotes initiative and independent learning, with pupils being able to make inform-ed judgements about when and where to use ICT to best effect, and to consider its implications for home and work both now and in the future* (DfEE/QCA, 1999, p. 96).

The current programmes of study for ICT are organised into four sections. These are:
- Finding things out
- Developing ideas and making things happen
- Exchanging and sharing information
- Reviewing, modifying and evaluating work as it progresses

The parallels between these headings and the interest which geographers have in people, places and cultures in a rapidly changing world are quite clear. Primary geography provides a rich and purposeful curriculum context for pupils to develop their ICT capabilities. At the same time, geography is enhanced considerably by the use of ICT, and teachers who plan for ICT in their scheme of work and integrate it well generally reap the benefits in terms of the quality and value of their teaching and their pupils' learning.

ICT and geography

The current statutory position is that ICT is a fundamental part of developing geographical skills. At key stage 1 children are expected to be using CD-Roms, photographs and videos. At key stage 2 the programme of study states that 'pupils should be taught ... to use ICT to help in geographical investigations' (DfEE/QCA, 1999, p. 112). This is expanded further:

> *Geography provides opportunities for pupils to develop the key skills of ... IT: Through using CD-Roms and the internet selectively to find information about places and environments, using e-mail to communicate and exchange information with people in other places, using spreadsheets and databases to handle and present geographical data, and developing IT skills specific to geography [for example, geographical information systems (GIS) and remote sensing]* (DfEE/QCA, 1999, p. 9).

Every primary teacher is therefore a teacher of both geography and ICT; the two subjects complement each other. The Geographical Association and the National Council for Educational Technology (GA/NCET, 1995) have identified five ways in which ICT can help pupils' geographical learning in the primary school. These are by:

ICT offers valuable support by helping teachers:

■ to keep abreast of developments in geographical subject knowledge and pedagogy, for example through websites from national bodies and reputable organisations, such as QCA (www.qca.org.uk/geography) and the Geographical Association (www.geography.org.uk);

■ to save time by providing varied resources through the creation of differentiated texts, and finding up-to-date data and information. (There is a growing number of teacher resource exchange websites);

■ to provide a flexible resource, for example through using digital images and maps which have value in a variety of contexts;

■ to provide a stock of resources for use with pupils, for example through adapting text, maps and images that can be downloaded from the internet and once filed electronically can be adapted for pupil use;

■ to keep in touch with other professionals through the use of e-mail; or by joining electronic newsletters or by contributing to a teachers' forum such as that found on the Staffordshire Learning Net (www.sln.org.uk/geography);

■ with whole-class or large-group teaching, for example through the use of a data projector connected to a computer to display maps and images; interactive whiteboards have also become increasingly familiar in classrooms;

■ with their administrative tasks, as planning, assessment records and data can be stored and updated as required.

Figure 1 | *The contribution of ICT to teaching geography.*

■ enhancing enquiry skills;
■ developing geographical knowledge;
■ understanding patterns and relationships;
■ providing images of places;
■ investigating the impact of ICT.

Enhancing enquiry skills

Most geographical enquiry-based work can be purposefully enhanced with ICT (Figure 1). For example, the use of databases, spreadsheets, data-logging equipment or the internet will certainly be an advantage where pupils are asking geographical questions, making observations, or recording and investigating data from fieldwork and secondary sources. These tools can also enhance the study of a local issue such as traffic and parking, investigations into shopping or leisure, the weather, or a local river study. If you use open-ended or generic software it will help pupils to organise the information they have collected, look for patterns and produce graphs. Another option is to use photographs, taken with a digital camera, to illustrate their work.

Geographical enquiries invariably involve creating, using and interpreting maps. Again, ICT can provide relevant support. For example, a wide range of maps can now be found on the internet (see websites such as Multimap, StreetMap and Get a map). Specialised software is also available to help pupils to generate maps.

Pupils can communicate the results of their enquiries in the form of text, pictures and graphs, with a word-processing or presentation program. It is often effective to use some ICT-generated work, alongside materials produced using other media. A worthwhile project would be for pupils to produce a 'tourist' guide or an 'eyesores and treasures trail' of their local area, illustrated with photographs combined with text. If the photographs are taken with a digital camera, the images can be transferred directly to a computer. Alternatively you can have your prints saved onto a CD-Rom by

Photo | *John Halocha.*

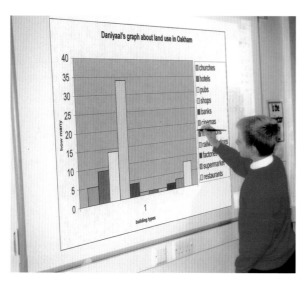

Photo | Kate Russell.

your photographic developer, thus transforming them into a more flexible resource.

Many schools now have their own website, which is used to publish information about the school. This may include information about the locality so there is an opportunity here for pupils to contribute material and information from their geographical enquiries. Electronic mail (e-mail) links now enable pupils to communicate with schools in other localities, and the rapid response time makes it possible for them to enter into a dialogue with other pupils. Much of this dialogue could focus on finding out about features of the other locality, what it is like to live there, exchanging photos and other resources. Reciprocal links like this are mutually beneficial as 'my here becomes your there' and may lead to exchange visits.

Developing geographical knowledge

In recent years the wide range of traditional sources of information available to geographers has been complemented by the rapid development of the internet and the use of e-mail. The problem now is often one of 'information overload' and pupils need to apply their literacy skills to access the information, select the most relevant and reject the rest. With the increase in use of the internet, CD-Roms have declined in significance, but may still be useful as a storage facility for digital images. There are also CD-Roms which focus on geography topics, and contain videos, animations, sound and other information, although, like books, they can soon become dated. A tape recorder is a very useful tool for collecting material from interviews, e.g. conversations with visitors about their jobs or life in another locality, or for recording sounds during a 'sensory walk' in the local area. Even young pupils will be able to operate the machine for both recording and playback.

Understanding patterns and relationships

As pupils develop their geographical understanding they need to be learn how to recognise spatial patterns. A programmable toy, such as a floor turtle, can help illustrate the notion of direction, especially if it is programmed to move around a large floor map or maze. While many computer 'games' have little to offer educationally, it is worth checking to see if they develop spatial and other geographical skills. Examples include games where use is made of an on-screen map or where players have to develop a mental map or make decisions. Beware, however, of programs that promote stereotypes or reinforce bad practice.

Databases, spreadsheets, some simulations and some websites can also help to provide an insight into geographical relationships such as weather patterns or causes and effects of water pollution. Multimap, for example, has the facility for overlaying a see-through map on top of an aerial photograph of virtually anywhere in the UK. InfoNation (provided by the United Nations) has a child-friendly means of displaying comparative data from all countries.

Providing images of places

Primary geography aims to develop pupils' awareness and knowledge of the culture and

character of a variety of contrasting places. The use of ICT tools such as fax, e-mail, CD-Roms and the internet can provide pupils with photographs, video, sound and other first-hand data. Some publishers are also producing materials on CD-Rom to complement textbooks and television programmes. Photo compact disks are another innovation which enable schools to build up a collection of photographs of the locality and which could be exchanged with schools elsewhere.

Investigating the impact of ICT

One requirement of the ICT national curriculum is that pupils at key stage 2 should investigate and compare the uses of ICT inside and outside school. Geography provides a natural focus for this work. For example, pupils can look at ways in which ICT has influenced communications, leisure and employment. They can also investigate how computers and other related devices are used by travel agents and the retail industry and how satellites help to forecast the weather.

During a visit to an airport, one school focused on how ICT and control systems were used there. The airport staff were very helpful and showed the pupils the check-in procedures, the seat allocation system, the luggage security scanners and many other ways in which ICT has had an impact on the workings of a modern airport. A visit to a local shopping centre is another option. Here pupils could observe and locate the uses of technology and control and monitoring systems such as automatic doors, cash dispensers, bar-code readers and supermarket stock control systems. They might also find out about 'store loyalty cards' and how the information they collect is used. The impact of high street video surveillance cameras, speed cameras and traffic and pedestrian crossing lights could also be considered.

Planning

The key to success in using ICT to enhance geography lies in careful planning. This should be a process in which several people are involved – the geography co-ordinator, the ICT co-ordinator and the relevant class teachers. The planning process will be a cyclical one, undertaken periodically when geography activities change or new ICT resources are acquired.

Curriculum planning is critical to ensure purposeful activities reflecting continuity and progression of skills and activities. For a coherent approach the geography and ICT co-ordinators should work together to identify opportunities for using ICT in geography and to plan meaningful learning experiences. Planning should start with the geography curriculum, reviewing areas which will naturally be enhanced by ICT activities. Try to introduce new activities slowly. If the changes are manageable then they are likely to be more successful. Any gaps can be filled in over time – do not attempt to make too many changes at once.

A good place to start in the planning process is to review current practice. This can be done by monitoring planning and pupils' work and by talking to teachers and pupils. The next step is to consider if you could make better use of ICT in your geography curriculum. Among the issues to be addressed are staff confidence and expertise, and access to hardware and software. If teachers lack the appropriate skills and confidence, plans need to be made for suitable training, either within school or through a centre-based course. If additional resources are required, then representation needs to be made to the head teacher and governors and

Figure 2 | *ICT resources checklist – does your school have these ICT tools?*

Although there has been an increase in the number of computers available in primary schools over recent years, there is an ongoing need to replace ageing equipment and to increase pupil access to up-to-date ICT resources. However, ICT does not just mean the use of computers, it also includes the use of other technological equipment such as programmable toys (Roamer, Pip and Pixie), tape recorders, cameras, scanners, fax machines, data-logging equipment, automatic weather stations, internet and CD-Roms. Does your school have these ICT tools? The resources which a teacher could reasonably expect to have access to in school are as follows:

Hardware and peripherals

- Computer system (with CD-Rom) with internet access and a printer in each classroom, preferably linked to the school network (if there is one)
- Larger schools are finding the benefit of computer suites for whole-class teaching, but this will often depend on the availability of accommodation
- An e-mail address (one for each teacher and each class as a minimum)
- A data projector, preferably linked to a laptop, for displaying presentations, digital slideshows, and a range of sources including websites
- Interactive whiteboard or Smart board
- Scanner
- Digital cameras
- Programmable toy(s) – there are several to choose from
- Data-logging equipment and/or automatic weather station and appropriate software

Software

Generic software suitable for the age of pupils including:

- Word processing
- Desktop publishing
- Presentation software
- Graphics or drawing
- Programming software (such as Logo)
- Database (or graphing software for very young pupils)
- Spreadsheet
- Internet browser
- E-mail

Geography-specific software

- Mapping software. My World is suitable for very young pupils but Local Studies (SoftTeach Ltd) has built-in progression and helps extend pupils' work through building up interactive maps with hyperlinks
- Simulations or adventures which involve 'mental maps'
- CD-Roms providing images and information about places including encyclopaedias and atlases

Towards the future – these are becoming more commonplace

- Portable computers for teachers and pupils
- A school website (geography can contribute to a local area study)
- Video conferencing
- Digital video camera
- An e-mail address per pupil
- Wireless controlled networked laptops which can be used in different classrooms

incorporated into the School Development Plan (See Figure 2).

It is important to ensure that pupils experience a broad and balanced range of ICT activities over a key stage. These need to be carefully worked out with clear regard to progression and differentiation, and be properly resourced. A few well-planned activities are better than several ad hoc ones with unclear objectives. They also need to be written into both the geography and ICT schemes of work.

Once the activities have been planned you need to refer to the relevant curriculum policy documents and national guidance, to show the links between the subjects. It is also valuable to review the activities after they have been undertaken and, if necessary, modify them in the light of experience. Figure 3 shows some of the ways ICT can be used to support good practice.

Figure 3 | *Suggestions to support good practice.*

1. Include some ICT-generated work, preferably by pupils, in your geography displays.
2. Set high standards of presentation and research in your professional work to model what you expect of pupils.
3. Develop locality photopacks of digital images using a digital camera or by having your photos stored on a photo-CD (a service available at many high street photo processing outlets). Once they have been created the images can be used flexibly in texts, printed out or displayed through a data projector using appropriate software.
4. Build up ICT resources alongside other resources for your topics (e.g. datafiles, a library of maps and images, a list of websites to support study of particular themes and places). It is worth taking time to 'file' these resources in an organised way so that they can be retrieved easily for future use.
5. Contribute some of your work in geography to the school website – celebrate pupils' work among a wider audience; particularly appropriate will be work about the local area, which will be of value to other schools too.
6. Keep abreast with your ICT skills and pass them on to your pupils – but also be open to being helped by pupils!
7. Reinvent the slideshow with PhotoJam, which enables you to present a slideshow of images (from a digital camera, scanned images or pictures saved from websites) in one of a range of styles; PhotoJam can be downloaded and is available for a free trial period or for purchase at a very reasonable price from the Shockwave website www.shockwave.com.

Working with pupils

1. Allow pupils to draft, redraft and edit work for final presentation at the computer – do not use a computer as an electronic typewriter to re-type handwritten work.
2. Aim for each pupil in your class to use ICT in at least two significant geography activities during the year, but preferably in each unit that they study.
3. Create the structure of a database, spreadsheet or base map in advance so that pupils can enter the appropriate information and analyse the results or add it to a base map.
4. Remember that some ICT activities are better suited to individual or collaborative work, especially word processing, and can initially be time consuming. However, this is compensated by the higher standard of work that results.
5. Encourage more able pupils to extend their work through additional research and independent learning.
6. Data handling is usually a group (or even whole-class) activity as pupils take it in turns to collect and enter data and then analyse it. Lower ability pupils may work better with support from their peers.

Resources

- Many internet sites and CD-Roms are valuable as sources of information for the teacher as well as for pupils.
- Ensure ICT equipment (especially printers which are a common source of angst) is in working order and that classrooms have access to the appropriate software and disks.
- Take health and safety considerations seriously; ensure computers are on suitable furniture and that the screen is at an appropriate height for children. There should also be sufficient space around the computer for working space and for books.

Catering for differentiation

All pupils, whatever their ability, benefit from ICT as it encourages motivation, develops skills and improves access to the curriculum. For pupils with mild learning difficulties, especially those in mainstream schools, the starting point should be to create activities for individualised learning using familiar equipment, rather than considering the need for different resources. However, pupils with moderate or severe learning difficulties may benefit from a range of specialised equipment, including voice recognition systems and software that reads text out loud. There are also several subject-specific learning programs which offer activities for a range of individual needs. More able pupils should be encouraged to use either a wider range of software or to use more of the functions of a particular program. They can also take more responsibility for their own work, including making decisions on the most appropriate tools for a task.

Photo | *John Royle.*

Using ICT to learn about mountains

There are many opportunities for incorporating ICT in a unit of work on mountains. Pupils could use maps and images from the internet as part of their research and produce climate graphs using a database or spreadsheet. They might also prepare a PowerPoint or PhotoJam presentation or devise a web page to communicate their findings to the rest of the school.

Useful websites
Mountain and mountaineering resources
Explore the mountain ranges of the world; a site for rock climbers, mountaineers, hikers, and backpackers who love mountains. www.peakware. com/encyclopedia/index.htm

Everest
There are many Everest sites commemorating the 50th anniversary of its conquest (2003).
Scholastic Publications: http://teacher. scholastic.com/activities/hillary/index.htm
Unlocking the Archives RGS: www.unlockingthearchives.rgs.org/themes/everest/default.aspx

Himalayas
Mogens Larsen has been trekking four times in the Himalayas – twice in India and twice in Nepal. The photos you see on this site are all from these treks: www.himalayas.dk
Nepal (from Jeroen Neele photo gallery): www.home.zonnet.nl/jneele/argentina/index.htm
Nepal and the Himalayas on the web: www.vic.com/nepal/related.html

Mountain Regions
Rocky Mountains, Canada: www.canadianrockies.net
Mountain Nature: www.mountainnature.com
Snowdonia National Park: www.eryri-npa.co.uk
Peru (Paddington Bear): www.mape.org.uk/Paddington/index.htm
Tiki the Penguin visits Peru: www.oneworld.net/penguin/peru/peru_home.html
The Austrian and Swiss Alps: http://alps.virtualave.net
Peakware World Mountain Encyclopaedia: www.peakware.com
Iceland: www.sln.org.uk/geography/SLNgeography@Iceland.htm
Virtual Montana: www.virtualmontana.org (This site is particularly good for the French Alps, Banat Mountains in Romania, and Snowdonia in North Wales. It is the result of collaboration between several groups including Liverpool Hope University College and the Field Studies Council.)
Global Eye Focus on Mountains: www.globaleye.org.uk/primary/focuson/index.html (This is one of a series from Worldaware's Global Eye with features on Everest, and Machu Picchu in the Peruvian Andes. Mountain Factfile might help the children to get going!)

Websites

Websites come and go; some change their addresses and it can be frustrating to see an error message saying a website has not been found. These sites are given in good faith at the time of writing; they are among the most useful and most flexible and mostly come from well regarded and well known sources. A good search engine such as Google www.google.com will help you find others, but searching the internet can be a time-consuming process. I recommend providing pupils with a few sites to look at for specific purposes in a lesson otherwise they may waste time searching and not have enough time for research!

Distant places
Action Aid: www.actionaid.org
Global Eye (Worldaware): www.globaleye.org.uk
Global Gang: www.globalgang.org.uk
InfoNation: www.cyberschoolbus.un.org/infonation3/basic.asp
On the Line (Oxfam): www.ontheline.org.uk
One World: www.oneworld.net
Oxfam for Kids: www.oxfam.org.uk/coolplanet/index.html
Save the Children: www.savethechildren.org
UNICEF: www.unicef.org
Vision Aid: www.vao.org.uk
Water Aid: www.wateraid.org
World Info Zone: www.worldinfozone.com
World Vision: www.worldvision.org.uk/main.htm

Photo | *Kate Russell.*

Guidance and advice
Geographical Association: www.geography.org.uk
Becta: www.becta.org.uk
ICT advice: www.ictadvice.org.uk
Innovating Geography: www.qca.org.uk/geography
National Curriculum: www.nc.uk.net/home.html
National Curriculum in Action: www.ncaction.org.uk
Ofsted: www.ofsted.gov.uk
Primary Geography Pages: http://homepage.ntlworld.com/anthony.pickford2
QCA: www.qca.org.uk
QCA Scheme of Work: www.standards.dfes.gov.uk/schemes3

Images
Aerial survey UK: www2.getmapping.com/home.asp
BBC: The Day/Week in Pictures: http://news.bbc.co.uk
Free Foto (great for UK): www.freefoto.com
Geo Images: http://geogweb.berkeley.edu/geoimages.html
Geography Photos (Ian Murray): www.geographyphotos.com
Images of England (English Heritage): www.imagesofengland.org.uk
Images of the World: www.imagesoftheworld.org
Jeroen Neele Visiting the World: www.home.zonnet.nl/jneele/argentina/index.htm
Physical landscapes: www.regolith.com
Picture Search: www.picsearch.com
Webshots: www.webshots.com
Window on our World: www.sln.org.uk/wow

LEA sites
CLEO (Cumbria, Lancashire): www.cleo.net.uk
Devon: www.devon.gov.uk/dcs/geog
Kent: www.kented.org.uk/ngfl/prigeog/index.html
Staffordshire: www.sln.org.uk/geography
Suffolk: www.slamnet.org.uk/geography
Wakefield: www.gowilder.org.uk/geog/
Hertfordshire: www.thegrid.org.uk/learning/geography
Kent NGfL: www.kented.org.uk/ngfl

London Grid for Learning: www.lgfl.net
North West learning Grid: www.nwlg.org
Shropshire IT4L: www.it4l.org/links/geog.htm
South West Grid for Learning: www.swgfl.org.uk

Maps

Get a map (OS): www.getamap.co.uk
Interfact atlas: www.gridclub.com/info/atlas/intro.shtml
Multimap (for maps and aerial photos): www.multimap.com
Old maps: www.old-maps.co.uk
Ordnance Survey: www.ordnancesurvey.co.uk
Ordnance Survey MapZone: www.mapzone.co.uk/main.html
Outline maps: www.eduplace.com/ss/maps
Streetmap: www.streetmap.co.uk
World atlas: www.graphicmaps.com

Weather and climate

BBC weather site: www.bbc.co.uk/weather
CNN weather site: www.cnn.com/WEATHER
Met Link: www.metlink.org
Met Office: www.meto.gov.uk
Staffordshire school weather data: www.amingtonheath.staffs.sch.uk/intro.html
Weather at Keele: http://pangaea.esci.keele.ac.uk/weather
World climate: www.worldclimate.com
Yahoo: http://weather.yahoo.com

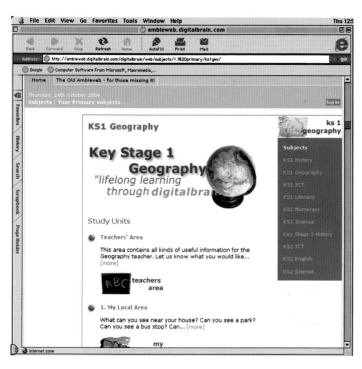

School websites

There are many exemplary school websites which are well-designed, feature pupils' work and have interesting information, including photos and maps of their localities. It is difficult to single out sites for specific mention, for fear of insulting the others. However, the following are sufficiently well-established and offer support and guidance to a much wider audience so warrant inclusion here:

- Ambleside Primary School (Award-winning site, especially good for key stage 1 geography) http://ambleweb.digitalbrain.com/ambleweb/web

Figure 4 | Sources of help in the use of ICT in our geography teaching

1. The ICT co-ordinator in your school (he/she will be delighted if you take an interest in ICT!).
2. LEA advisers (ICT and geography) and your local ICT centre.
3. The Geographical Association Annual Conference (this is held around Easter and includes an exhibition of new materials as well as ICT workshops).
4. Regional conferences on primary geography.
5. *Primary Geographer* the termly magazine published by the Geographical Association.
6. Educational shows and exhibitions, especially the BETT Show (in January) the Education Show (in March) and GA Annual Conference (at Easter).
7. Publications and support materials from The British Educational Communications and Technology Agency (Becta) (details on their websites www.becta.org.uk and www.ictadvice.org.uk). Includes a newsletter and on-line CPD.
8. Local support in your school cluster (liaison between secondary schools and primary schools is important – ensure ICT is on the agenda).
9. NAACE is the professional association for those concerned with advancing education through ICT (www.naace.org)
10. QCA and other national guidance (e.g. National Curriculum in Action: www.ncaction.org.uk from QCA which gives examples of pupils' work including ICT work).
11. The Primary Strategy has produced guidance for integrating ICT in geography.
12. *Technically Geography*: Lessons combining geography and ICT (Bowden and Copeland, 2004-) sets out lessons based on the schemes of work.

■ Snaith Primary School (especially for sections on Seaside, Water, Paris, India)
   http://home.freeuk.com/elloughton13/index.htm
■ Perton First School (especially for the travels of Barnaby Bear)
   www.perton-first.staffs.sch.uk/index.html

Who can help me?

ICT is essentially a practical tool, requiring a 'hands-on' approach. While it is hoped that this chapter may have provoked ideas and provided a context for development, you may still be left wondering 'Yes, but how?'. Figure 4 suggests some other sources of help.

References and further reading

Becta (2003) *Using web-based resources in Primary Geography.* Coventry: Becta.

Bowden, D. and Copeland, P. (2004) *Technically Geography: Lessons combining geography and ICT for Y1/2.* Sheffield: Geographical Association.

Bowden, D. and Copeland, P. (2004) *Technically Geography: Lessons combining geography and ICT for Y3/4.* Sheffield: Geographical Association.

Bowden, D. and Copeland, P. (2005) *Technically Geography: Lessons combining geography and ICT for Y5/6.* Sheffield: Geographical Association.

DfEE/QCA (1999) *The National Curriculum: Handbook for primary teachers in England.* London: DfEE/QCA.

DfEE/QCA (1999b) *Geography: The national curriculum for key stages 1 and 2.* London: DfEE/QCA.

DfEE/QCA (1999c) *Information Technology in the National Curriculum.* London: DfEE/QCA.

DfEE/QCA (2000) *Geography: A scheme of work for key stages 1 and 2 (Update).* London: DfEE/QCA.

DfES (2004) *Information and Communications Technology In Schools In England: 2002 (Provisional).* London: DfES.

DfES (2004) *Primary Strategy: Learning and teaching with ICT.* London: DfES.

Geographical Association (termly) *Primary Geographer.* Sheffield: Geographical Association.

GA/Becta (1998) *Planning for ICT* (leaflet). Sheffield/Coventry: Geographical Association/Becta.

GA/NCET (1995) *Primary Geography: A Pupil's entitlement to IT* (leaflet). Sheffield/Coventry: Geographical Association/NCET.

NCET (1995) *Approaches to IT Capability, Key Stages 1 and 2.* Coventry: NCET.

Ofsted (2002) *Geography in Primary Schools: Ofsted subject reports series 2001/02,* (e-publication) www.ofsted.gov.uk

Rodgers, A. and Streluk, A. (2002) *Primary ICT Handbook: Geography.* Cheltenham, Nelson Thornes.

SCAA (1997) *Expectations in Geography at Key Stages 1 and 2.* London: SCAA.

Postscript

ICT should make an active rather than a passive contribution to learning, and as such it is possibly the most difficult aspect of the national curriculum to organise and manage. It is also expensive in terms of both resources and the time required for training and management. Because ICT itself is constantly changing, teachers sometimes feel threatened by it. The fact that pupils may have greater expertise and confidence in using computers than they do does not help. However, ICT is exciting and motivating and nearly every pupil will take to it with enthusiasm, irrespective of their age and ability. Finally, and most important, the fact is that ICT has enormous potential for enhancing and supporting the geography curriculum. Provided you start in a manageable way, and follow the guidelines set out here and in the many other support publications now available, your chances of success are high. Good luck!

Our local area

IN THIS CHAPTER YOU WILL FIND KEY IDEAS ON
DISPLAYS • ENQUIRIES • ENVIRONMENT • GOOD PRACTICE • INSERVICE TRAINING
• NATIONAL CURRICULUM • PLANNING • SCHOOL JOURNEYS

Using the school locality

The school locality is a rich resource for geographical enquiry. It is highly accessible for fieldwork, can be studied in a great variety of ways and helps pupils to develop their sense of place. The school and its surrounding area provide a meaningful context for developing geographical skills and studying national curriculum places and themes. Its potential for practical, first-hand work, building on existing knowledge and experiences, gives pupils of all abilities the opportunity to develop their confidence and achieve success both in school and in adult life.

By exploring the underpinning geographical principles and adopting good practice strategies for geographical teaching and learning in this area, teachers will become increasingly confident in evaluating and making more effective use of the school locality.

Curriculum requirements

The curriculum guidance for the Foundation Stage recognises the importance of allowing pupils to explore their environment and learn from practical, first-hand experience (DfEE/QCA, 2000). All six 'areas of learning' identified provide opportunities for developing geographical understanding. One of these areas, 'knowledge and understanding of the world', is especially relevant to local area work (see Chapter 2).

The national curriculum specifies how pupils' understanding of geography can be developed through the study of localities as they progress through the primary school (DfEE/QCA, 1999). At key stage 1 pupils are required to undertake a study of their school locality. Similarly, there are designated localities for key stage 2 – one of which has to be in the UK and could focus on the local area.

The term 'locality' is used in the national curriculum in a technical sense. At key stage 1 it is defined as 'the immediate vicinity, including school building and grounds and the surrounding area within easy access' (p. 111). At key stage 2 it is more extensive, being 'an area larger than the school's immediate vicinity' which, we learn, 'will normally cover the homes of the majority of the pupils in school' (p. 114).

It is valuable to think of the locality as three separate areas (Figure 1) as suggested by the National Curriculum Council (NCC, 1993). These are:

A. The school buildings
B. The school grounds
C. The surrounding area

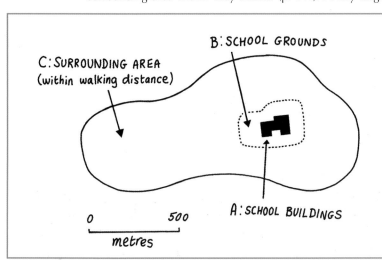

Figure 1 | The locality is best thought of as three separate areas. After NCC, 1993.

You might also consider views of more distant features, such as hills, church towers and factories, if they contribute to the character of the area. As a school you need to decide what boundaries you think are appropriate for the studies you undertake.

The enquiry approach

Structuring geographical learning around the enquiry process (see Chapter 7) enables effective investigational work to be undertaken in the immediate environment. Teachers and pupils should identify a place, theme or issue to explore and then consider appropriate questions to structure their investigation. Pupils will then be involved in planning, collecting, analysing, evaluating and hypothesising. Ideally, the enquiry should focus on a real issue: for instance, 'Why is there a problem with litter in the playground?' or 'Where shall we site a new litter bin?'. The school grounds and local streets can be used as a focus for learning about the environment through the acquisition of knowledge, learning through the environment as a resource for study, and learning for the environment to consider issues, values and attitudes.

Outdoor work is an important aspect of the curriculum, offering opportunities for both independent learning and for co-operation with others; pupils will respond to these forms of learning in different ways. With infants, such work is best done little and often. Short periods of 10-30 minutes outside are best – although with younger pupils it can take more time to get them ready to go out than they actually spend working outside! Careful timing, allowing pupils to complete observations without getting bored, is essential. Unfortunately such work is always weather dependent and you need to learn to make the most of every possible learning opportunity. Remember that you can use the winter as well as the summer months. Pupils are used to going out in the playground in all temperatures – it is often the teachers who complain of the cold!

Planning

Any work outside the classroom demands careful planning and preparation. If more than one adult is available, and the activity has been explained beforehand, it will be possible to work outside with a whole class. However, with younger pupils especially, the ideal arrangement is a small group with a dedicated adult. A number of small groups can cover different aspects of the study in a controlled learning context, which allows individuals to exploit their own interests, with each group returning to the classroom when their tasks are complete. Different aspects of the study will need to be discussed beforehand both with your pupils and helpers. Careful preparation of helpers will help to ensure that all pupils are supervised, but each group is allowed some independence. Ensure that all pupils know exactly what they are doing, have chosen a suitable resource or activity, can operate any necessary equipment and are physically comfortable. When pupils return to the classroom they must know where and how to store their findings, and time should be made available for groups of pupils to share their findings with others.

If you are well organised you will be more confident and better prepared to take advantage of any spontaneous opportunities which will inevitably arise when working outside and which frequently provide the high points of a study. All outdoor work must, of course, follow your school and LEA policies for fieldwork. Conducting an appraisal of your locality will also form an essential part of your planning. As teachers we are often guilty of failing to explore the

1. What are the key geographical features in your local area? Make an inventory of particular aspects of your locality such as:
- different areas in the classroom, school buildings and grounds
- viewing points for map and compass work
- the layout of the site
- local journeys and routes
- physical features such as hills and streams
- changes and new developments
- local issues and concerns

Display your findings on a large master base plan of your district. Remember to update this on a regular basis so that you can see how your locality changes over a period of time. Consider if pupils would identify the same features.

2. What types of equipment are needed for local area fieldwork? You need clipboards, collecting and measuring equipment, and field study boxes, which any member of staff can use with their pupils. They should contain everything pupils will need to work outside successfully; small, self-sealing, plastic boxes are ideal.

3. Do you have an emergency fieldwork survival kit? This should allow you to stay outside no matter what! It might include everything from a pencil sharpener to a plaster or spare clothing.

4. Have you considered issues to do with safety and inclusion? Is there a school policy defining organisational procedures and requirements? Do you offer safety advice to colleagues? Are you able to cater for pupils of different abilities (see Chapter 24)?

5. How could effective use be made of other adults? Do you involve other adults in in-service training? Is there a training programme for new helpers? Are teaching assistants involved?

6. What vocabulary can be developed and how? Look at activities planned in the school grounds at both key stages. Make lists of key directional and place related vocabulary. Is there increased sophistication of language and vocabulary developed through fieldwork and active learning outside the classroom?

7. What resources do you need? Do you have access to recent as well as older photographs of your school building and local area? Are there any plans, maps and memorabilia? Do you have large-scale Ordnance Survey maps and aerial photographs of the school and its surroundings? Have these been used to devise an A4 simple base plan of your school and grounds that everyone can use?

8. Which parts of the national curriculum can you cover in your local area work? Make copies of the programmes of study and ask colleagues to highlight appropriate sections at both key stages. Are you covering these effectively? If you follow the QCA schemes of work, are you making the most of the opportunities for local area work?

9. Is there scope for developing geographical skills and cross-curricular links? Consider if you could make better links to literacy, numeracy, citizenship and ICT as you explore your locality.

10. How does your local area work help to develop a sense of place? Can you identify an issue or enquiry question to explore with colleagues? How can they plan to ensure active learning throughout the school? What are the opportunities for assessment? What records do you want to keep?

Figure 2 | *Questions to consider as you conduct an appraisal of your locality.*

school grounds and locality on foot ourselves; the result is that we know less about them than our pupils do! If possible try to identify a school in-service training day when you can clarify your approach to local area work. Figure 2 provides a checklist of questions to consider.

Using the school and its grounds

Pupils learn best when activities revolve around their own interests and concerns. Younger pupils are particularly curious about their own sphere of operations – the building and grounds of their school – and primary school practitioners have long recognised the value of the school and its immediate environment to their pupils' geographical learning. In their reports, HMIs have consistently identified the value of geographical work carried out within the immediate locality of the school, where practical fieldwork observations can most easily take place.

Within the school grounds, investigations can involve simple map-making, early observations 'in the field' and talking to people about their way of life and the world of work. Pupils can also be introduced to the inter-relationships between human and physical processes. Their developing environmental awareness can be harnessed by encouraging them to take responsibility for their school environment. In this way the three key dimensions of primary school geography – place, space and the environment – can all be developed to provide pupils with memorable, vivid, active and relevant outdoor learning experiences.

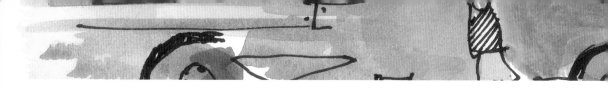

Photo | *Stephen Scoffham.*

Exploring the local area

Using a small area within easy walking distance has many advantages. It is a small-scale, relatively safe environment to which short visits can be fitted easily into school sessions. As well as being easily accessible, the school grounds also provide a safe environment from which to study the immediate locality, e.g. to undertake observational drawings of shops or houses across the road. All schools will have some distinctive features, inside and out, which are worthy of study. Using the school grounds enables pupils and teachers to explore micro changes in their environment and respond to any immediate changes that may occur, e.g. the arrival of workmen on the school site or flooding after heavy rain.

The following activities will help you to develop a greater geographical understanding of the school and its grounds. Some of them can be built into your curriculum plans for geography, others might provide the focus for an in-service training day. Such activities also provide an opportunity to convince others of the benefits of first hand experiences and practical work.

1. Sensing

One of the most meaningful types of local environmental work for younger pupils is through using their senses: identifying what they can see, hear, smell and feel. The fifth sense of taste, for obvious reasons, is best avoided! Observation – learning to look carefully – is an important geographical and cross-curricular skill to learn and develop. Here are some ideas you might like to try:

- Spotting shapes and features
- Identifying likes and dislikes
- Brainstorming and composing word pictures
- Drawing a memory picture or mental map
- Going on a sensory walk
- Blindfold activities, requiring pupils to feel and use language to describe and identify features in the school grounds. (Their sense of smell might also be used to help them identify where they are.)

2. I-spy

Organise simple five-minute activities such as 'I-spy' around key vocabulary associated with the school building and grounds. Using a word bank, dictionary or alphabet display – particularly important when helping pupils to gain access to secondary sources – you can

focus on features associated with particular letters or sounds through the use of an index. If you set up a database to store the data you have collected, it could become a starting point or a reference point for pupils involved in future local enquiries.

3. Listening and hearing

Sounds and noises make a significant contribution to the quality of the environment. Some sounds, like bird song and running water, have a peaceful quality; others can be disturbing, irritating or even frightening. What sounds can the pupils hear? Do the sounds differ from place to place? Can they record the sounds they hear and explain what they are?

4. Making rubbings

Give younger pupils the opportunity to record their findings in a variety of ways. Have you tried allowing them to make rubbings using large sheets of paper and chubby wax crayons? Try rubbing various features and then locating the rubbings on a large base plan, marking their correct positions.

5. Field sketches

Making sketches is one of the best ways of getting pupils to engage with their surroundings. Younger pupils find it hard to record whole scenes but will quickly gain confidence if directed to small details such as a door handle, an individual plant, or a small section of wall. Encourage them to add a title and make simple notes of the features they think are important.

6. Numbers walk

The school grounds can be used for practical work such as simple collecting, counting and measuring exercises, e.g. counting or plotting the position of doors, windows, drainpipes, covers, grates, steps and flags; or in a weather topic to read a thermometer, measure rainfall, observe cloud formations and movement, calculate wind speed and direction or even to take a closer look at soil.

7. Boundaries and barriers

Boundaries and barriers are a common feature of any school environment. Pupils will be familiar with the rules which affect them but may never have considered the reasons why they are there. Security, safety and organisation are usually involved in one way or another. Make a survey of boundaries and barriers around your school. Why are they there? Are they physical or do they depend on the interpretation of notices and signs? Why do we need them?

8. Weather watch

The weather affects everyone – particularly children. The school grounds provide a real context in which pupils can experience this phenomenon and make first-hand investigations and observations, such as:

- When do pupils need to wear a coat outside? Why?
- Why does the school roof leak?
- Why do some window frames need to be re-painted or replaced?
- What is the weather like today?
- How does it change?
- What seasonal differences occur?

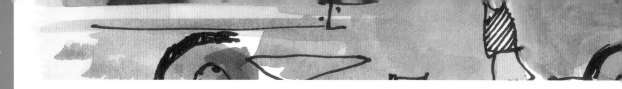

9. Building materials

Take a close look at the building materials used in your school. What are they? Where did they come from? Why were they chosen for that part of the building? Use sensory activities to help pupils explore the textures, pattern, size and shapes of the materials. Set up a display table showing different examples.

10. Photographs

Organise a short walk or route round the school building or grounds linking different points of interest. Get the pupils to take photographs of key features using a digital camera or print film. Encourage each pupil in the group to select one item to record (not their friends!) and display the pictures with captions round a map or plan in a later lesson.

11. Looking up, down and between

Learning to 'look down' is vital to the development of the plan view concept and mapwork skills. You might organise a walk with stopping places where pupils can record what they see when they look down on the ground and up in the sky. Can they sketch what they see? Extend the work by looking for views through archways or between walls and buildings. If the pupils draw and cut out the pattern of roofs and chimneys against the sky you will be able to set up a wall display showing local silhouettes.

12. Direction games

The playground can be a useful space for directional games. You might use giant directional arrows to reinforce key vocabulary such as up, down, eye-level, forwards, backwards, right and left, overground and underground. A simple base plan of the playground can be used for activities and trails. Pupils could consider which parts of the school grounds are natural/built/green. What sort of habitats have been created in this place? Which creatures live there and why?

13. Compass points

Use the playground and school grounds to give pupils first-hand experience of compass points. What can they see to the north, south, east and west? Get them to record their findings in words or pictures. Use a local map and atlas to find out what lies further away in each direction. Which direction would they need to take to reach the Sahara Desert or the United States for example? See if they can plan a route round the grounds involving compass directions and 90° turns. This will be useful practice for orienteering work at some later stage and could be re-inforced by using a programmable toy or toys.

14. Surveying

You can use the school building and grounds to help pupils to develop simple survey techniques and learn to plot symbols on a base plan. For example, a simple land use survey could be carried out using a colour code system. Why do pupils think certain activities take place in particular locations? Are there better alternatives? Can you use the school building and grounds for practising questionnaire and interview techniques? For sampling specific locations for intensive small-scale study, use a PE hoop to delimit a sample area and focus the pupils' attention on a particular area for study.

Photo | Tina Horler.

15. Mapping

Mapping is a unique geographical skill, and every opportunity should be taken to practise and develop it. In the school building and grounds you have the most accessible and meaningful learning context for mapping skills. Using a familiar environment helps pupils relate their direct experiences to map making. Simple orienteering skills can be developed, based on trails using pictures or symbols as clues, or by following string to collect objects. Maps can also be used to follow and record routes and to identify locations, features and positions.

16. Ordnance Survey maps

Where possible you should use large-scale Ordnance Survey (OS) maps of your school building and grounds from an early stage. They are an excellent teaching resource and pupils often enjoy looking at them to spot information and find out more about their locality. Use OS maps to help you devise maps of the area, which pupils can use in their fieldwork activities. You can also enlarge selected areas for wall displays. Simply make an OHP transparency of the relevant section, project the image on a sheet of clean paper pinned to the wall and draw round the outline. You can vary the magnification by moving the projector backwards and forwards.

17. Aerial photographs

Aerial photographs are an excellent source of information about your locality and a useful way of complementing maps and plans. Research shows that children as young as three years old can make sense of aerial photographs of a familiar area. Older children can use the photographs to identify features (start with the school and their own house), trace routes between places and find out about changes and developments in their local area.

18. Eyeballing

Eyeballing combines observational and recording skills. Select a particular part of the school building or grounds and give pupils a photograph to focus their observations.

Can they go to the spot and make an observational drawing of a specific feature, e.g. a door or window?

Can they copy examples of writing in their environment?

Can they make a field sketch or conduct a census to answer a key question?

Can they find any date-stones or other clues which give historical information?

What can they see from different viewing points?

Can they add information to base plans or drawings to record where they have found the clues?

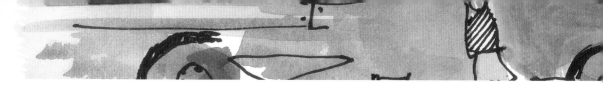

Photo | John Halocha.

19. Changes in the environment

Most schools have been altered and changed over the years. Some may even date back a century or more and have gas fittings, old fashioned furniture and datestones. Pupils enjoy looking for clues that indicate change and quickly adopt the role of detectives. Challenge them to locate a range of features working from photographs you have prepared beforehand and to mark them on a base map. See if they can devise a changes trail or walk of their own as an extension exercise.

20. Scavenger hunt

Explore the school grounds to find examples of pollution and other environmental problems. Think about the following questions to help you structure the work.

What items can they collect? (Use plastic bags as gloves.)

Where did they find their items?

Can they mark the locations on a base plan of the school?

What do the items tell you about your school grounds as a place?

Did other groups find similar or different things? Why do you think that is?

Can you find out more about the items you collected by talking to others?

21. Environmental quality

You can explore pupils' environmental concerns by asking them questions such as:

What do you think about the local environment? Are there any places which you think could be improved and who might make the changes? Once they have conducted their appraisal pupils can be encouraged to identify a project and set out to make the improvements themselves. Locating routes for new pathways, working out the plan for a wildlife area or designing extra play equipment are the kinds of projects which motivate pupils and harness a range of cross-curricular skills.

Using local streets and buildings

Local streets and buildings are invaluable teaching resources and you should take every opportunity to develop outdoor work both in geography and in other subjects. The benefits in terms of pupil motivation, providing meaningful learning experiences and illustrating ideas in practical ways are immense. Recent scientific discoveries make it clear that outdoor work provides a rich, multi-modal environment in which our brain is likely to flourish (Scoffham, 2003). It is no coincidence that in a recent survey Ofsted noted how first-hand experiences, including visits both locally and further afield, enriched the curriculum in the most successful primary schools (Ofsted, 2002).

The following suggestions indicate some possibilities for local area work. They include a mixture of topics covering physical, human and environmental geography. The opportunity for applying and using skills such as mapwork and ICT will be apparent throughout. The need

for a large-scale (1:1250) map of the area and other resources such as aerial photographs should also be noted.

1. Landscapes
Consider the landscape in your area and the way that people have responded to it. You might begin by recording hills and slopes on a base map of the local area. Are there any areas, e.g. sports grounds, which are particularly flat? Are there any cliffs or beaches near your school? You might also consider local rocks and soils. Set up a display table showing interesting samples from local homes and gardens. Some of these materials may have been brought from other areas so you will begin to discover things about the surrounding region as well as your own locality as you embark on this work.

2. Weather
Pupils can record the direction and the force of the local wind, and the amount of rain in terms of the time it falls and the water it provides. Encourage pupils to find places around the school which are especially wet or dry. What causes these variations in micro-climate? Are there any places in the immediate environment which tend to flood in heavy rain? Do back-copies of local newspapers carry reports of unusual weather and its effects on local people?

3. Houses and homes
You might begin a study of houses and homes by identifying different building materials, e.g. tiles, bricks and timber. The way these materials are used and combined has a significant impact on the character of an area. Pupils can study front doors, windows, and the construction of the buildings, e.g. terraced, semi-detached or detached. They should be given information about the age of buildings so that they can be aware of the growth of their local area. Identifying any listed buildings and plotting them on a map is another way of revealing historical patterns. An alternative approach is to identify or visit current construction sites. This will show how your settlement is changing and developing. Thinking about the future is an important part of geography.

4. Shops
Visits to local shops lead naturally into an investigation of the goods available and questions about where they originate. There may be opportunities for making a visit to a local supermarket to find out how it is organised and the work that people do there. Try to consider the issues that lie beneath the surface. How much of the produce on sale comes from local sources? What impact do warehousing, transport and distribution have on the environment? How much do producers get paid and does the shop stock 'fair trade' goods?

5. Community facilities
Leisure and community centres provide places where people can meet and they add to the quality of local life. What facilities exist in your area and who do they cater for? Examples might include mother and toddler groups, youth centres and old people's homes. Most schools are also within walking distance of a church. If you can arrange a visit and talk to the vicar about their work it will add to the pupils' understanding of the area. What facilities do pupils think are missing? You might identify a suitable project and draw up a proposal to discuss more widely in the school and, perhaps, the local community.

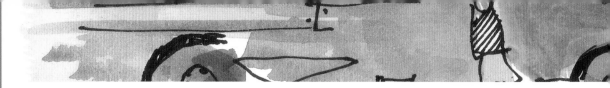

River under threat from treated sewage release

Housing growth without jobs could spell 'disaster'

'Shabby station is letting area down'

'Greenfield land not the place for business park'

Dismay as yew tree gets the chop

Bid to set new Walking Bus record in Kent

Intrepid fund-raisers reach their peak
Rotarians from the Forest of Blean club have hiked up the highest peaks in Scotland, England and Wales.

Turn that waste into compost

Historic building is saved from oblivion

Figure 3 | *Media reports often deal with geographical issues. These headlines all came from one issue of the* Kentish Gazette.

6. Transport

Pupils can record the number of cars, lorries, bikes etc. passing their school or another location and make maps and diagrams of railways, tramlines, cycle routes and footpaths. Are there any problems such as traffic congestion which need to be addressed? Are the public transport facilities adequate and does traffic have a negative effect on people's lives? You could bring the work back to a child level by recording the journeys and methods of transport which the pupils use themselves. Are all these journeys necessary and could they be made by other means?

Caring for the environment

The pupils should be encouraged to engage in practical projects such as caring for gardens and pets, using water, energy and other resources wisely, and understanding that waste can be recycled and not just 'dumped' (Figure 3). Charities such as the Groundwork Trust may have projects in your area which are well worth learning about. Alternatively you might invite professional planners, architects and landscape designers to talk to the pupils about how their work involves an environmental perspective. Conduct an audit or your own school and local environment. What demands does it place on the environment and natural resources? How could this impact be reduced?

Making a trail

Trails can be an excellent way of structuring local area work and serve to focus attention on specific themes or issues. You might devise a trail of your own (perhaps as an in-service activity) or get the pupils to make one themselves. Either way there are a number of issues to consider:

■ Decide whether there will be a focus on a theme (e.g. changes) or a topic (e.g. transport), or if you want the pupils to select points of interest along the designed route

■ Decide whether you will observe the surroundings continuously along the walk or stop at set sites to undertake a number of studies

■ Decide exactly where you are going to cross roads and access any other safety hazards

■ See if you can include interesting or unexpected elements, e.g. by selecting a route down paths and alleyways

■ Do not make the route too long and decide whether you want to create a circular walk

Photo | *Paula Richardson.*

- Provide pupils with a map to work with
- Set up questions and activities that have a clear educational purpose
- Keep the text brief so that participants do not spend too long reading the script while they are outdoors
- Try to include a variety of recording techniques, e.g. recording information on a table, adding labels to a drawing, making sketches, drawing maps and so on
- Take photographs of different features along the route which can be studied in detail back at school

Conducting research

As they work as 'geography detectives' and find out about their local area, pupils will need to use a range of sources. It is important to teach them how to conduct their research. You can help them by identifying suitable research questions and issues and discussing the approach they are going to adopt. Some of the options include:

- Reading books and pamphlets about the local area, e.g. town guides, civic society appraisals, historical guides
- Reading newsletters and reports, e.g. from the local council, local clubs, parish churches
- Meeting with local residents, who have key information about the area
- Looking at photographs and artwork of the local area, both historic and current
- Watching videos which indicate views on and activity in the area
- Visiting the local library and museum
- Contacting the local authority planning department to find out about new schemes and proposals
- Gathering information from the internet, e.g. from local websites
- Reading the local papers, including past articles
- Studying maps, plans and aerial photographs
- Investigating data files, e.g. local census statistics
- Taking part in fieldwork outside the classroom

When pupils have completed their investigations it is helpful if the resources they have collected are stored in a resource bank. Future classes will then be able to access this for their research.

Recording findings

When the enquiry sessions have been completed it is important that pupils record the observations and information they have collected. Try to get them to collaborate and share ideas among themselves. Classroom discussions can help to raise questions and identify the issues that have emerged. Get pupils to draw their own maps and devise graphs, diagrams and graphical displays. Here are some other ways of presenting findings:

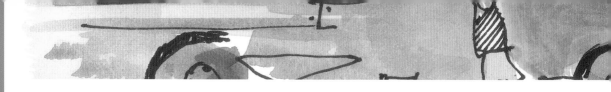

Photo | *Lois Gunby.*

■ Design diagrams with annotations and labels

■ Write a report summarising findings

■ Create a mini pamphlet

■ Write answers to prepared questions

■ Write personal thoughts about key issues

■ Draw a picture using set art skills for a display

■ Design a postcard

■ Design a map showing findings

■ Print out digital photographs and write an explanation about each one

■ Set up a display which includes sample items and display models

As they present their findings pupils will gain confidence in using and studying local maps. They should also be given opportunities to draw maps of their own. These might show important streets, compass directions, routes between places, specific sites and other geographical features. If you can provide a range of different maps of the local area – tourist maps, maps prepared by the council, bus maps and so forth – it will give pupils a range of examples to emulate as well as extra information about your locality.

Finding out about the local area is an important component of geographical work from the nursery through the primary and secondary school to GCSE, A-level and beyond. Blyth and Krause (1995) suggest that geography can be divided into two main components: 'Big geography' starts from distant places and considers the world as a whole space and environment at a national and continental scale; 'Little geography' fits within this framework and involves studies of particular places and their immediate surroundings. Pupils need to study both 'little' and 'big' geography at all ages. However, as they grow older and move through the curriculum their perspectives broaden and deepen. Contrary to popular belief children do not grow out of local area work as they get older. Rather they study it in a more sophisticated way. That is why learning about the locality is so important.

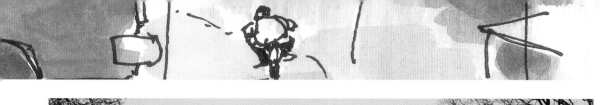

Photo | *Lois Gunby.*

References

Blyth, A. and Krause, J. (1995) *Primary geography: A developmental approach.* London: Hodder and Stoughton.

DfEE/QCA (1999) *The National Curriculum: Handbook for primary teachers in England.* London: DfEE/QCA.

DfEE/QCA (2000) *Curriculum guidance for the foundation stage.* London: DfEE/QCA.

NCC (1993) *An Introduction to Teaching Geography at Key Stages 1 and 2.* York: National Curriculum Council.

Ofsted (2002) *The curriculum in successful primary schools.* London: Ofsted.

Scoffham (2003) 'Teaching, Learning and the Brain', *Education 3-13*, 31, 3, pp. 49-58.

Photo | Stephen Scoffham.

IN THIS CHAPTER YOU WILL FIND KEY IDEAS ON
ASSESSMENT • DISTANT PLACES • ENQUIRIES • EUROPE • ISSUES • MISCONCEPTIONS •
NATIONAL CURRICULUM • PLANNING • PROGRESSION • QCA UNITS

Contrasting localities

T his chapter gives guidance about UK and other European localities schools might study in the context of the programmes of study for key stages 1 and 2. Supporting resources and appropriate teaching strategies are considered and guidance given on using QCA units, planning and assessment.

> *The development in children of a "sense of place" is a fundamental aim for geography teaching – to help children to develop a feeling for the unique identity of a place; what the places they already know mean to them and what it might be like to be in places that are as yet remote (Wiegand, 1993).*

Children live in a world that is full of relevant and significant places. These include not only their home and school environment, play areas and holiday destinations, but also places that they learn about from television programmes, films and books. Each place has its own 'character' which is determined not just by human and physical features and processes but also by our individual responses to them. Consequently, a central question for pupils to answer when studying any place, especially those they have not visited, is 'What would it feel like to live in this place?'. This question is relatively easy to answer when pupils are investigating their own environment but more complex when the place is further away and, therefore, unfamiliar to most or all of a class.

What is certain is that children are fascinated from an early age by places that are unfamiliar and distant. However, their ideas about these places may, understandably, involve inaccuracies and stereotypes. Place studies give teachers an opportunity to investigate and, through their teaching, challenge and modify pupils' initial ideas. Thus, before starting any place study it is important to explore pupils' existing knowledge and understanding. The results allow subsequent activities to proceed from the known to the unknown.

What ideas do young children have about unfamiliar places?

What notions do children who live in cities have about the nature of English villages? Before undertaking a study of a Surrey village, pupils in a year 4 class in an urban school were asked to complete a questionnaire about their ideas. These are some of the things they said:

- English villages are 'quite poor, not a rich place'.
- The people wear clothes that are 'torn because they don't have much money'.
- You might see 'girls wearing long black gowns and white aprons'.
- Work includes having 'to go into the forest to collect wood to light the fire with it' and being a 'shoe-shiner'.
- You would see 'pots where the food has been cooked' and hear some 'wild animals'.
- Life for people living in a village in England would be 'like African people they beg for food and water'.

In general, most pupils in this class saw villages as poor, impoverished and technologically backward places. Their ideas about village life were largely negative, historical and wildly inaccurate. Some years ago Graham and Lynn (1989) noted how children tended to associate

life in the developing world with primitive, hunter-gatherer images. It may well be that as children's ideas about distant places develop, they pass through a stage when they confront their deep-seated fears of an unknown, hostile world.

By contrast, the small number of pupils who actually had first-hand experience having been to an English village had much more accurate and positive impressions. One pupil, for example, recounted visiting his grandmother in a Somerset village, describing it as:

> *A place where there are shops, houses, not much traffic and with mostly old people but with some children. People work in a Somerfield supermarket and there is also a police station and the children there play football. The village is sometimes quiet and sometimes, not usually, noisy.*

As children make sense of the world, they rely on what Wiegand (1992) calls either direct experiences obtained through first-hand encounters, or indirect, second hand experiences. While direct experiences may be most influential and 'real' for children in informing their ideas and opinions about places, they are more regularly affected by indirect influences such as accounts of places visited by friends or family, through stories and, probably most significantly, television programmes.

Investigating pupils' existing ideas is simple and can be done in a number of ways. One approach is to gather evidence through writing, annotated drawings, word banks and word lists. Another option is simply to ask questions. This can be particularly revealing as it exposes children's existing knowledge and personal experiences as well as their opinions. With modification the following question set could be applied to any locality:

- What is a village?
- Have you ever been to a village?
- Do you know anyone who lives in a village?
- What would you expect to see/hear/smell there?
- What kind of work do people do there?
- What do you think it is like to live in a village?
- How do you think village life is changing?
- What issues do you think will affect the village in the future?
- Would you like to live in a village? Why/why not?
- In what ways is the village similar to/different from our place?

What are the opportunities and guidance for studying European localities in the curriculum?

The geography national curriculum specifies the places that children need to study beyond their local area (DfEE/QCA, 1999). At key stage 1 pupils have to investigate a contrasting locality either in the UK or overseas. At key stage 2 they are required to study (a) a locality in the UK

Photo | *Paula Richardson.*

and (b) a locality in a country which is less economically developed. At both key stages pupils also have to develop geographical enquiries and skills at a range of scales from the local to the national.

What does this mean in practice? At key stage 1 teachers have a free hand and can select any place they like. At key stage 2 schools are told that their UK locality study might be either the school locality itself or a locality elsewhere in the UK. While the curriculum may be flexible there are strong arguments for studying both the school locality and a locality elsewhere in the UK. Covering both will provide a wider and more coherent approach to developing children's sense of place, defined as 'the "personality" of a place, together with a sense of the range of diversity of places around the world' (DES, 1990).

Guidance for other European place studies is less explicit. However, at key stage 1 pupils could certainly select a European locality other than the UK as their contrasting locality study. At key stage 2 the curriculum specifically requires pupils to study different parts of the UK and Europe in their study of localities and themes. It is also important to know that as a member of the European Union the UK, along with other member states, has been committed since 1988 to incorporating a European dimension into appropriate areas of the curriculum such as geography.

Whichever places you choose, the scale of the study is most important. At key stage 1 the contrasting locality should be 'an area of similar size' to that of the school's 'immediate vicinity, including school buildings and grounds and the surrounding area within easy access' (DfEE/QCA, 1999, p. 111). At key stage 2 the locality chosen should be 'an area larger than the school's immediate vicinity' and will normally include 'the homes of the majority of the pupils in the school' (DfEE/QCA, 1999, p. 114). This enables children to make valid comparisons and contrasts with their own place.

What can I do to support a locality study?

Since the inception of national curriculum geography in 1991 most schools have taught about distant places using resources from one of the following:

- A photopack produced by an educational publisher
- A school-produced locality pack
- TV programmes
- Story books
- A twinning arrangement with a school from a contrasting locality
- A residential school visit

Professionally-produced resources are widely used and they have the advantage of being conveniently available 'off the shelf', are often written by professional authors, tend to be well presented and not too expensive to buy. These resources are likely to remain the best option for many schools. However, geography subject managers might feel that it is time to change the focus of their UK locality study, in which case they can either change the resource pack or, preferably, devise some materials of their own. The great advantage of the second option is that fieldwork can be incorporated into the teaching programme, providing an invaluable extra dimension to pupils' depth of understanding about a place that simply cannot be achieved in classroom study alone.

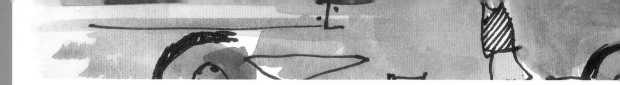

How can a school produce its own resource pack?

A locality in the UK that contrasts with the children's own locality does not have to be very distant or very different. The majority of UK primary schools are probably no more than a short way away from a place that has characteristics that would allow children to find both differences and similarities with their own place. Moreover, the processes involved in producing a photo pack are largely similar for all localities. Here are some ideas to get you started.

First, select a place that is contrasting but easily accessible, to enable resources to be gathered easily and to allow fieldwork. Next carry out some research. A visit is essential but the internet may be useful in finding both written and pictorial information. If possible, make contact with a primary school in the place you have selected to see if you can establish links or gain extra information from local teachers. Additionally, you may be able to find a family who will co-operate in representing a 'typical' picture of everyday life.

The geography photopacks currently in use in your school will provide ideas for appropriate resources and teaching strategies but the most important task is to decide on suitable photographs. These will be more influential than anything else in informing pupils' ideas about the place and are likely to be the resource from which they will gain most enjoyment. If possible take your photographs on a bright day, preferably in the summer when the content of gardens, open spaces and so on is most interesting. You should also take more pictures than you intend to use and later select the twenty or so that provide a balanced and authentic perspective. While taking the photographs remember it is also possible to gather other valuable information and resources such as maps, old photographs and reports of current issues and events.

In general you should try to take photographs covering the following themes:
- physical characteristics, e.g. fields, hills, rivers, seashore, weather, rocks
- human characteristics including (depending upon the nature of the place) the variety of homes (cottages, flats, bungalows, terraces), types of shops, significant buildings (town halls, churches, police stations) etc.
- places where people work (factories, office blocks, farms)
- places for recreation (playgrounds, sports grounds, cinemas)
- environmental and planning issues to illustrate the concept of change and conflict

Ideally your images will also include:
- a primary school, including its interior and grounds
- a 'focus' family in their home with their children who attend the primary school
- historic prints of the place showing aspects of change and continuity

It is important for people, preferably named people, to appear in some of the photographs as this gives an invaluable 'personal' link with the place that primary children find particularly appealing. Digital cameras make the job of taking a large number of photographs relatively cheap and easy. You can use a computer later to manipulate the size and content of the images you have chosen. Good quality colour reproduction is possible with modern colour printers. Enlarging the photographs to A3 size is helpful for infants, but A4 is suitable for juniors. Laminate the photographs to make them more durable.

An invaluable resource that gives a unique overview of any place is a vertical aerial photograph. Aerial photographs taken across the UK are available from Getmapping (www.getmapping.co.uk) and Multimap (www.multimap.com). They can also be viewed at Streetmap (www.streetmap.co.uk) for London boroughs only.

A range of maps is essential for any place study. For places beyond the pupils' own locality small-scale maps allow them to locate the place (Where is this place?) and to consider the interconnectedness of places (How is this place connected to other places?). Larger scale maps encourage exploration of the place (What is this place like and why is it like it is?) and recognition of change in a place (How is this place changing?).

More detailed exploration of places is best served by OS 1:2500 maps (1cm represents 25m), available for rural areas, or 1:1250 (1cm represents 12.5m), available for urban areas. Such large-scale maps show small areas in great detail.

The Ordnance Survey is an excellent source for a range of maps but the internet has sites that offer free maps such as Multimap and Streetmap (see above). Additionally, ICT can be used to help children make their own maps. A good map-making programme is Local Studies (from SoftTeach, see page 203).

You should also try to include case studies as these are very effective in helping younger pupils to explore, in context, the effect of change in a place, including environmental issues. Issues arise when there is conflict over how the land is used. Find out if there is an issue facing your chosen place. It might be a proposal for a new housing estate, the building of a superstore or a by-pass, for example. Studying issues encourages not just an understanding of why change happens and how conflict arises ('How is this place changing?'), it also provides opportunities for pupils to consider their own and others' feelings about change ('What do I, and others, feel about this change?').

In posing and answering these questions you can meet teaching objectives within both the cognitive and affective domains. The cognitive domain 'is essentially concerned ... with knowledge and skills' while the affective domain is concerned with 'the learning of values and attitudes' (Bale, 1987). Using issue-based case studies encourages 'confluent education' – the linking of the cognitive and affective domains – towards enabling each pupil to become a 'feeling/thinking person' (Fien, 1983).

Why and how can fieldwork be incorporated into a UK locality study?

The disadvantage of using a published resource for locality studies is that it is unlikely that the places covered are accessible for fieldwork, and such work is essential for expanding and refining pupils' knowledge and understanding. Depending on the nature of the place, activities might include field sketching, land use and traffic surveys, map making, using all senses to describe feelings, tape recording sounds, listing work done in the place and taking photographs. Pupils gain a much broader view of a place when they study it directly and are also likely to appreciate some sense of those intangibles that contribute to its 'personality' and character. A feature of all places is that they are dynamic and ever changing. Recognising and acknowledging this dynamism can be problematic unless regular information updating takes place. Thus, it is important to try to make annual visits to the locality so that you can build up additional resources over time. These might particularly include new photographs that demonstrate change.

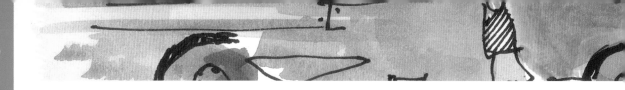

How can schools use the scheme of work units in UK locality studies?

Within a whole-school framework for geography all schools should be able to use or amend the QCA schemes of work. Unit 3: An island home and Unit 13: A contrasting UK locality – Llandudno (DfEE/QCA, 2000) focus on contrasting UK locality studies and are both supported by a range of published resources.

Unit 3: An island home

This unit links geography and literacy. The fictitious island of Struay featured in the Katie Morag stories (Hedderwick, 1997, 1998) is based on the real Scottish island of Coll. An authentic geographical enquiry needs photographs, but Coll's remote location has made this difficult, prompting the publication of several Coll resources. These include *Discover Coll: The Real Struay* photopack (Graham and Walker, 2003, see panel below) and *Isle of Coll map* (Geographical Association, 2004). Using these resources with the Katie Morag stories should expand the potential and range of the Unit.

The Isle of Coll

Together with a colleague, I have compiled the contrasting UK photopack about Coll mentioned above. Physical geography features prominently on Coll so photographs were taken of hills, beaches, lakes and cliffs. The human dimension was represented by images of the school, various homes, shops and the island's only hotel.

In selecting the photographs compromises had to be made. There was no room, for example, for interesting events like the Coll summer show. Additionally, intangibles like the friendliness of the islanders, the peacefulness of the beaches, the strength of the wind and the sometimes unrelenting rain cannot be properly captured in words or pictures. Ultimately there can be no substitute for fieldwork.

Photo 1 | *A beach on the island. Photo: Douglas MacDowall. © Bluesky International.*
This photograph is of Feall Beach, typical of many around Coll. There is evidence of the tidal nature of the sea and there are sand dunes and cliffs in the distance. Beaches are common on Coll and even on a sunny day in early summer, there are no people here. A photograph such as this could allow key stage 1 pupils to talk about their own experiences of visits to the beach and to consider comparisons and contrasts.

Photo 2 | *Ben, Ellie and their father by some rock pools.*
Photo: Douglas MacDowall. © Bluesky International.
This photograph is of three members of a tourist family amongst the rock pools on Rubh 'a' Bhinnein beach on Coll. It shows not just the physical characteristics of the coast, but most importantly visitors using and enjoying the place. Additionally, the children are finding out about sea creatures such as crabs and shrimp. The fact that the children are named provides a 'personal' dimension to a secondary source of information for pupils using the photograph.

Photo 3 | *Vertical aerial photograph of Arinagour.*
Photo: Getmapping plc. © Bluesky International.
This photograph, at a scale of 1:5000, shows the only settlement of any size on Coll, the village of Arinagour. Such photographs are a wonderful source of information about places. This image shows buildings, including piers as well as homes, a hotel, farm and school, and the physical characteristics of water and land. Additionally, it shows patterns in the land such as the varying depth of the sea, green fields and the rather barren, rocky land that covers much of the area. Used alongside a similar image of the pupils' own locality, fascinating opportunities are offered for comparison and contrast of both physical and human features.

Photo 4 | *Highland cattle.*
Photo: Greg Walker. © Bluesky International.
This appealing image shows the unfamiliar breed of highland cattle found on Coll. These cattle are ideally suited to the wet and windy climate with their double-thickness coats. They also have an environmental link with the island's vegetation, as they help distribute seeds from the plants as they graze.

The Royal Society for the Protection of Birds (RSPB) site in the south-west of Coll provides a case study of an environmental issue. This area is a breeding site for many birds, including the globally endangered corncrake, as well as the most suitable part of the island for farmers to grow crops. Corncrakes nest late and on the ground and when farmers cut their crops early in summer it disrupts breeding. A compromise over this conflict has been reached between the RSPB and the farmers who now cut later in the season and cut from the field centre outwards, allowing any birds still on the ground to escape. This example enables pupils to understand not only why the conflict existed on Coll but also to see that through communication between the two parties, and seeing the other's point of view, it has been possible to reach a compromise. Simultaneously, pupils have a context to understand important concepts of conservation and sustainability.

Unit 13: A contrasting UK locality – Llandudno.

Two published packs offer quite different approaches to teaching Llandudno as a contrasting UK locality. *Llandudno: The study of a contrasting seaside locality* (Rowbottom, 2002) provides a wide range of maps and photographs with suggested teaching strategies, while *Storylink: Llandudno* (Dryden and Hare, 2001) combines fictional stories with factual case studies featuring real people from Llandudno.

A problem with this unit arises where fieldwork is required. There are two main solutions:

1. Adapt the unit questions to acknowledge that the study will be entirely classroom-based (Unit 10: A village in India provides ideas for this).

2. Apply the ideas in the unit to a closer locality enabling fieldwork and perhaps creating your own locality pack as part of the fieldwork.

For UK and other European place studies in key stage 1 you could also combine Unit 5: Where in the world is Barnaby Bear?, using the excellent Barnaby Bear books and associated resources (Jackson, 2000-2003; Lewis, 2001-2003). This engaging bear has been to a range of places including Edinburgh, Norway, Brittany and Dublin. The associated teacher's notes are more than adequate to fulfil the national curriculum requirements. A strong case also exists for introducing Barnaby Bear into the Early Years curriculum giving nursery and reception children an enjoyable introduction to unfamiliar places and providing them with an early opportunity to expand their knowledge and understanding of the world.

Opportunities also exist in key stage 2 for European studies. In the thematic studies a European example might provide an interesting river focus, a European settlement could be one of the examples of settlements of different size and character and it might even be possible to find a European Union context for an environmental issue. Up-to-date Europe-based locality resources appropriate for key stage 2 are not common. However, it is possible to devise resources for a place in Europe using school links, holidays, tourist information and the internet. If you do not have your own contacts the Central Bureau for Educational Visits and Exchanges (see page 203) will help put you in contact with an appropriate school.

How can my school plan for progression and assessment?

Schools are at liberty to organise whole-school geography plans according to their own circumstances. You may choose to follow the QCA suggestions by:

1. Using Unit 3 or a Barnaby Bear book in year 2, enabling children to draw on knowledge, skills and understanding of their own locality gained earlier in key stage 1 when investigating the contrasting place.

2. Using Unit 13 in year 5 with progression achieved through 'increasing the range and complexity of geographical questions, skills and contexts' (Sweasey, 1997, p.44).

3. Using a European locality, in support of thematic studies, in any year in key stage 2.

When assessing pupils' achievement it is important to use a wide range of tasks. Many of the teaching activities in the QCA units support assessment and the learning outcomes provide suggestions for assessment criteria. You can view valuable guidance on assessment at www.ncaction.org.uk through varied examples of pupils' work.

What makes an effective conclusion to a study of contrasting UK and European localities?

Before completing the study you should find out how pupils' ideas have changed by asking them the same questions you used at the beginning. When this was done with the pupils in the inner city school described earlier it transpired that their ideas were much more accurate and authentic. The new ideas reflected knowledge and understanding gained from the resources and activities used in the village study, especially the visit. The pupils had greater confidence in showing their knowledge and understanding orally and in their writing and drawing. Additionally, they had a much more positive attitude towards life in the village and they valued what it had to offer those who lived there. Significantly, most pupils were now positive when asked if they would like to live in the village, and the influence of the field trip seems to have been the main factor in changing their attitudes.

When pupils are taught about localities in the UK and mainland Europe using a wide range of activities and resources, including fieldwork, they will be getting a real opportunity to refine their growing 'sense of place' and increase the depth of their feeling and thinking. Teachers should also be optimistic that they will be able to answer, with confidence and accuracy, that vital question: 'What would it feel like to live in this place?'.

References and further reading

Bale, J. (1987) *Geography in the primary school.* London: Taylor and Francis.

Bowles, R. (ed) (1999) 'Raising Achievement in Geography', *Register of research in geography.* London: Rachel Bowles.

Department for Education and Science (DES) (1986) *Geography for Ages 5 to 16.* London: HMSO.

DfEE/QCA (1999) *The National Curriculum: Handbook for primary teachers in England.* London: DfEE/QCA.

DfEE/QCA (2000) *Geography: A scheme of work for key stages 1 and 2 (Update).* London: DfEE/QCA.

Dryden, I. and Hare, R. (2001) *Storylink: Llandudno.* Sheffield: Geographical Association.

Fien, J. (1983) 'Humanistic Geography' in Huckle, J. (ed) *Geographical Education: Reflection and action.* Oxford: Oxford University Press, pp. 43-5.

Geographical Association (2004) *Isle of Coll Map.* Sheffield: Geographical Association.

Graham, J. and Walker, G. (2003) *Discover Coll: The Real Struay.* Coalville: Wildgoose.

Graham, L. and Lynn, S. (1989) 'Mud huts and flints; Children's images of the third world', *Education 3-13*, 17, pp. 29-32.

Hedderwick, M. (1997) *Katie Morag and the New Pier.* London: Red Fox.

Hedderwick, M. (1997) *Katie Morag and the Two Grandmothers.* London: Red Fox.

Hedderwick, M. (1998) *Katie Morag Delivers the Mail.* London: Red Fox.

Jackson, E. (2000-2003) *Barnaby Bear* big books. Sheffield: Geographical Association.

Lewis, L. (2001-2003) *Barnaby Bear* little books. Sheffield: Geographical Association.

Rowbottom, L. (2002) *Llandudno: The study of a contrasting seaside locality.* Coalville: Wildgoose.

Sweasey, P. (1997) *Studying Contrasting UK Localities.* Sheffield: Geographical Association.

Wiegand, P. (1992) *Places in the primary school.* London: Falmer Press.

Wiegand, P. (1993) *Children and primary geography.* London: Cassell

Contact information

Central Bureau for Educational Visits and Exchanges: http://www.centralbureau.org.uk

SoftTeach: http://www.soft-teach.co.uk

We found out about rainforests

lovely
magical
wonderful
valuable
special
beautiful

STORYBOX
THE PARROT THAT TOLD THE TRUTH

VING THE
NFORESTS

Why do we have
DESERTS AND
RAINFORESTS

THE
GREAT GREEN FOREST
Paul Geraghty

Photo | Kathy Alcock and John Collar.

IN THIS CHAPTER YOU WILL FIND KEY IDEAS ON
DISPLAYS • DISTANT PLACES • LOCATIONAL KNOWLEDGE •
MISCONCEPTIONS • PLANNING • QCA UNITS

The wider world

1 Uses and develops their interest and natural curiosity about places.

2 Provides opportunities for them to explore ideas and skills.

3 Develops their existing knowledge and understanding of places, environments and cultures.

4 Helps them to examine and clarify their existing experience and awareness of places.

5 Develops spatial awareness towards a global scale.

6 Helps them to recognise their interdependence with the rest of the world.

7 Builds positive attitudes towards other people around the world.

8 Builds a global perspective that extends their present perspectives.

9 Helps them to value diversity in places, environments and cultures.

10 Combats ignorance, partiality and bias, thus helping to avoid stereotyping and the development of prejudice.

***Figure 1** | Ten good reasons for studying other places with young pupils. Adapted from Catling, 1995.*

***Photo** | Stephen Scoffham.*

Why study distant places?

Geography helps children to extend their world and to move towards a better, richer understanding of real places and people. Studying places beyond their local area provides great opportunities to capitalise on children's natural interest in and curiosity about people and places, and this alone can make it a very satisfying and enjoyable experience for both teachers and pupils. But there are many other good educational reasons for studying other places, including overseas, with young children, as Figure 1 identifies.

Indeed, the reasons for teaching children about distant places and cultures from an early age appear compelling, as research suggests that they often develop attitudes to other people and countries before they have appreciable knowledge of them (Marsden, 1976; Aboud, 1988; Wiegand, 1992). Without intervention, infants are liable to accept uncritically the bias and discrimination they see around them. Stories of war, famine and disasters in the media may distort their perceptions as may stereotypes promoted in advertisements. The influence of parents and, later, peer group pressure, may also serve to confirm negative views. However, a classic UNESCO study (cited in Carnie, 1972) indicated that between the ages of seven and eleven children become increasingly favourably disposed towards peoples of other lands. If children have a balanced teaching programme and are given access to appropriate information and ideas they are willing to abandon their previous ideas and adopt new, more positive, images (Scoffham, 1999). They are more likely to see people as equals, whatever their race or nationality. By the time children reach secondary school, stereotypes become much harder to shift because they become lodged in children's emotional life and are not easily dislodged by logical argument or factual evidence. It is vital, then, that we introduce primary age pupils to a global perspective before prejudices and stereotypes become established and ideas entrenched. The inclusion of a world dimension in the geography national curriculum should serve to promote international goodwill and multi-racial tolerance – one of the key issues for the future.

How big should my 'distant locality' be?

When deciding which locality to study, it is important to choose a small area that children can compare and contrast with their own school locality and other localities they may have studied. It also needs to be manageable to understand and easy to relate to. A distant locality study is not, therefore, the study of an area as large as a country or region; rather, it is likely to be a village, part of a town or city, or perhaps a small island. However, in all locality studies, pupils need to become aware of how the locality they are studying exists within a broader geographical context. When a distant locality is being studied, pupils need access to maps, globes and other sources to locate the position of the chosen locality within a country and continent, and to enable them to view it alongside their own home/school locality. At key stage 2 it will be relevant to include some locational knowledge, as identified in the examples of significant places and environments in the geography national curriculum.

It is essential to have some background information about the country in which the locality is set so that you can discuss the main physical and human features (e.g. its climate and the location of its capital and major cities) and how the chosen locality relates to or is influenced by these features. In addition, we should build on pupils' own experiences of different places and tap into the knowledge and experience of adults associated with the school. If you make an audit of the places where people in your school have lived you may find you can make use of their knowledge and experience.

How do I choose a 'distant' locality?

At key stage 1, studying a locality overseas is optional within the national curriculum so schools can select a 'distant' locality in the UK, rather than overseas, as long as either the physical and/or human features of the chosen locality contrast with those of the locality of the school. However, geography co-ordinators should encourage colleagues to include an overseas locality in their planning, to widen pupils' perspectives and introduce them to other people and cultures. There are also opportunities for introducing knowledge and understanding of environmental change and sustainable development – all places are changing and there are problems and issues to consider. The recent increase in, and improvement of, resources for overseas studies at key stage 1 make this an interesting and exciting option both for teachers and their pupils. The *Barnaby Bear* series of big books (Jackson, 2000-2003) and little books (Lewis, 2001-2003) is an excellent example.

When considering which locality to choose, it is valuable to consider the following questions:

Photo | Elaine Jackson.

- What distinctive features does the locality have?
- How does the chosen locality provide a contrast with our locality?
- Do we want to visit the locality? If so, how/when?
- Could I use the local school as a host base?
- Will there be any opportunities for fieldwork?
- How appropriate is the locality for undertaking a geographical enquiry?
- Which localities are being studied at other key stages and years?
- What resources are available to sustain and develop the study?

Figure 2 | *Some advantages and disadvantages of sources of information for locality studies.*

Source	Main advantages	Main disadvantages
Artefacts	Interesting, 'real' evidence. Promotes discussion. Can enhance a display.	Insufficient on own. Can be difficult to 'extract' geography.
Account or talk by other adult, e.g. person from a distant place	Recent information and a good starting point. Interesting, particularly if combined with photographs, slides, artefacts, etc.	Often little follow-up material available. Doesn't always focus on one distinct locality. Talk may not be pitched at right level or include sufficient geography.
Personal knowledge of the locality chosen	Your first-hand experience can bring the locality 'alive'. Photographs, slides and artefacts from the locality can enhance the study.	You will need to prepare appropriate activities for the pupils to undertake. It takes time to put together.
Commercially-produced pack (usually includes photographs and some activities)	Can reduce planning time. Usually cheap as a relevant resource.	Often more material than is required, so need to be selective about which parts you use. Not always video or slides intended for key stage 1. Activities may not be geographically focused and lead to 'topic drift'.
Slides of a locality	Whole class or group can view at one time. Useful as a focus for discussion with the teacher. The most interesting slides are your own.	Needs setting up. Teacher needs clear information about slides. It is difficult to talk about other people's slides. Not readily available for all localities. A 'static' view of a locality.
Video of a locality	Usually an interesting addition to a study. Whole class can view at one time.	Talk not always pitched at right level. Doesn't always focus on geography requirements. Difficult to find – often has a country or regional approach, not locality.
Photographs of a locality	Allows individual pupils/groups to 'see' the locality in some detail. If good quality, then can be very interesting. Easy to organise.	Selective and therefore may indicate bias and stereotyping.
Locality in the news	Interesting, topical, relevant. May have different accounts of the news of a locality, with some photographs.	Not always appropriate for key stage 1. Often insufficient detail for the study. May present an unbalanced view. Focuses on the unusual, rather than everyday life.
Book/story	Can be a good starting point. Can easily refer back. Usually high quality photographs. Useful for reference or as an additional source.	Few focus on a locality – usually a country or regional approach. Not always intended for key stage 1. Selective material, sometimes subject to bias/stereotyping in pictorial evidence. Usually an overview giving insufficient detail. Single copy in class unmanageable.

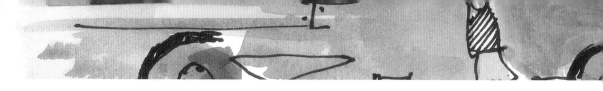

Photo | *Paula Richardson.*

■ What potential is there to link the skills and themes of geography with the locality?

■ Can I use my own experiences in the study?

■ Have any children or adults in the school visited this place?

■ Which localities am I, as the teacher, interested in?

At key stage 2, pupils have to study two localities. One of these has to be in the UK (either the local area or a locality elsewhere), and the other a locality in a country that is less economically developed. This country could be in Africa, Asia (excluding Japan), South America or Central America (including the Caribbean). As it is unlikely that your pupils will be able to visit the overseas locality at key stage 2, it is essential to consider what sources are available to you before you undertake the study, as these will affect the extent to which you can address the geography programmes of study. The ethnic composition of your school will also be significant. There is a growing range of resources available to schools, some of which are more useful than others. Figure 2 sets out the main advantages and disadvantages of using the different types of sources for locality studies.

What about using a television programme?

There is no doubt that television programmes can significantly increase interest in a place. The moving image is attractive and our pupils live in an age where it is pre-eminent. As more programmes become available we need to consider the following:

■ Does the programme focus on a specific locality and aspects within it which are relevant to your studies? (The programme may not have been made with these in mind.)

■ Does the programme have a positive commentary? (Children are not sophisticated viewers – the commentary needs to be exact, precise and true.)

■ Does the programme show extremes or exceptions, highlighting the unusual or different? (Powerful images stay with you – the commentary is often forgotten.)

■ Does the programme consider issues such as change and environmental problems? (Generalised descriptions of places can present a misleading portrait by, for example, only focussing on the positive or negative aspects of life in that place.)

■ Are there suitable resources which can be used to follow up the programme?

Children today are bombarded with images and information, but do not have the skills to appraise them critically. Although a programme may be used to introduce a topic, it is often better used along the way. In any event, always record the programme so that the pause facility can be used, and always view the programme beforehand to check its suitability.

How do I plan for my chosen locality?

When planning activities for your children, combining the three elements of geography – skills, places and themes – is the ideal, and represents good geographical practice. The three geographical themes at key stage 2 (water and landscapes, settlements and environmental issues) may be studied in the context of the local area, if you choose. However, as you should also include work at regional and national scales you need to be sure that over the key stage your planning allows work at a range of scales and in a range of contexts, including the UK, the European Union and the wider world. Several of the QCA units provide good opportunities to do this, e.g. Unit 18 'Connecting ourselves to the world', Unit 16 'What's in the news' and Unit 10 'A village in India'. At key stage 1 Unit 22 focuses on the overseas locality study of the village of Tocuaro in Mexico and provides another option.

As you plan the work, you need to consider opportunities for the development of ICT capability and the role of ICT in geography. Geography offers an increasing number of opportunities for pupils to be involved in ICT. There are increasing numbers of useful secondary sources for information, such as CD-Rom, e-mail, CDi, and mapping packages, in addition to open-ended software such as word processors and spreadsheets. These need to be included at the long-term planning stage to ensure appropriate coverage.

Some possible options for consideration when planning are set out below:

- You may decide to build some of each theme into every locality study over the key stage.
- Your school could 'adopt' a river, such as the Rhine or Nile, stopping off at a number of settlements along the way. This approach would allow opportunities to develop much of the geography work in an integrated way.
- Each locality chosen will potentially provide work about settlements. You may decide to look at a range of sizes and types of settlements over the key stage, comparing settlements in the UK, Europe or in more distant locations.
- Consider choosing a locality where weather data, particularly rainfall, is available or can be obtained. This can be compared with world-wide data printed in some national daily newspapers, or data from the television. Ideally the places chosen would be those with which the pupils are familiar, perhaps through news stories, holiday visits or cultural background.
- Will your approach to planning mean that you will be able to capture the pupils' interest and imagination?

What should I consider when teaching about distant places?

The study must develop pupils' knowledge and understanding of places. The pupils need to be taught about the main physical and human features that give the locality its character; how the locality is linked with other places; and issues and problems that may affect it. These features will include, for example, rivers, landscape, hills, woods, weather, housing, land use, where people shop and what people do. At key stage 2 it will extend to how and why places change and the wider geographical context. It is easy to overlook some of the different elements of geography. The compass rose diagram, originally devised by Birmingham Development Education Centre, is an invaluable checklist for a balanced study (Figure 3).

One of the difficulties in studying distant places is the very real risk of presenting an inaccurate, biased or over-simplified picture. Geographers have an important role in providing opportunities to promote understanding, and in challenging ignorance and prejudice. A focus on a distinct locality, rather than a country or region, can help to avoid misleading generalisations which may be the basis of many prejudices. A locality study is a manageable size and so easier to comprehend. It can also be about real people with real lives,

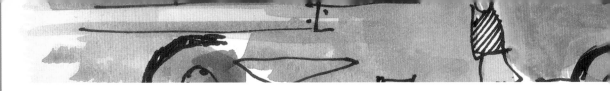

Natural
These are questions about the environment – energy, air, water, soil, living things, and their relationships to each other. These questions are about the 'built as well as the 'natural' environment.

Who decides?
These are questions about power: who makes choices and decides what is to happen, who benefits and loses as a result of these decisions and at what cost?

Economic
These are questions about money, trading, aid and ownership.

Social
These are questions about people: their relationships, traditions and culture, and the way they live. They include questions about how, for example, gender, race, disability, class and age affect social relationships.

Figure 3 | *The Development Compass Rose.*
Source: Birmingham DEC, 1995.

therefore engaging pupils' interest. However, it is equally important to consider the locality within its country-wide context, otherwise pupils may have a distorted view of that country, assuming that it is all like the locality studied. This can be a particular problem if the resources that you are using focus on a single village in, for example, Africa or Asia.

Pupils need a broad, balanced picture, rather than a one-sided viewpoint, so consideration should be given to rich as well as poor, urban as well as rural, women as well as men. Beware of over-emphasis on poverty accompanied by simple and dogmatic explanations of the causes of poverty abroad. We must help our pupils to appreciate and value other cultures, ways of life and ideas. A distant place study should stress that spiritual values are more important than material values for millions of people in the poor – and in the rich – world.

Even before they reach school age, children will have attitudes, beliefs and values about other peoples and places. Some of these may be positive but others may be negative and critical. Indeed, we each have our own attitudes, beliefs and values about other peoples and places which we have not visited. Ideas and images come from a range of media sources such as television, articles in newspapers, holiday brochures, travel literature and films. Unfortunately these can, and usually do, present a very distorted view because of their selection for communication. They present unbalanced and incomplete images which may be reinforced within the home environment. Schools have a very important role in countering prejudice and racial discrimination and this is clearly reflected in the statement of values which underpin the national curriculum. Schools should ask themselves:

- Will our personal images of a place give the right messages to our pupils? For example, describing a place as 'primitive' or 'simple' is unhelpful and biased.
- Do we misinform because of our own ignorance of particular places, or because we use terms incorrectly? An example might be using the word 'country' to describe Africa.
- Do our texts in school describe some places as 'problem areas'? For whom are they said to be problematic? For example, in the Amazonian rainforests, the indigenous populations have lived there for centuries in equilibrium with their environment, but problems for these areas have come since the relatively recent settlement by Europeans.

It can be helpful to consider distant localities alongside your own locality. How would your locality appear to a visitor? What would you want a visitor to see? How would you present your locality to a visitor from abroad? You might also consider exploring visitors' ideas about other parts of our country – plenty of stereotypical images abound still. Consider also our traditions and culture in relation to how people from abroad might view them – Bonfire Night is an interesting one! Of course, all images have their limitations and in overcoming one set of stereotypes new ones may be created. However, by using a wide variety of images and providing plenty of opportunities to question and discuss them, it should be possible to build up a more rounded picture of life in a distant locality. The important thing is to be aware of these pitfalls so that we can make sure we do not fall into them and so that we can help pupils to develop an informed and balanced view of other people, places and cultures and the richness and diversity of our own heritage.

How do I study another place?

Pupils need to use a range of geographical skills and enquiry questions when they learn about a distant locality. The questions they need to ask are the same as when they study their own place:

- Where is this place?
- What is it like?
- Why is it like this?
- What is the landscape like?
- What goods and services are available?
- Where do people go to shop? What work do the adults do? How do people travel?
- What is it like to live there?
- What links does it have with other places?
- What is the weather like there?
- How is the place changing? How do the people feel about the changes?
- How is the locality similar to/different from our locality/other localities we have studied?
- How may this place change in the future?

How should I start my locality study?

Some options:

- Identify the locality on maps and globes. How would you get there? Consider various options – land, sea, flights, direct/following a particular route.
- Use a story from the country. What does it tell us about that country?
- Exchange ideas about what we know (or think we know!) about the country where the locality is situated. Discuss how we can find out about a country or place.
- Look at some photographs of the locality. Show a photograph with some of the picture hidden and ask what might be in the unseen part. Uncover the area and show the pupils the whole photograph. Are they surprised? As a whole class devise questions which will provide starting points to help them to find out more about the locality.

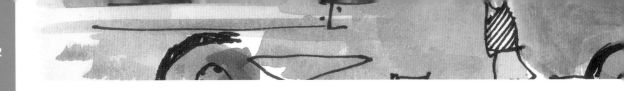

■ Show the pupils an artefact. Choose something which will capture the pupils' imagination. What is it? What is it used for? Who uses it? Where has it come from? What is it made from? What can it tell us about the place and people it belongs to?

If you start from the locality, remember that you will need to set the locality within its country, its continent and within the world, using maps and globes.

Can I design my own locality study?

Yes! Ideally, the chosen locality should be one which you have visited, or one that particularly interests you and for which you have collected plenty of relevant information and other materials. The information needs to be:

■ recent

■ accurate

■ relevant

■ interesting

■ unbiased

If possible, collect information and resources relating to the following features/aspects of the chosen locality:

■ the location (include local maps)

■ landscape

■ weather (include local data if possible)

■ the people and their way of life

■ homes

■ work

■ access to services (water, for example)

■ transport

■ health

■ recent and current changes

Figure 4 | *Jobs and schools provide good comparisons. Photos: Stephen Scoffham.*

You may also wish to collect information about other aspects of the locality such as religion, customs, food, the local school and leisure pursuits (Figure 4). Although these are not essential to meet the programmes of study for geography, your pupils are likely to ask you questions about all of these and more. The information needs to be collated in a way that enables pupils to gain access to it easily. If the locality cannot be viewed at first-hand by the pupils, then visual aids are vital. A range of clear photographs (prints or slides), showing the landscape, housing, the people and their lifestyle would be particularly helpful, as would a video, even an amateur one, to make the place 'come alive' for pupils.

What maps will I need for a locality study?

Suitable maps and plans of the locality are essential. If these are too detailed for younger pupils to use, you could draw simple outline maps to include only the information that is necessary, and in a form which relates to the level of the pupils' mapping skills. Your map should have a title, an indication of direction (perhaps an arrow pointing north) and a scale (even if only approximate). A simple key may also be included. Ideally a range of maps should be used for a locality study, including the following:

- Map of world – continent marked.
- Map of continent – country marked.
- Map of country – locality and significant features of country marked.
- Map of locality – main features marked.
- Plans of any important buildings (for example, named person's home, school, farm).

All small-scale maps involve some element of distortion (area, size, etc.) so it is essential to use a globe to identify the position of the locality in the world. To enhance the study, you could make a display of artefacts from the country, e.g. stamps, maps, food and products, coins, items of clothing, the national flag, a newspaper, stories, travel brochures and

Figure 5 | *Classroom display of artefacts. Photo: Stephen Scoffham.*

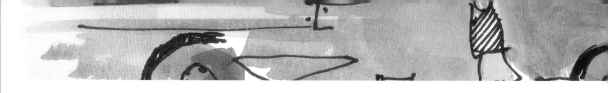

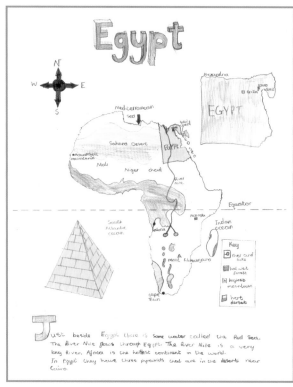

Pupils' work | *Y4 Abbey Primary School, Leicester.*

other pictures of the country. Even bus tickets, receipts and paper bags can be valuable sources of evidence which pupils can interrogate. Artwork from the locality, and recordings of local music are an excellent way of adding colour and atmosphere to the study (Figure 5). If the pupils can dress up in local costumes, eat some traditional food or say a few words in the local language such as 'good morning' and 'my name is ...' they are likely to engage even more whole-heartedly.

Should we link with our overseas locality?

There are obvious benefits in making a direct link with an overseas locality. Forming overseas links has the potential to develop in pupils a sympathetic and caring attitude towards other peoples and ways of life, as well as a sense of responsibility for the environment, both locally and globally. It also helps to counteract prejudice. Your pupils may wish to contact people in the chosen locality and so may ask you whether they can write to pupils in the local primary school, for example. This may not always be easy, especially if the locality is in a remote part of Africa, Asia or South America. The language differences may be overcome, but unless there are fax or e-mail links it may take weeks or even months for replies to arrive. If the locality you have selected is the subject of a commercial photopack, then many schools, perhaps hundreds, may be finding out about it and this may be a potential problem for the locality school. Where links are very successful it is usually where the chosen locality is unique to a school and where there is a long-term commitment to the link by teachers in both schools.

References and further reading

Aboud, E. (1988) *Children and Prejudice.* Oxford: Blackwell.

Birmingham Development Education Centre (1986) *Hidden Messages: Activities for exploring bias.* Birmingham: DEC.

Birmingham Development Education Centre (1995) *Development Compass Rose – A consultation pack.* Birmingham: DEC.

Carnie, J. (1972) 'Children's attitudes to other nationalities' in Graves, N. (ed) *New Movements in the study and teaching of geography.* London: Temple Smith, pp. 121-34.

Catling, S. (2002) *Placing Places* (third edition). Sheffield: Geographical Association.

DfEE/QCA (1999) *The National Curriculum: Handbook for primary teachers in England.* London: DfEE/QCA.

Jackson, E. (2000-2003) *Barnaby Bear* big books. Sheffield: Geographical Association.

Lewis, L. (2001-2003) *Barnaby Bear* little books. Sheffield: Geographical Association.

Marsden, W. (1976) 'Stereotyping and third world geography', *Teaching Geography,* 1, 1, pp. 228-31.

Scoffham, S. (1999) 'Young Children's Perceptions of the World' in David, T. (ed) *Teaching Young Children.* London: Chapman, pp.125-38.

Weldon, M. (1997) *Studying Distant Places.* Sheffield: Geographical Association.

Wiegand, P (1992) *Places in the Primary School.* London: Falmer.

Acknowledgement

The author is grateful to Stephen Scoffham for his comments on earlier drafts of this chapter.

seesied in afrika – Danielle

i went too brietn – Elise

Pupils' work | *St John's Primary School, Redhill.*

Geography and the global dimension

'Only when the last tree has died and the last river been poisoned and the last fish been caught will we realise we cannot eat money'
(Cree proverb).

What is the global dimension?

This Native American proverb gets to the nub of the global dimension. It is in essence about social justice and living sustainably: how we choose to behave towards each other and towards Earth itself. In educational terms the global dimension can be applied to teaching and learning of any subject. However, it is inextricably linked with geography, not just because the words 'geography' and 'global' have an obvious connection, but because of the contribution that the global dimension can make to geography.

When geography is taught through the global dimension it has the capacity to promote meaningful spiritual, moral, social and cultural development. It also has the potential to encourage people to live more sustainably and gain a lasting curiosity, respect and concern for the planet. The global dimension is about developing critical thinking and gaining a better understanding of how the world works. It involves challenging racist, stereotyped and discriminatory views and promoting greater understanding and appreciation of different issues, places and people in the world. According to official guidance (DfEE, 2000) there are eight key concepts underpinning the global dimension. These are:

■ Citizenship
  Gaining the knowledge, skills and understanding necessary to become informed, active, responsible global citizens
■ Sustainable development
  Understanding the need to maintain and improve the quality of life now without damaging the planet for future generations
■ Social justice
  Understanding the importance of social justice as an element in both sustainable development and the improved welfare of people
■ Diversity
  Understanding and respecting differences and relating these to our common humanity
■ Values and perceptions
  Developing a critical evaluation of images of the developing world and an appreciation of the effect these have on people's attitudes and values
■ Interdependence
  Understanding how people, places and environments are all inextricably inter-related and that events have repercussions on a global scale
■ Conflict resolution
  Understanding how conflicts are a barrier to development and why there is a need for their resolution and the promotion of harmony
■ Human rights
  Knowing about human rights and understanding their breadth and universality

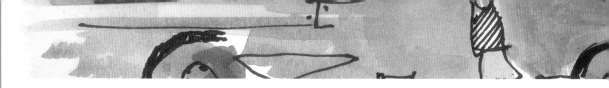

Photo | Indymedia,
http://www.indymedia.org.uk/

These terms cover a great deal and are complex, so really require some further thought. Regarding 'interdependence', for example, Graves (2002) makes the point that '... without discussion, the term suggests a level playing field or co-dependence; that each country is equally powerful and able to make choices without regard to external pressures.' Clearly, this is not how the world actually is.

Nevertheless, the concepts which the DfEE identify are central, not only to the global dimension, but also in related initiatives such as global citizenship, development education and education for sustainable development. They also stand at the heart of good education and are clearly acknowledged in the national curriculum (DfEE/QCA, 1999), particularly in the section on values, aims and purposes:

■ Education is 'a route to equality of opportunity for all, a healthy and just democracy, a productive economy and sustainable development.' (DfEE/QCA, 1999, p.10)

■ The curriculum 'should enable pupils to think creatively and critically, to solve problems and to make a difference for the better … pass on enduring values, develop pupils' integrity and autonomy and help them to be responsible and caring citizens capable of contributing to the development of a just society.' (DfEE/QCA, 1999, p.11).

In order to translate the above rather broad and general aims into real-life classroom practice, it is helpful to consider (a) the knowledge and understanding, (b) the skills and (c) the values and attitudes that are required to enable pupils to understand the world around them and to make sense of their own place within it. Oxfam (1997) has devised a framework which shows how to integrate the global dimension into the curriculum (Figure 1). Many of these key elements relate directly to the material in the geography orders for key stages 1 and 2.

Why is the global dimension important in geography?

It is important to teach geography through the global dimension:

… because we live in an interdependent world and we have responsibilities towards each other.

We are affected every day by geographical issues that highlight the interdependence of our world. Each day we make choices about where we go and how we get there. We choose what to read, listen to, or see on TV, and, perhaps, think about how much of it to believe.

Knowledge and understanding about:
■ Social justice and equity
■ Diversity in all its forms
■ Globalisation and interdependence
■ Sustainable development
■ Peace and conflict resolution

Skills of:
■ Critical thinking
■ The ability to argue effectively
■ The ability to see and to challenge injustice and inequalities
■ Respecting people and things
■ Co-operation and conflict resolution

Values and attitudes of:
■ Having a sense of identity and self-esteem
■ Empathy
■ Commitment to social justice and equity
■ Appreciation of diversity
■ Concern for the environment and commitment to sustainable development
■ Holding a belief that people can make a difference

Figure 1 | *Key elements of global citizenship.*
Source: Oxfam, 1997.

The decisions we make affect others across the globe. Buying Fairtrade goods that have been produced within an agreed ethical code has a direct bearing on the welfare of the producers. For instance, the Day Chocolate Company pays small-scale producers in co-operatives £100 extra per tonne for social welfare benefits. With Fairtrade goods there is also a minimum price ceiling so that, unlike in non-Fairtrade situations, producers do not find themselves subject to price fluctuations that can have a devastating effect on their lives. In all areas of consumerism we are having an impact on others on a daily basis:

■ When buying new clothes are we supporting sweat-shop labour where human rights are being violated?

■ Are we aware of the ethical operating practices of the pharmaceutical or fuel companies that we patronise?

■ Are we supporting local producers and initiatives?

■ Is our bank lending money to those making landmines?

Of course, you may well be acting unwittingly, but the information is available if you want it: try *Ethical Shopping* (Young and Welford, 2002) for a start, or many of the journals listed at the end of this chapter. These are complex issues, and it will take time to do the research, but be assured, you are making an impact on someone, somewhere, whatever you do. Given the modern technology available to us in the UK, the excuse that we don't know what is going on is simply not plausible: if we are concerned about the impact we are having on other people and places, we need accurate information to inform our choices. Many young people acknowledge this: 81% of young people in a MORI poll conducted for the Development Education Association (DEA) believed that 'young people need to understand global matters in order to make choices about how they want to lead their lives' (DEA, 1998).

... because of the need to address the discrimination present in our society.

We all hold responsibility as educators and as human beings to address injustice in our society. The national curriculum (DfEE/QCA, 1999) also charges us with promoting equal opportunities and tackling prejudiced views (p.11). In addition, the amended Race Relations Act requires public authorities 'to take the lead in promoting equality of opportunity and good race relations, and preventing unlawful discrimination' (CRE, 2002). Vigilance against racism is especially important in light of the continued activity of the British National Party within some local councils, frequent reports from helplines of racial harassment and the ongoing public hostility towards asylum seekers and refugees.

Institutional racism is an on-going and deeply-rooted problem. A study of the treatment of 3000 prisoners made by Professor Jeremy Coid of the Royal London School of Medicine highlights huge differences in treatment based on ethnicity (Coid, 2003). In Britain today black people are six times more likely to be sent to prison than white people. So, while almost a quarter of the prison population is from a minority ethnic background they only represent approximately 5% of the population of Britain. Other studies present a similar picture, suggesting that very little seems to have changed since the publication of the MacPherson Report into the death of Stephen Lawrence (MacPherson, 1999).

... because we need to counter misinformation and stereotyped views about each other.

It is most important for children to learn about people and places with whom they are

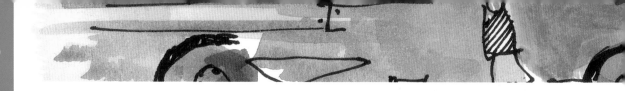

unfamiliar. As an illustration, here are two descriptions by pupils of countries they have never visited. Can you identify them?

Country A
There are no poor people
There are no black people
The policemen wear red and black uniforms
They eat frogs and snakes
(Source: Ritson, 1996)

Country B
They have very tall mountains
People who live there travel in big buses
They have beautiful coins
There are lots of industries

In fact both the descriptions are of the same country – England! The first list (A) was compiled by Kenyan children and the second (B) by Greek children. Both show how those who are unfamiliar with England can have very distorted perceptions. What about how similarly inaccurate our views of other places can be? It is equally important for educators to be aware of potential misunderstandings, generalisations and assumptions. Without an understanding of the issues involved, such work can lead to stereotyping of individuals or whole nations and actually serve to promote mistrust and intolerance. As educators, we have a responsibility to ensure that pupils gain an accurate and fair image of different people and places. We can do this by presenting them with up-to-date, realistic information and by challenging views that are unfair and inaccurate.

… because there are so many inequalities in the world, which are caused by the way the world works.

One way of showing inequalities around the world involves comparing statistics for different places. For instance, on 11 September 2001 approximately 3000 people died in the USA as a result of a terrorist attack. Aggregated figures show that around the world on the same day approximately 2700 children died of measles, 6020 people died of diarrhoea and 24,000 people died of hunger (New Internationalist, 2001). This quiet carnage has continued each day since then but goes largely unreported, whereas the events in the USA rocked the world. How can this be if all human lives have the same intrinsic value? Or are we saying that they don't?

The other shocking point to be raised here is why, in this day and age with the technology that is available, are thousands of people dying from preventable causes? Answer: for reasons related to the unequal distribution of resources, unequal access to technology, and greed and self-interest on the part of those people in the world with power – the majority of whom live in the so-called developed world.

… because it is imperative that we live more sustainably.

Education for sustainable development is an explicit part of the geography component of the national curriculum. This gives those teaching geography an important task. However it is not necessarily an additional task, because sustainable development can be brought into all areas of the programme of study, and all the QCA learning objectives. As stated at the beginning of this chapter, sustainability is about more than environmental matters – it is about how we live and interact with others and whether or not we are promoting social justice in our everyday lives. Having said that, there are some easy ways of living more sustainably and reducing the impact we have on the planet. Turning the tap off when cleaning your teeth is one way. Basing

decisions to buy wooden products on whether or not they have come from a sustainable forest reserve is another. Recycling as much as possible is a third. Currently Britain only recycles 11% of its household waste whereas our Northern European neighbours recycle around four times this amount (Brown and Hencke, 2003). We clearly have a long way to go in using resources more responsibly.

... because we can affect what happens in the future.

Professor David Hicks (2001) states that research shows that young people in many different countries are very interested in what sort of world they are likely to inherit. He suggests that it is important to encourage children to think about the kind of future they want and to help them move towards it. Certainly, an unpublished survey done by Oxfam Education showed that children of primary age are interested in global issues: Fair trade, global warming and war were all mentioned. However, many also realised that they could make a difference through their personal actions, and suggested various ways of doing so, e.g. by learning more about the world, lobbying the council, wasting fewer resources or joining groups of like-minded people. We all have power to affect the future, and should take heart from the number of campaigns started by individuals or groups, which have championed particular causes and had a significant impact. If geography is about creating a sense of awe and wonder about the world, alongside that goes an opportunity for us to say what we think about its treatment in a variety of ways.

... because many schools have developed a link with another school.

Many schools and some ITE colleges in the UK have a link with, or arrange visits to, comparable institutions in another country – often a country in the 'developing' world. In order for these experiences to be mutually beneficial, it is of utmost importance that those involved in such linking are aware of the wider issues relating to this kind of relationship. For instance, are the expectations on each side equal and realistic? What are the perceptions of the schools/communities towards each other, and do these need to be challenged? What happens if the key link person moves schools? There can be many benefits all round to such exchanges, but only if these issues are addressed.

... because teaching approaches which incorporate the global dimension are of interest to pupils and can promote learning.

Several research projects have shown that active teaching methods such as drama and simulation games which are integral to the global dimension are highly effective in promoting learning (Fisher and Hicks, 1987). In addition, research by Lynn Davies (1999) at Birmingham University showed that involvement of pupils in schools councils where the school ethos supports democracy and equity can lead to a drop in exclusion levels. In addition, Stephen Scoffham (1999) suggests that well-planned units of work can promote positive attitudes to the 'developing' world. The effectiveness of such programmes reduces considerably as pupils grow older and their attitudes harden, so it is important that they are undertaken in the primary phase.

Ten strategies to promote the global dimension in geography

So, if we feel that it is important to teach through the global dimension, the question to address is: How can we do that? Here are some ideas.

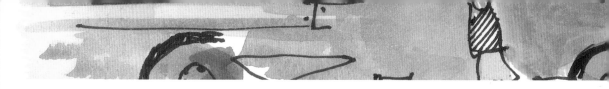

Figure 2 | *The epitomy of Englishness?*
Photo: Mary Young.

1. Give an all-round view of a place or country.

When studying any place or a country, it is important to present an all-round view otherwise incorrect assumptions can be made or reinforced. As an example, look at the photo in Figure 2, taken in England. If this was the only photo you used to illustrate life in England to someone who had never been to the country before they would, under-standably, get a rather distorted view of it. In order to avoid giving a one-dimensional picture, think in terms of opposites: urban-rural, traditional-modern, young-old, leisure-work, rich-poor, women-men. You really need a minimum of ten photos to give any sort of representative view. Try out this approach with a country you know well, then apply the same criteria when you put up a display on a country with which you are unfamiliar.

2. Don't generalise

It is nonsense to make sweeping generalis-ations about people, places or countries; yet it often seems to be done. For instance, take this recently overheard remark: 'Africans live in mud huts'. Well, yes, some people in some countries in the African continent do live in mud huts (which are often a very appropriate form of housing) but not every single person lives in one, in the same way that not every person in England lives in a stately home. Generalisation removes individuality and fuels stereotyping, but can be countered by presenting exceptions and focusing on particular individuals and instances.

3. Look for similarities before differences

When comparing people's lives around the world, focus first on how people across the world are similar rather than different. It is, of course, also important to acknowledge the differences between people and their lives. We are different colours, we have different experiences – but to look at similarities first is one way to stress our common humanity and to promote empathy. To find links, try this simple activity – it works equally well with pupils or in staff development sessions. Give each group a photograph of someone from a distant locality (see Figure 3). Ask them to find as many links or connections as they can between their lives and the life of the person in the photograph. It is helpful if you have some knowledge about the people's names and the places in the pictures, although this is not always possible. You may be surprised to find that it is always possible to make links, even if they are sometimes as tenuous as 'We all breath the same air'.

Figure 3 | *Two women collecting water near Lilongwe, Malawi, Central Africa. Photo: Mary Young.*

Some possible links between their lives and mine are:

■ They look pleased to be having their photo taken and I like having my picture taken too.

■ I have some material at home similar to that piece the woman on the right is wearing.

■ They look like friends, and I have friends.

■ There is a hill like that near my house.

4. Compare like with like

Avoid asking questions which will inevitably lead to answers which present one place or country as superior to another. It is very difficult to compare a British city with a village in India without coming to the conclusion that in terms of apparent wealth, services, modern technology and access to health care the British city is 'better'. (Of course, if you use a different yardstick, e.g. living sustainably, looking after the elderly or cleanliness of the air, it may well be the village comes out on top.) However, if Delhi and Edinburgh are compared, there will be many similarities in terms of traffic, cafés, shopping areas and places of cultural and religious significance. With comparisons, stick either to rural or urban scenes.

5. Be prepared to challenge any discriminatory views that may arise

With all work exploring people and places, if you intend to promote equality and fairness it is essential that you are aware of and can tackle any prejudiced or discriminatory views that pupils may express. As an example, when students on a PGCE course were shown a photograph of life in a country town in Pakistan one of them commented 'They all look a bit primitive to me'. If such a comment is made, it is essential that you challenge it. To allow prejudiced comments to go unchallenged is in effect to condone them. Dealing with prejudicial remarks must be done sensitively, with an explanation given as to why the comment is unacceptable. You may choose to have a further quiet discussion with an individual after the session, but always let it be known that something unacceptable has occurred, in order to give a clear message to the whole class. Here are some ways in which prejudiced comments can be addressed:

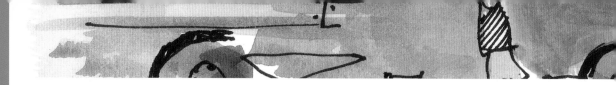

- Ask the whole class if they think a particular comment is fair or not – and why. Then discuss ways it could be re-phrased to make it fairer. A good test would be to ask if the pupils would feel comfortable to make the comment in front of someone from that place or country.
- Talk about the power of certain vocabulary. Explain that some words carry negative connotations or are offensive to some people, and that in consequence they should be avoided.
- Encourage pupils to realise that something can be offensive to another person even if the person saying it didn't mean it to be – it is how the comment is received that is important.
- Do a circle-time activity based on things that make us feel good and things that upset us. Link this to how we talk about and treat others.

If possible, avoid reacting crossly in telling the pupils a comment is unjust, or it may become something they purposefully repeat. For similar reasons always discuss why certain remarks are not acceptable.

6. Look for reasons and explanations

When looking at any problem or disaster anywhere in the world, encourage pupils to think about the underlying reasons rather than simply what has happened. World events may be complex, but their causes are very often to do with basic issues such as greed, unfairness, lack of sharing, and selfishness – basic concepts that children understand and often experience on a daily basis. Furthermore, don't be afraid of contentious issues. Geography is about the real world and we do children a disservice if we present them with a bland and cosy picture that they know to be untrue.

7. Think of solutions and not just problems

It is important when difficult issues arise in geography, for instance, the destruction of the Brazilian rainforest, that you enable the pupils to feel that they can do something positive about it. Leaving pupils feeling powerless can lead to them 'switching off' completely. Equally dangerous is to encourage pupils to think that simple solutions will fix things, e.g. 'The Brazilians should just stop cutting the trees down'. However, there are many things pupils could do if faced with this difficult situation. For instance, they could find out what is happening in the rainforest and why; read about the achievements of the rainforest campaigner Chico Mendes; or find out about how Fairtrade and other initiatives can help the indigenous people. You might encourage the pupils to think about how to conserve places local to the school – perhaps those which have high landscape value or are rich in flora and fauna. An alternative is to set up an area of high biodiversity within the school grounds (perhaps Eco-schools could help). As the slogan which emerged from the 1992 Rio Convention put it: 'Think globally, Act locally'.

8. Draw on the richness of the school and community

Often, there are pupils within the school whose experience and expertise can be drawn on in many ways, e.g. in teaching pupils another language, or in giving first-hand accounts of different places. Many children have multiple identities, have family in more than one

country, speak several languages, are both Muslim and Welsh or feel both Tibetan and British, and so forth. Rather than overlooking this complexity, geography teachers should help pupils work out who they are by both understanding ourselves and by being understanding of the pupils.

9. Find the right information and pictures to help you

There are many places where you can find up-to-date information and pictures to help you teach geography through the global dimension – just ensure you understand whose view is being presented! Development Education Centres have resources to help you, and there are many good websites, e.g. Oxfam's Coolplanet and Global Eye and Global Express. Find time to keep abreast of current affairs – some radio and TV programmes as well as newspapers can be helpful in this respect, and The *Guardian Weekly* is a particularly good digest of news. It is useful to keep a file of newspaper cuttings arranged either under thematic headings such as weather, landscapes, cities, transport and the environment, under different regions of the world. Good magazines and journals include *Orbit, New Internationalist, Resurgence, Developments, Ethical Consumer* Magazine and *CEE-Mail.*

10. Avoid being tokenistic

Sometimes schools see the global dimension of geography as enabling pupils to experience 'different cultures'. This work often includes a topic such as 'Food from round the world' or encouraging pupils to try on saris, or arranging a visit from African drummers. All of these things can work very well, but only if they are undertaken within the context of anti-racism, where issues of diversity have been explored by the teachers and pupils. Otherwise these activities may serve to reinforce stereotypes. For example, it is easy for pupils – especially in mono-culturally-white areas – to explain that rice is eaten in India, while forgetting that they have it at home too! Missing this crucial link that brings us together across the world as human beings with many shared experiences means that the differences again prevail, and that stereotyped views and assumptions are not challenged. The same issues are true with the drummers, often billed as African (but from which country in that huge continent?). Without preliminary work and a commitment on the part of the school to promote the global dimension, pupils' comments about the 'exotic' costumes, the 'difficult' accents, the, 'Oh that's what they all do in Africa' go unnoticed, uninvestigated and unresolved.

Schools linking

Schools linking tends to be seen as the remit of geography, often acting as a major stimulus for collecting information about your own locality, whether in maps, text, sketches, photographs or numerical data. A link can be set up with a school in the same country, another country economically similar to your school, or a country in the Majority ('developing') World. Wherever the link is, schools linking is an activity that has the potential to lead to a greater understanding of different cultures and can be a powerful way of finding out about children's lives, hopes and aspirations. However, it can also be tokenistic and promote inequality. Before embarking on a school link it is important to address some key questions to do with understanding different cultures, equality and the exchange of information:

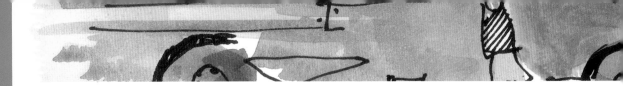

Promoting greater understanding and respect for different cultures

■ How will you enable the children in each school to gain the sort of information about each other that will be likely to lead to greater understanding rather than to compounding of stereotypes?

■ How will you avoid promoting attitudes of superiority or pity towards your link school in terms of how you deal with any differences of wealth, resources, or opportunity in different pupils' lives and in the details of their daily routines?

■ How will you tackle dispelling any stereotypes the partner school pupils might have about this society, e.g. everyone is rich/there is no crime?

Issues to do with equality

■ Will staff visits be reciprocal?

■ What are the expectations of each school, and how will these be met?

■ Will you fundraise for your link school? How will you balance the issues of helping your partner school financially or materially and seeing the two schools as equal?

■ How will you use the knowledge you gain from the professional experience, knowledge and perspectives of your colleagues/peers in your partner school? Similarly, what will they gain from you?

Issues to do with the exchange of information

■ Might there be any technical problems in exchanging information?

■ What effect would this have?

■ How could it be resolved?

■ Will the link be a whole-school commitment for both schools? If not, what will happen if the key staff member leaves the school?

The intention behind this chapter is to encourage you to think about the global dimension in your teaching. The issues may be challenging, but there are plenty of examples of good practice as to how geography and the global dimension are being linked. And remember, the global dimension is based on the premise that in order to teach about geography, or indeed anything at all, it is necessary to be anti-discriminatory, essential to be able to think critically, and important to be well-informed. What would be the point of education without these things?

References and further reading

Ashmore, B. (2000) *Mapping Our World*. Oxford: Oxfam.

Brown, P. and Hencke, D. (2003) 'Britain's abysmal record' in Donellan, C. (ed) *Waste Pollution*. Cambridge: Independence Education Publishers.

Coid, J. cited in Osman, S. and Harris, P. (2003) 'Black people more likely to be jailed than white', *Guardian Weekly*, 2-8 January, p.10.

Council for Racial Equality (2002) *The Amended Relations Act*. http://www.cre.gov.uk/ duty/index.html

Cree proverb http://www.geocities.com/quote garden/arbor.html

Davies, L. (1999) *Schools Councils and Pupil Exclusions*. London: Schools Council UK.

Day Chocolate Company website: http://www.Dubble.co.uk

DEA-commissioned MORI poll (1998) *Children's knowledge of global issues, a research study among 11-16 year olds*. London: MORI.

DfEE (2000) *Developing a global dimension in the school curriculum.* London: DfEE.

DfEE/QCA (1999) *The National Curriculum: Handbook for primary teachers in England.* London: DfEE/QCA.

Fisher, S. and Hicks, D. (1987) *World Studies 8-13.* Edinburgh: Oliver and Boyd.

Global Express website: http://www.dep.org.uk/ globalexpress.

Global Eye website: http://www.globaleye .org.uk

Graves, J. (2002) 'Developing a global dimension in the curriculum', *The Curriculum Journal,* 13, pp. 303-11.

Hicks, D. (2001) *Citizenship for the future: A practical classroom guide.* Godalming: WWF-UK.

MacPherson, W. (1999) *The Stephen Lawrence Inquiry: Report of an inquiry by Sir William MacPherson of Cluny.* London: HMSO.

Oxfam (1997) *A Curriculum for Global Citizenship.* Oxford: Oxfam.

Oxfam Coolplanet website: http://www.oxfam.org/coolplanet

Ritson, C. (1996) *Speaking for Ourselves, Listening to Others.* Leeds: Leeds DEC.

Scoffham, S.(1999) 'Young Children's Perception of the world' in David, T. (ed) *Teaching Young Children.* London: Paul Chapman Publishing

Young, M. and Commins, E. (2002) *Global Citizenship: The handbook for primary teaching.* Cambridge: Chris Kington Publishing.

Young, W. and Welford, R. (2002) *Ethical Shopping.* London: Fusion Press.

Journals and contact details

Development Education Centres and DEA membership. Tel: 0207 490 8108; E-mail: devedassoc@gn.apc.org

Guardian Weekly. Tel: 0870 066 0510

Orbit via VSO. Tel: 020 8780 7500

New Internationalist. Tel: 01865 728181; Web: http://www.newint.org

Resurgence. Tel: 01208 841824; E-mail: subs.resurge@virgin.net

Developments via DfID. Tel: 0845 300 4100; E-mail: enquiry@dfid.gov.uk

Ethical Consumer Magazine. Tel: 0161 226 2929; Web: www.ethicalconsumer.org

CEE-Mail. Tel: 0118 950 2550; E-mail: enquiries@cee.org.uk

Eco-schools. Web: http://www.eco-schools.org

Acknowledgement

Thanks to Dick Palfrey and Lance Lewis for comments on the draft.

Section 4

Themes and topics

Chapter

Auntie Mabel flew her aeroplane in the pouring rain.

Hannah

Oliver W.

pouring

thunderstorm

The water cycle

A cloud is millions of drops of water.

Clouds are blown by the wind.

Water evaporates from trees and plants.

Water runs into rivers.

...ater is heated ...the sun and ...aporates from ...e sea.

Rivers flow into the sea.

computer

RM

Photo | Kathy Alcock and John Collar.

Weather and climate

This chapter shows how observing and keeping regular weather records from the earliest classes can enhance understanding of the impact of weather and different climates on both people and the environment.

Weather affects every aspect of our lives – clothes, food, buildings, the way we feel, what we see and hear, and so on. Daily variations in the weather provide material for greetings and small-talk. The results of accurate observation and measurement are distilled and transmitted daily at regular intervals into our living rooms by weather forecasters, in language we can all understand. In short, weather is a cross-curricular topic, combining elements of English, mathematics, science and geography, and presenting opportunities in music, art and drama and other subjects.

The geography curriculum does not require pupils to study weather and climate as a separate theme. However, the weather is a key aspect of physical geography and will always feature in locality studies. If your school follows the QCA schemes of work there will be opportunities for learning about the weather in the 'continuous' units (Units 5, 16,18 and 24) as well as when you explore connections using e-mail links with other schools (DfEE/QCA, 2000).

There are five aspects to learning about the weather:
1. Observing and recording weather conditions and the effects of these on human activities
2. Making weather instruments and recording data
3. Observing seasonal patterns which make for differences in climate
4. Finding out about weather conditions and patterns around the world
5. Considering the causes and effects of unusual weather conditions

Weather studies for key stage 1

The first question in class discussion time is often 'What is the weather like today?'. The answers can be used to prompt pupils to think about daily weather changes, as well as introducing them to an enquiry approach. You could start by focusing upon what a toy bear or other creature is, or should be wearing, to develop observation skills and to remind pupils that they too need to wear clothing which suits the weather. Making a daily weather diary is an ideal activity for encouraging pupils to observe, question and record, and to communicate ideas and

Monday	Tuesday	Wednesday	Thursday	Friday	Saturday	Sunday	
							Get up
							Lunch
			Forgot				Evening Meal
					Forgot		Bedtime

Pupils' work | *Claire McAlpine.*

information. Personal diaries and stories are also valuable in helping pupils to select appropriate vocabulary to describe both the quality and quantity of sun, rain and wind. By year one, pupils should be able to make practical measurements and use charts and graphs to note changes during the day, between days and also weeks, months and eventually years.

Observing and recording the weather on a regular basis from the earliest moments of a child's school life provides an important foundation for developing awareness about the wider world and is vital to an understanding of hot and cold places. The use of secondary sources such as pictures, stories and photographs is also important in this respect and illustrates how such things as clothes and vegetation can provide clues to different conditions. Often, story-book pictures illustrate an environment as a whole. For example, *A Balloon for Grandad* (Gray and Ray, 1991) shows the change from well-watered green Britain to the drier places of the Mediterranean, thus providing an opportunity to discuss how weather affects the environment.

Weather is one of those experiences you can focus upon as it happens. The pupils could take photographs of, for example, the pond frozen over, or an unwatered tub of flowers looking like a desert after a dry holiday, and attach them to their weather record for that day or week. You can use direct evidence of this kind when looking at other places, in particular places which contrast with the home locality. By the end of year two, pupils should also be able to understand the difference between the seasons and how seasonal changes are reflected in the landscape. They also need to be able to relate consequences to causes and processes. For example, a flooded playground is the consequence of a heavy rainstorm. From observing and recording a local weather incident such as this, pupils could progress to looking at the water cycle.

Stories and pictures

Nursery rhymes, e.g. 'Incey Wincey Spider', 'Dr Foster', 'Rain rain go away', sayings, e.g. 'Red sky at night ...', and stories, e.g. Postman Pat (Cunnliffe, 1983) and Caribbean Tales (Salkey, 1980) contain a wealth of inspiration for continuing discussions on the daily weather and what it feels like on different days and in different places. As well as considering the way daily weather affects the clothes we wear, pupils can look at what is happening to plants and trees. They could bring in fruits, leaves and flowers, noting how each season has distinctive colours, often related to temperature (e.g. red leaves are a reaction to cold, white blossoms appear as the temperature rises, an abundance of green reflects regular rainfall, yellows reflect harvest and dry grass). The variations from day to day and week to week can be used as starting points to develop a vocabulary and literary repertoire. Gradually repetition and change will become apparent.

Photo | *Anna Gunby.*

Year 1

Observing and talking about a familiar place on a regular basis emphasises the idea that weather changes from day to day. Once a class is settled, more systematic observation and ways of recording can be developed. Take specific

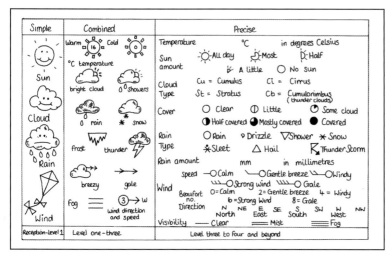

Figure 1 | *Progression in weather symbols.*

weather types, e.g. warm weather, cold weather, wet weather, windy and foggy weather. In the autumn term a late warm spell (Indian summer) can link with holidays past and to come. As the weather is changing, you might get the pupils to devise and use symbols (Figure 1). The symbols can be matched to pictures for each kind of weather. Have ready clothes suitable for dressing a cardboard figure or teddy for each season, to help emphasise that the weather influences what we wear. Pupils' understanding of changes in weather can be assessed by matching the clothes to the symbols at the end of the week.

Start using a thermometer by year 1 – there are specially large ones for infants. Look at the daily newspaper report to see if any other places had the same temperature. In Britain, the north cools down more quickly than the south in autumn, and spring temperatures are correspondingly later in rising. In winter the west is usually warmer than the east and vice versa in summer. These ideas can be developed by checking via e-mail with distant relatives if snowdrops, apple blossom or other flowers are blooming; looking at children in different climates and talking about the weather with other schools via fax or e-mail to send images and drawings; and deciding what to wear in hot and cold places (see Kindersley and Kindersley, 1995). Locate these places on a globe and relate them to the Poles and Equator.

As pupils become more secure in their number work they can begin to make measurements with more precision and ask questions about the element measured. Collect rain in a wide container, then measure it in a tall, narrow container – this maximises the smallest shower! Use consistent, non-standard measures, e.g. one jam jar per day. Decide which way the wind is blowing and use the playground compass or wind vane to name the direction from which the wind is coming (i.e. the opposite direction to which the arrow is pointing towards). Stand with backs to the wind and look at the way hair and clothes are blowing. Most winds are from the west or south-west so the most sheltered place in the playground will be on the east side. Shadows help – they point to the west in the early morning, to the north at midday, and to the east at home time. At first, use terms such as 'in front', 'behind' and so on to describe location, then casually introduce the four compass points. Practise location language to describe the wettest, driest, hottest and coldest places at school, locally, and in the world.

We experience the elements, such as sun and wind, with some or all of our senses and react to them in different ways: snow and frost are bright, cold and crunchy, and snow crystals make patterns. The following questions can be used to direct pupils' learning about the various elements:

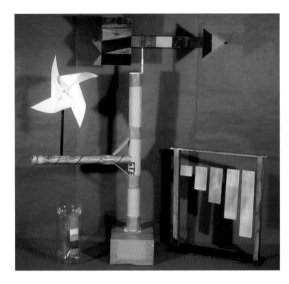

Photo | Stephen Scoffham.

Rain and snow

- Why do we protect ourselves (and plants and buildings) from rain and snow?
- What is waterproof, and what waterproof materials are there?
- What happens to rain after it has fallen? Compare trays of sand/soil, one covered with plastic to represent tarmac.
- What happens to puddles? Watch what happens to water in a saucer placed on a sunny window sill.
- Why are roofs often different shapes in wet/dry places? Look at sloping and flat roofs.

Wind

- In what ways is the wind useful? Can it push things - us, sails, boats, mills? Make a toy windmill.
- What does wind do to litter in the streets?
- In what ways is the wind harmful?

Sun and shelter

- What is shelter? Use a terrarium and a box model to consider the differences between greenhouses and ordinary houses.
- Why do we wear hats and sun cream?
- How do animals and plants adapt to the sun and heat?

Elements
breeze, clouds, fog, frost, gale, hail, ice, lightning, mist, rain, raindrops, snow, sun, thunder, wind.

Seasons
Summer: boiling, bright, calm, cloudless, downpours, drought, dry, fresh, hot, humid, light, pouring, shine, showers, storms, sunny, warm
Winter: chilly, cold, cool, crystals, dull, floods, foggy, freezing, frosty, gales, overcast, rainy, severe, snowy
Spring and *Autumn:* changeable, damp, dew, drizzling, frosty, gentle, mild, puddle, spell, spitting, splash, strong

Terms
anemometer, chart, diary, direction, gauge, heat, moisture, period, polar, rainfall, record, rose, season, shade, shelter, sock, symbol, temperate, temperature, thermometer, tropical, vane, weather, wind

Sayings
bitter wind, cats and dogs, crunchy snow, dawn, dusk, mackerel skies, mares tails, rainbow, squalls, thunderbolt

Year 2

In year 2 you might develop a seasonal topic each term – these lend themselves to cross-curricular work and can be revisited with increasing detail and complexity according to the development of the class. Figure 2 indicates the kind of vocabulary needed and possible by the end of year two. A weather word book or weather word bank could be compiled using computer software.

Figure 2 | *Weather vocabulary appropriate for the end of year 2.*

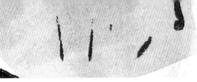

Level	Rain	Temperature	Wind	Air pressure
1/2	Collect in wide bowls, measure in narrow, tall jars, one per day.	Thermometers should all be alcohol filled. Giant ones make reading easy and develop good recording habits.	Use home-made wind vanes showing four compass points, and wind speed indicators which give higher readings the stronger the wind.	
3	Collect in a straight-sided, level-bottomed jar with a scale in millimetres facing inwards. Record the reading in mm each day at the same time.	Maximum-minimum thermometers (from good hardware stores or garden centres) show temperature is an ongoing change.	Hand held anemometers and ventimeters (more robust) measure wind speed. The best of three readings should be recorded.	Use a sensitive home-made barometer with a scale graduated between high and low pressure.
4	Collect in a jar placed inside a cylinder sunk into the ground (the cylinder, e.g. tall coffee tin, acts as a shield for the rain collection jar). The rain is collected via a 12.5cm diameter funnel resting on the top of the jar and measured in a standard science measuring cylinder, as with official gauges.	Wet and dry bulb thermometers allow humidity to be measured – the more humid the air the more equal the readings, and the more uncomfortable the air! Cheap thermometers/ hygrometers from garden centres give similar results.	Anemometers can be connected to a digital recorder for constant readings. Link wind speed readings, taken at the same time as other observations, to the eight points of the compass and the Beaufort Scale (Figure 4). Record on a wind rose.	Use an aneroid (banjo) barometer and link the readings to Radio 4 shipping forecasts. The concept of rising and falling pressure becomes interesting when considering storms at sea and the lines of equal pressure (isobars) shown on weather maps.
graph	Bar *Discontinuous* data	Line *Continuous* data	Circular *Directional* data	Line *Continuous* data

Figure 3 | *Progression in measuring techniques.*

By year 2 pupils can use home-made equipment to make more detailed and regular measurements of the weather to identify rain and temperature patterns (Figure 3). Plan to make reasonably accurate measurements of rain, temperature and wind for at least one week, three if possible, in more than one season.

It is helpful if there is a whole-school policy about the kind of weather symbols the pupils will use, and the progression of those symbols (see Figure 1). You also need to decide on progression in the use of measuring equipment. Finally, agree when detailed weather observations will take place and the audience for whom they will be useful.

Here are some suggestions for activities at this level:

■ Link weather observations with myths and legends (e.g. 'Persephone and the wind' from Hadley, 1983 or weather poems, e.g. Wright, 1982)

■ Make picture maps

■ Role play as weather forecasters

■ Use different words to describe moisture, from dew, fog, mist, rainspots, showers, drizzle to downpours, storms and blizzards (thus making links between English and geography)

■ Combine and extend weather symbols (see Figure 1)

■ Use grid plans of the playground and neighbourhood to locate the best places for sun, shade and shelter, and refer to the four compass points

■ Use fair tests to show evaporation from puddles or containers left outside

■ Show how steam from a kettle condenses on to a cold surface

■ Use two plants, watering one only, to show the importance of rain to growth

■ Link wind with clouds – are all the clouds moving in the same direction as the wind (usually this is so)?

■ Use a toy windmill to measure winds of different strengths

■ Translate the Beaufort Scale into pictures (Figure 4). Pupils could make felt shapes from the pictures, then sew or glue them onto card as part of a display about the weather.

Force No	Name	Wind Speed			Effect
		knots	kmph	mph	
0	Calm	0	0	0	Smoke rises straight up
1	Light air	1-3	1-5	1-3	Smoke drifts. Windvane does not turn
2	Light breeze	4-6	6-11	4-7	Leaves rustle. Wind vane moves
3	Gentle breeze	7-10	12-19	8-12	All leaves move. Flags flutter
4	Moderate breeze	11-16	20-28	13-18	Small branches move. Litter blows
5	Fresh breeze	17-21	29-38	19-24	Waves on water. Small trees sway
6	Strong breeze	22-27	39-49	25-31	Large branches move. Wires whistle
7	Near gale	28-33	50-61	32-38	Trees bend. Umbrellas useless. Walking difficult
8	Gale	34-40	62-74	39-46	Branches break. Difficult to walk any distance
9	Severe gale	41-47	75-88	45-54	Tiles blown off. Fences down
10	Storm	48-55	89-102	55-63	Trees uprooted. Buildings damaged
11	Violent storm	56-63	103-117	64-70	Widespread damage (rare in UK)
12	Hurricane	64+	118+	71+	Devastation to everything (not in UK)

On weather maps a circle is shown with a line pointing in the direction from which the wind has blown. On the line are 'feathers' to show the strength. Half a feather = 1 and a full feather = 2 A Gale = 4 feathers

Current UK weather reports give wind speed in miles per hour (mph) (where 1 knot is equivalent to 1.15 mph) or in names on the Beaufort Scale; Wind measure scales are in mph, kmph or metres per second.

Figure 4 | *Beaufort Scale.*

Weather studies for key stage 2

Weather studies form a valuable component of geographical work at key stage 2. In national curriculum terms there are direct links to geographical enquiries and skills, the understanding of places, and work on patterns and processes. Several of the QCA units also involve weather studies. The most obvious is Unit 7: Weather around the world.

Weather studies at key stage 2 will involve pupils in making more precise observations than at key stage 1 and lead them to analyse evidence and recognise patterns. Pupils can collect evidence in the school grounds, as part of a study of a contrasting locality, and through using a fax and other ICT such as e-mail and video. What is important is that the evidence is integrated with the study of places and other themes such as environmental change.

In the context of locality studies, weather can be identified as an important contributor to the character of a locality and therefore as a characteristic to be used when comparing one place with another – in this place, rain falls mainly in daily thunderstorms throughout the year; in that place, droughts happen for months rather than weeks; and so on. In turn, these characteristics influence human activities and can, in the case of prolonged drought for instance, lead to profound changes.

Weather information about places around the world can be gathered from daily world reports in newspapers and used to help develop pupils' understanding of climate. Newspapers and other news media are also useful as sources of information about world events which may be directly or indirectly related to weather conditions. So, for example, a volcanic eruption or earthquake, while not being directly related to weather, raises questions concerning shelter and warmth for victims of the events. Similarly, pupils' appreciation of seasonal weather patterns could be fostered by asking them to think about what clothes an aid worker might take when visiting places where disasters have occurred.

An insight into the ways in which micro-climate may influence land use in a locality can be developed through work in the school grounds. For example, measurements taken of the amount of rain, sun and wind in different parts of the playground can lead to discussions about the siting of vineyards on south-facing slopes, or of forests or ski slopes on north-facing ones.

Year 3/4
Record for three or four weeks, at the same time each day.
With a young, inexperienced class start with symbols.

Figure 5 | *Progression and differentiation in recording at key stage 2.*

Week beginning			Mon	Tue	Wed	Thu	Fri	Sat	Sun
Temperature	°C ⌂ °F								
Rainfall									
Cloud cover									
Wind speed									
Direction									

Progress from pictogram graphs to linear
(temperature), bar (rainfall) and circular
(cloud and wind) graphs.

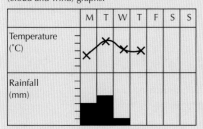

	M	T	W	T	F	S	S
Temperature (°C)							
Rainfall (mm)							

Begin gathering clues for wind speed
from a Beaufort Scale (see Figure 4).

Years 3 and 4
The lower junior pupil should be able to recognise that weather not only changes from day to day but also during each month and year. Combining more precise observations serves to show that sunshine does not always mean high temperatures and that rainy days can be mild – depending upon which way the wind is blowing. You can adapt the daily recording sheet (Figure 5) to take account of these variations as appropriate.

With weather data from around the world available in the daily papers, on Ceefax/ Oracle, and the internet (see below) you can devolve the responsibility for collecting data for several places to 'roving reporters'. In addition, pupils might discuss the different symbols that are used. Weather forecasting involves pupils in thinking about why people in different occupations need to predict the weather. Design technology can be used to make reasonably accurate measuring equipment and science can contribute towards under-standing weather phenomena through simple experiments (see Chapter 12 for more ideas on the application of ICT).

Wind
Wind is air moving from one place to another; it brings heat and moisture from the place it has just left. Consequently winds are named after the direction from which they are travelling. The west wind usually brings cool, wet weather to the British Isles, picked up over the Atlantic, often in a series of low pressure systems. The north wind brings cold, clear or stormy conditions from the polar regions regardless of the time of year. Easterly winds bring snow, frost and fog in winter, coming as they do from a continent that is very much colder than our islands. In summer the position is reversed and east winds bring long spells of hot weather and even heat waves in July and August. Sometimes the British Isles are engulfed in warm, humid air – very welcome in winter, unpleasant in summer. This air has originated over the Tropics, in the region of the Azores. One of the reasons Britain's weather is so variable is that warm air from tropical regions is continually mixing with cold arctic air in our latitudes.

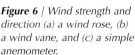

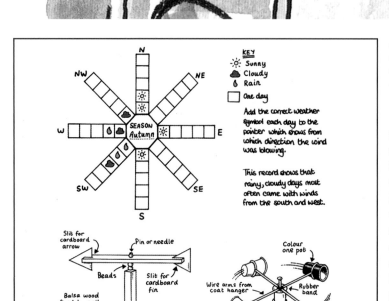

Figure 6 | *Wind strength and direction (a) a wind rose, (b) a wind vane, and (c) a simple anemometer.*

In teaching about wind it is important to discuss how the source of wind affects its character. Support discussions about television and radio weather reports by looking at the globe to see where the air has originated. Use science experiments to help explain why land and water cool and warm up at different rates. Discover the wind direction by getting pupils to observe the action of dust, litter, leaves or tissue streamers in the playground and record the results on a wind rose (Figure 6). Use the UK map to mark on the wind directions in relation to your school. Do the winds cross land and sea? Could they have become drier before reaching you? Are winds from the west always wet?

Wind speed is a measure of the strength of the wind (see Figure 4). You could make several different wind measurers, such as a yoghurt pot anemometer (Figure 6), and refer to an official ventimeter (school supplies or ship's chandlers) or anemometers (school scientific suppliers) and adjust the scales so that all measurements relate (calibrate). Discuss how wind affects certain people, e.g. pilots, sea captains and athletes. Wind-generated electricity is a topic which has links with 'energy' in science, and raises environmental considerations in relation to siting wind farms in scenic areas.

Rain

Wind brings moisture, usually in the form of rain. Fog and dew may occur when warm humid air meets colder conditions, as when the ground has cooled during the night. Clouds are the result of the same process at higher levels – which is why the tops of mountains are often shrouded in

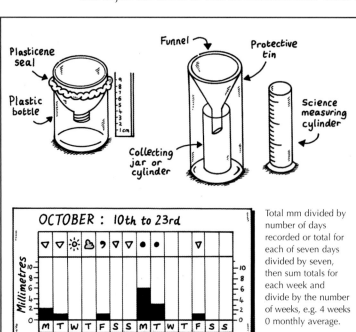

Total mm divided by number of days recorded or total for each of seven days divided by seven, then sum totals for each week and divide by the number of weeks, e.g. 4 weeks 0 monthly average.

Figure 7 | *Measuring and recording rain.*

Language	Description	Type and abbreviation		Height	Weather
Wispy, mares' tails, curls	High streaks fanning out ahead	Cirrus	Ci	High	Often more wind; usually storms after good weather
Sheet, layers, blanket	Blanket of low grey cloud which can reach the ground	Stratus	St	Low to high	Drizzle, fog which can become steady, wet, rain
Billowy, fluffy, heaps	White, flat-bottomed clouds which look like cotton wool	Cumulus	Cu	Low	Small – fair and sunny. Large – showers
Dark, towers, castles	Large, dark clouds from ground to the highest level	Nimbostratus	Ns	Medium	Heavy rain or snow
Dark base, billowy towers	Marked, dark base which gets darker as rest of white cloud billows up to anvil shape	Cumulo-nimbus	Cb	Low to high	Very heavy showers with hail, thunder, lightning. Move away at speed

Figure 8 | *Cloud characteristics.*

mist, while valleys are clear and sunlit. Condensation on a glass of iced water illustrates the formation of dew and fog, and clouds can be created by blocking a glass jar containing hot water with an ice cube. The water vapour from rising, warm, moist air cools to form a cloud under the ice.

Rain falls when the droplets of condensed moisture are large and heavy enough to drop through the rising air beneath the cloud. They can be small (drizzle) or as large as hailstones. All the different forms of 'rain', including snow, are known as precipitation.

Measuring rain can be frustrating and disappointing – even when it rains heavily only a small amount will collect in a gauge (Figure 7). This is why the longer the observation period the more probable it is that significant data will be collected. Gauges can be positioned at contrasting sites round the school (such as under trees, at the foot of walls sheltered from rain-bearing westerly winds, in a puddle area, in a 'dry' area) and the results noted, together with possible explanations. You could extend the work by checking on the accuracy of national forecasts and putting up a display of photographs about droughts and floods at home and abroad.

Clouds

The appearance of clouds, and the patterns they make, are good indicators of weather to come and there is a rich vocabulary associated with their description (Figure 8). When recording cloud types and cloud cover it is important to look through the same window each day and to estimate the cover either as quarters or eighths (oktas). If your are learning about cloud types, choose a week when low pressure systems or 'unsettled weather' are forecast and make it an end-of-morning or afternoon game. Connect cloud observations with forecasting. Can pupils forecast the afternoon weather from the morning observations?

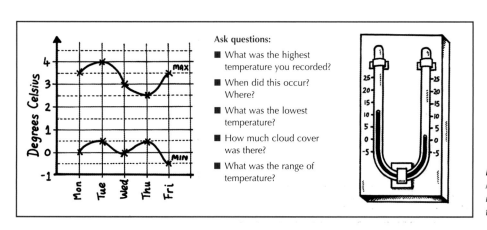

Ask questions:
- What was the highest temperature you recorded?
- When did this occur? Where?
- What was the lowest temperature?
- How much cloud cover was there?
- What was the range of temperature?

Figure 9 | *Measuring and recording temperature.*

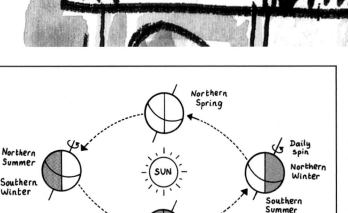

Figure 10 | *What causes the seasons? Demonstrate the movement of the Earth round the sun with a central light and small globe on its stand (this keeps the Earth's tilt).*

Temperature

When studying 'hot places' and 'cold places' it is important that pupils understand that temperature varies in these places both between day and night, and between seasons. For example, even in hot deserts the soaring heat of the daytime may be replaced with sub-zero temperatures by the end of the night, and polar regions are sometimes as hot by day in the summer months as places much nearer the Equator.

Some specialist thermometers record temperature variations over time. For example, maximum-minimum thermometers 'save' the lowest and highest temperatures (Figure 9) and are used regularly by gardeners to monitor changes between night and day temperatures. Using these thermometers will help pupils to understand why 'tender' plants from Mediterranean areas, such as geraniums and busy lizzies (Impatiens), have to be kept indoors until late May or June when all danger of night frost is past. The link between overnight temperatures and cloud cover can also be explored. Was there a blanket of cloud preventing the loss of warm air, or could the stars be seen? Is there a link between morning hoar frost or heavy dew and night-time weather?

Elements and terms
air masses, anticyclone, atmosphere, atmospheric pressure, cloud names, cold front, condensation, continental air, cyclone, depression, drought, evaporation, high pressure, humidity, insolation, isobar, isohyet, isoline, isotherm, knot, latitude, longitude, low pressure, maritime air, mass, maximum and minimum temperature, precipitation, prevailing wind, radiation, saturation, season, solstice, storm, storm tracks, sun hours, synoptic chart, tornado, tropics of Cancer and Capricorn, transpiration, tropical storm, tropical air, visibility, warm front, wind chill

Instruments and scales
Anemometer, aneroid barometer, Beaufort wind scale, maximum and minimum thermometer, ventimeter

Climates
Arid, equatorial, desert, maritime, Mediterranean, monsoon, mountain, polar, temperate, tropical

Figure 11 | *Vocabulary appropriate for the end of year 6.*

Years 5 and 6
Weather and climate

The conditions that prevail at a particular moment in a particular place make up its weather. The pattern of weather over a period of time is known the climate. In general, the distance from the Equator is the key factor in determining climate, along with proximity to open water and the influence of the Earth's tilt and orbit (Figure 10). Upper juniors should be introduced to some of the different climates around the world, and make comparisons between their own locality and other places using appropriate geographical vocabulary (Figure 11). If pupils have regularly observed and recorded the weather earlier in their school careers they will be in a good position to undertake this work. A good school atlas will be an invaluable resource for extending the study and contains sample graphs for rainfall and temperature in representative locations around the world. By studying these, pupils can identify the hottest, wettest, coldest and driest months. You may also be able to obtain a satellite view of the world which shows the major areas of natural vegetation and discuss how these are related to climate conditions.

Pupils need to understand that climate statistics are based on average weather conditions. Use your own records to demonstrate average temperatures and rainfall for a week, or month

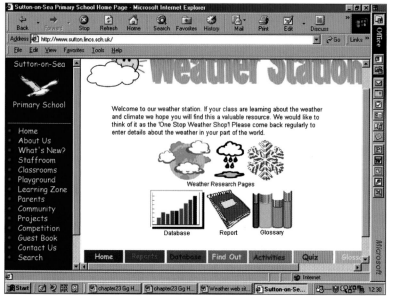

Figure 12 | *Sutton-on-Sea Primary School's weather website pages.*

if possible. You may also be able to gather together information on precipitation and temperature for your region for a whole year (remember to archive on floppy disk or CD-Rom). Daily newspapers will provide weather data. Other sources of data are the local weather centre, the local reference library (which keeps back-issues of most newspapers), the Meteorological Office and local secondary schools. Discuss whether the hottest, wettest and other extreme readings are obvious. Make a climate graph for your own locality and, if possible, for contrasting localities that you have studied.

You might invite individual pupils to act as foreign correspondents who report back on weather conditions in their part of the world. This works particularly well if their locality or its region is prone to extreme conditions such as tropical storms (hurricanes), monsoon flooding, severe snow falls or long droughts. The reports could also refer to the lack of spring and autumn in tropical lands, and the twelve-hour day (in these areas, where the sun rises around 6.00am each morning and sets at 6.00pm, the seasons are identified by differences in the amount of rain that falls rather than by changes of temperature).

Microclimates

The way weather conditions vary around the school grounds has already been mentioned. Upper juniors can compare at least four sites and relate them to different projects and developments such as new seats, flower tubs, animal hutches or a pond. You may be able to broaden the study into the local area and consider possible sites for improvement projects, e.g. a sheltered corner where elderly people can rest, a greenhouse on allotments or a new bus shelter to keep out the wind and rain. The temperature, amount of sun, rainfall, and speed and direction of wind will all have significance. At this level, pupils should be taking accurate readings from instruments and encouraged to be as precise as possible. This will enable you to make comparisons with official and other automatic readings. Connect with school networks, e.g. Sutton-on-Sea Primary School's weather station website pages (Figure 12) are part of an online Montage Project, sponsored by the British Council, for schools around the world. This site has lots of weather-related data available, and it is possible to submit your own data which can be added to a larger weather databank.

Using ICT

Since 1997 the internet has enabled schools to set up interactive projects that involve exchanging weather data (The GLOBE project; 4Seasons; OZ-Teachernet). The GLOBE project alone has over a 1000 school links – mostly in the US but also in Europe and Japan,

Photo | *Tina Horler.*

including over 300 primary schools in the UK. In addition, there are several internet sites which provide webcam pictures or transmit instant weather images.

Many current projects emphasise the core curriculum but incorporate geography in order to establish the significance of place in the worldwide pattern of weather and climate. This means that the enquiry process becomes an important tool. For instance, pupils might ask the following questions when considering places for a holiday: How hot is it? Shall I need warm clothing for the evening? Is it likely to rain? Will the rain be heavy? Will it be windy? What games are played there? To answer these questions, pupils can visit a number of weather information sites such as BBC, CNN and Yahoo. Compare the information available from each – some are more comprehensive and child-friendly than others.

This chapter has emphasised ways of observing and recording the weather. When it comes to offering explanations for weather phenomena, primary school pupils often run into difficulties and end up giving confused and contradictory answers. In recent years there has been some fascinating research into children's ideas. One of the most comprehensive reviews is presented by Henriques (2000) whose findings are summarised in Figure 13. There are plenty of issues here which deserve further research. As a teacher it is important to realise that the best way to challenge misconceptions is through rigorous attention to detail when observing, recording and analysing weather data from the earliest age. Some misconceptions will, however, only be overcome as pupils mature.

Keeping records for the future

Weather studies in the primary school offer many opportunities for developing and using skills in recording and observation. In addition, they encourage pupils to think about the reasons behind some of the similarities and differences between areas around the world, and to empathise with those whose lives are dependent on or directly affected by weather conditions. Gathering weather data on its own could be an arid activity – geographers help to give life to the data. Both people and nature have learnt to survive in the hottest, coldest, wettest and driest places in the world – and all places in between. The climates which are found in these areas have in turn had no small part in developing the cultures created by people who live there. In order to be good global citizens and to understand the 'rules' of sustainability young children need to appreciate the inconstancy of the one medium, weather, which controls the source of the most constant human need – water.

By keeping and maintaining weather records over long periods pupils are part of a long and valuable tradition of weather watching that goes back at least as far as ancient Roman and Greek times. While the increased detail and accuracy of modern weather observation has obvious benefits for us all, even simple methods, such as those we use in our schools, still enable us to make useful predictions. And our records, which may span many years or decades, are things of value which in future years will provide a rich source of information about past patterns and changes.

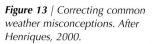

Figure 13 | *Correcting common weather misconceptions. After Henriques, 2000.*

Weather phenomena	Children's misconceptions	Comment
Water Cycle – Water evaporates from land and sea, the water vapour condenses to form clouds from which rain falls back to Earth, runs down slopes and flows into rivers and seas.	Children sometimes have an incomplete notion of the water cycle.	Children understand the concept of freezing before the more complex notions of evaporation and condensation.
Transpiration – Plants also contribute water to the atmosphere via transpiration	Water only gets evaporated from the ocean or lakes.	Possible source: poor textbook, poster and media diagrams which only show evaporation from the ocean.
Clouds – Clouds are created when water vapour condenses onto dust or other particles in the air.	Clouds are made of smoke, cotton-wool and so forth. Clouds (and rain) are made by God. Clouds come from somewhere above the sky.	Some of these explanations are found in myths and legends.
Cloud types – A visible cloud is primarily tiny water droplets and/or tiny ice crystals; it is not water vapour.	Empty clouds are refilled by the sea.	Observe and discuss different cloud forms – cirrus, cumulus and stratus
Wind – Winds are produced by the uneven heating of the Earth's surface.	Clouds block wind and slow it down. Low temperatures produce fast winds. Winds are caused by the spinning of the Earth.	All of these false associations can be disputed with observations in different seasons and referring to records.
Why clouds move – Clouds move when wind blows them.	Clouds move when we move. We walk and the clouds move with us.	Observe clouds move in the playground and local area.
Rain – Rain begins to fall when water drops in the cloud are too heavy to remain airborne.	Rain comes from holes in the cloud. Rain comes from clouds melting. Rain occurs when clouds are shaken. Rain occurs when clouds collide.	Demonstrate the effects of condensation using a fine net suspended over a kettle. Watch how the drops grow larger and eventually fall off.
Frost – Frost forms when water vapour comes into contact with very cold surfaces. The water vapour freezes directly instead of condensing into a liquid first.	Frost falls from the sky. Frost is frozen dew.	Use the minimum thermometer to measure night temperatutes. Discuss how freezers ice up when the door is left open!
Floods – Flooding occurs when there is more water than the ground or rivers can accommodate.	Flooding only occurs along rivers when the snow melts in the spring. Flooding only occurs after heavy rainfall.	Half truths; in towns the 'ground' can be blocked drains; in winter the ground can be frozen.
Droughts – Droughts occur when the normal rainfall fails to occur causing distress to plants and creatures.	Droughts only happen in deserts. Droughts are caused by God.	A drought is a disruption to an established pattern. Deserts are naturally arid and thus do not suffer from droughts.

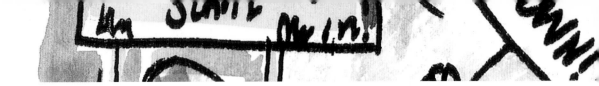

References and further reading

BBC weather website http://www.bbc.co.uk/weather

Chambers, B., Nelder, G., Paterson, K. and Wareing, H. (1989) 'Viewing the earth from space', *Primary Geographer*, 1, pp. 7-10.

CNN weather website http://www.cnn.com/weather

Cowcher, H. (1990) *Tropical Rainforest*. London: Picture Corgi.

Cunnliffe, J. (1983) *Postman Pat*. London: Hodder.

Dove, J. (1999) *Immaculate Misconceptions*. Sheffield: Geographical Association.

DfEE/QCA (2000) *Geography: A scheme of work for key stages 1 and 2. Update*. London: DfEE/QCA.

Fraser, A.B. (2004) 'Bad meteorology', http://www.ems.psu.edu/~fraser/BadMeteorology.html (viewed 18/2/04).

GLOBE project website http://www.globe.gov

Gray, N. and Ray, J. (1991) *A Balloon for Grandad*. London: Orchard Books.

Hadley, E. (1983) *Legends of the Sun and Moon*. Cambridge: Cambridge University Press.

Harrison, S. and Havard, F. (1997) *The Weather Book for Primary Teachers*. London: Simon and Schuster.

Henriques, L. (2000) '*Children's misconceptions about the weather: A review of the literature*', at http://www.csulb.edu/~lhe riqu/NARST2000.htm (viewed 18/2/04).

Kindersley, B. and Kindersley, A. (1995) *Children Just Like Me*. London: Dorling Kindersley/UNICEF.

Manley, G. (1952) *Climate and the British Isles (New Naturalist Series)*. London: Collins.

Morgan, W. and Bunce, V. (1992) 'Landscapes and weather, volcanoes and hurricanes', *Primary Geographer*, 9, pp. 6-7.

Morgan, W. and Bunce, V. (1995) 'Devastating Debbie (a hurricane)', *Primary Geographer*, 22, pp. 28-29.

Online Weather website http:// www.onlineweather.com

OZ-Teachernet website http://www.schools.ash.org.au/paa1/wlinks.htm

Russell, T., Bell, D., Longden, K., and McGuigan, L. (1993) *SPACE Research Report: Rocks, soil and weather*. Liverpool: Liverpool University Press.

Salkey, A. (1980) *Caribbean Folk Tales and Legends*. London: Bogle L'Ouverture.

SCAA (1997) *Expectations in Geography at Key Stages 1 and 2*. London: SCAA.

Primary Geographer (1997) 'Focus on weather'. Sheffield: Geographical Association.

Smeaton, M. (1991) 'Project Thailand: A study of a tropical locality', *Primary Geographer*, 6, p. 2.

The 4 Seasons Project website http://www.4seasons.org.uk/teachers

Warren, N. with Cambier, A. and Ranger, G. (1997) 'Geography meets French in the classroom: a contrasting locality study at key stage 2', *Primary Geographer*, 29, pp. 20-21.

Weeg, P. (2003) 'Resources for Educators', http://www.globalclassroom.org/resource.html#news (viewed 18/2/04).

Wright, K. (1982) *Hot Dog and Other Poems*. London: Puffin.

Yahoo weather website http://weather.yahoo.com/

Useful addresses

Advisory Unit Computers in Education, 126 Great North Road, Hatfield, Herts AL9 5JZ. Tel: 01707 26614. Website: http://www.advisory-unit.org.uk

Meteorological Office, Scott Building, Eastern Road, Bracknell, Berkshire RG12 2PW. Tel: 0845 300 0300 E-mail: education@metoffice.com Website: http://www.met-office.gov.uk/education/resources/ index.ht

SoftTeach, Sturgess Farmhouse, Longbridge Deverill, Warminster, Wiltshire BA12 7EA. Tel: 01985 840329. Website: http://www.soft-teach.co.uk/

Geopacks, 92-104 Carnwath Road, London, SW6 3HW. Tel: 08705 133168 E-mail: service@geopacks.com

Griffin and George. Tel: 01509 233344 for catalogue or Website: http://www.griffinandgeorge.co.uk/

Invicta Plastics, PO Box 9, Oadby, Leicester LE2 4LB. Website: http://www.invictagroup .co.uk/

Royal Meteorological Society, 104 Oxford Road, Reading RG1 7LL. Website: http://atschool.eduweb.co.uk/radgeog/metneteur/metneteur.html

TTS Ltd, Nunn Brook Road, Huthwaite, Nottinghamshire NG17 2HU. E-mail: sales@tts-group.co.uk Tel: 01623 447800. Fax: 01623 447999. Website: http://www.ttsgroup.co.uk

UKWebcameras, Buglehorn ltd, 250 Swan Lane, Hindley Green, Wigan WN2 4EY. Website: http://www.ukwebcameras.co.uk/Maps.asp

IN THIS CHAPTER YOU WILL FIND KEY IDEAS ON
DIFFERENTIATION • ENQUIRIES • FIELDWORK • NATIONAL CURRICULUM •
PLANNING • PROGRESSION • QCA UNITS

Rivers, coasts and the landscape

The delight all children have in playing by the water's edge provides an excellent starting point for teachers when planning work on physical geography. Remembering our own explorations as children can be a powerful planning tool for it is surely in our pupils' best interests that we begin by viewing the features we want them to investigate through their freshly enquiring eyes, and not our own.

The good news is that this may well be less difficult for a 'non-specialist' than it is for a more experienced 'geographer'. It is all too easy for the latter to slip into using specialist terminology and to start presenting pupils with complex ideas. Someone for whom the work is new, on the other hand, may be able to gauge the pupils' starting points more aptly, and share in any subsequent enquiry more sincerely. It is also comforting to non-specialists to realise that at this early stage using the enquiry process itself is at least as important as the knowledge gained from it.

Geography as a subject is essentially investigative. Physical geography is less a series of facts to be learned than a series of questions about the processes shaping the environment. This is recognised in the national curriculum which advocates direct experience, practical work, exploration, investigations based on fieldwork and, above all, the use of geographical questions. It is fortunate that these requirements make it essential that we undertake some work in the field, as this immediately raises the profile of geography in pupils' eyes!

Like all outdoor activity, investigation of the physical environment needs careful planning. Identifying opportunities for local fieldwork can be an excellent way of providing enjoyable continuing professional development for colleagues and building the confidence of non-specialists. Teachers planning fieldwork should take account of the risk assessment and adult-child ratio stipulations of their school and LEA. You also need to investigate local weather reports, tide tables, river levels or upland walking conditions (see useful websites, page 259), depending on the location.

This chapter is intended to support work relating to water in the landscape, especially in riverside and coastal locations. It therefore offers starting points in relation to QCA geography scheme of work units 4: Going to the seaside; 11: Water; and 23: Investigating coasts. However, it does not set out to promote physical geography as a separate subject but illustrates how physical features and processes are part of a holistic understanding of places and the way they have evolved. By offering a range of possibilities loosely linked to age groups the chapter hopes to encourage teachers to adapt QCA schemes (DfEE/QCA, 2000) and create their own field-based units. While there are some suggestions for progression and differentiation, teachers are left to match or adapt activities to the pupils in their class.

It is also acknowledged that for financial reasons school-based teaching will need to supplement, and sometimes even replace, field experiences. Some of the suggestions, especially for key stage 1, are therefore based within and around the school grounds. It is in key stage 2 that more focused studies of other landscapes become appropriate and can be

accomplished by developing work around one or two accessible field locations. An alternative is to build the work into a local or contrasting locality study, which should consider physical as well as human characteristics. The key to both curricular and financial economy appears to lie in making some judicious choices at whole-school planning level to ensure progression through well-designed holistic place studies.

Earliest encounters with the physical world

Where the school locality includes a river or stream there will be opportunities for site visits where very young pupils can talk about what they see and begin to think about the qualities of flowing water. You might also encourage them to think about how people make use of local rivers and why valleys are important routeways. Teachers sometimes take young pupils to the seaside as part of an integrated history/geography project. Using geographical terms such as beach, waves, cliffs, caves, dunes, and so forth falls naturally into such work while sand play, both in school and on the beach, offers many opportunities for early ideas about physical processes to be developed in ways accessible to the young child. There are also some excellent resources which support this work, such as the Wildgoose playmats. Shared literacy texts, if carefully chosen, can also be a great asset in building early physical concepts. For example, the big book *Waves* (Stott-Thornton, 1997) gives an excellent account of depositional wave action in the simplest of terms, while *Out and About* (Hughes, 1998) describes the full gamut of outdoor features and weather conditions and provides an endless source of geographical discussion points. Other relevant story books include Webster's Walk (Dow, 2001) which describes the effects of a drought on some ducks, and Be Quiet, Bramble (Blathwayt, 1989), the story of a brave cow who saves her friends when they are marooned when the river suddenly floods. Finally, Come Back Hercules (Lewis, 1988) is a delightful rhyming saga about a bored goldfish who escapes via the loo to his local river and eventually finds his way down to the sea, passing many features of geographical interest on the way. I have seen this story colourfully retold in infant drawings on a long thin wall 'map', detailing the sequence of the story.

The effects of weather on ourselves and our surroundings is a common key stage 1 science topic which can assist with geographical learning. Most children are aware that it is un-comfortable to walk in the rain and that we need special clothing to keep dry. Beginning with the question 'How does waterproof clothing work?' pupils can explore the properties of waterproof materials, and see how the shape of the surface directs water into creases and down slopes. This could lead to a detailed study of other items such as a sou'wester-style hat (whose shape directs water to flow down the back of the wearer), an umbrella whose spokes drip well away from the user, and plants whose leaves direct the water to drip onto the root base. Young pupils might even be encouraged to study their own faces and think about how the eyebrows form a barrier which directs water around the eyes, or how water flows over their body in the shower. It is a only short step from here to understanding the patterns created by flowing water on the uneven surface of a landscape.

Learning about coasts

The study of coastal areas in key stages 1 and 2 is an excellent way of introducing pupils to physical processes at work in the landscape. England and Wales have some 4425km of coastline in which it is rare to find the same kind of coastal scenery for more than 16-25km.

Over half of all counties have varied coastlines within fairly easy reach which schools can use in their fieldwork studies. Even if your school is inland, the seashore makes an excellent contrasting locality and will offer opportunities for work across the whole range of the national curriculum.

It is difficult to conceive of any area which bears more witness to human activities which have changed or destroyed or improved the environment than the British coastline. Its popularity, and use for leisure purposes, places it at the interface of an environmental conflict between demand for an increasing range of recreational facilities on the one hand, and a desire for the conservation of the natural environment on the other.

Beaches are constantly changing owing to the sudden effects of storms and periodic high tides. We need to be aware of this in undertaking pre-visit preparations. It may be more appropriate to consider the broad possibilities offered by the site rather than details which may change before the visit. Making a video to use later can be a useful addition to classroom resources. In particular it will be important to check tide timetables. The local coastguard has tide tables and can give exact times well in advance. Many of the ideas which follow focus on opportunities for lower primary classes. While it is envisaged that the work may well be part of a wider history/literacy/geography package the aim here is to support only the physical geography element with which teachers are often less familiar.

Beach play

For the youngest pupils the difference between the beach and other places they know will revolve around the different play opportunities it offers. Beach play is multi-sensory. It involves the sounds of wind, waves and seabirds; the touch of spray on the cheeks, of fine and sharp sand, of smooth pebbles, silky feathers and slippery, slimy rocks; the taste of salt on the lips; the smell of drying seaweed and fish; and innumerable new sights from sparkling water to distant ships. A short period of free play on arrival is valuable. First it allows for acclimatisation and enables the children to develop a 'sense of place'. It also gives the teacher a valuable opportunity for observing individuals at play, informing assessment of their needs, and it provides pupils with the evidence to support discussion of the key question 'What is this place like?'.

Beaches are potentially dangerous because natural sound levels are high and sudden warnings are difficult to hear. You need to set clear boundaries for pupils and establish clear rules of behaviour before setting out. For example, you might use a loud whistle or bell to tell the pupils to stop whatever they are doing and gather together. A high proportion of additional adult help is required on outdoor visits and at the free-play stage it helps if these assistants form a boundary cordon rather than being involved with the play itself. Then it really is free play and the starting points it provides are securely within the pupils' remit. It need not take long; five to ten minutes is enough.

It helps to have a large tray ready to receive the objects that pupils will inevitably bring to you and which might form the focus of initial discussions. It is also worth joining hands in a long line along the shore to stare at the sea, listen to the sound of the waves and observe the action they have on the beach. An activity which encourages pupils to concentrate and observe closely is to test the well-known hypothesis that every seventh wave is the largest. If you can be sure that an unexpectedly large wave will not soak their feet, it is a good idea to do this by listening with eyes closed.

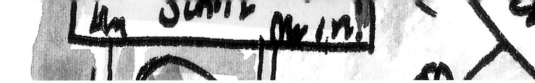

An excellent pre-visit preparation to enhance work with any age group is to make a short video of the waves. This will be most useful if it includes mainly close-up shots, with the sea breaking over rocks as well as on sand and pebbles, and if it shows the waves operating in contrasting weather conditions. Since physical geography particularly involves considering the power of the sea, it is worth considering an out-of-season visit, especially with older pupils, provided that there is some nearby indoor accommodation to allow them, and you, to thaw out.

Photo | Lois Gunby.

Building in sand

Sand is the feature of a beach which looms largest in children's imaginations and this will form the chief raw material of their play. Direct experience of sand, its qualities and physical properties, may also be gained by building sandcastles. Laying out the intended plan of the castle as a sand map is a useful way of practising simple cartographic skills, and photographs taken from different compass directions and looking down on the structure will demonstrate the basic idea of aerial views in a context which has personal significance for pupils.

With careful thought sandcastle building can contribute to pupils' conceptual development in other ways. For example, if you challenge them to build the walls as high as possible they will have to grapple with the way loose materials find their own angle of rest. As soon as the castle walls become too steep then catastrophe in the form of miniature landslides will strike. This activity not only illustrates the properties of materials it also provides first-hand experience of the nature of slopes. For older pupils it is but a short step from here to considering the instability of local cliffs or the fact that sand

How waves shape the coast

The principal agent shaping the coast is the sea, which reaches the shore as a never-ending series of waves of varying size, strength and frequency. Some waves are generated by storms. Known as 'swell waves' these generally diminish the further they get from the storm, as do the waves created by a passing boat. Most waves, however, are caused by the frictional drag of wind which disturbs the surface of the water as it blows. The size of these 'sea waves' depends on both the strength of the wind and the length of time (or distance) it has blown over that particular body of water. Hence waves reaching the west coast of Ireland which have had an uninterrupted passage across the Atlantic Ocean will be bigger and stronger than waves generated by the same wind on the west coast of Wales where the wind has only crossed the Irish Sea. This uninterrupted distance is known as the 'fetch' of the waves. It explains why spectacular feats of surfing can take place on some coasts, such as Bondi beach in Australia, even in good summer weather conditions.

Major exceptions to these two general types of wave include 'tsunamis', created by undersea earthquakes, and the tidal waves caused by suction of the sea into the intense low pressure of the eye of a hurricane as it comes ashore. Fortunately, we are most unlikely to meet either of these on our local beach!

The movement of waves is interesting, for while the eye tells us the water is moving forward, logically this cannot be so or the waves would keep on moving up the beach. Bear in mind they are still moving forward when the tide is going out. It is in fact the wave form that is moving, not the body of water. If we could track a single particle of water within that wave form we would find that it moves in a circular path, as shown in Figure 2. A cork or ball thrown out into the sea will bob up and down rather than travel directly towards the shore. We can demonstrate this general idea to children with a skipping rope. An up and down movement of the arm will create waves which progress laterally along the rope. A ribbon or bobbin fixed to the rope will emphasise that while the wave form moves forward, the marker bobs up and down, staying in roughly the same position.

Waves can be either constructive or destructive. Constructive waves occur in calm weather, are low in height and have a long wavelength. When they break, the energy dissipates quickly and the backwash is weak. This means that such waves tend to add new material to a beach. Destructive waves, by contrast, are stronger and larger. They are commonest in storm conditions and have a powerful backwash which can cause considerable erosion. They tend to attack headlands where the coast slopes more steeply.

Figure 1 | Knowing about the effects of waves and how beaches are formed can help you draw pupils' attention to appropriate enquiries.

dunes exhibit constant slope angles. This kind of open-ended activity focuses on processes rather than facts and stimulates progressive levels of thinking from infants or juniors. Whether we regard ourselves as 'geographers' or not, we can all have the confidence to raise and answer questions.

It can be particularly exciting to construct castles so close to the advancing sea that they will be attacked in the course of the beach visit. Guessing how much and which part of the castle will disappear next leads to penetrating observation and hence detailed experience of the erosion process, enhanced by emotional involvement. Allowing the children to battle in vain to save the structure with ramparts and ditches can only add to their eventual understanding of the inevitability of the sea's destructive power and makes a particularly good video sequence. This theme is explored for very young pupils in *Where is Barnaby's castle?* (Lewis, 2002) and *Barnaby Bear at the Seaside* (Jackson, 2001).

The effect of the waves

Understanding the basic effects of wave action (Figure 1) opens up another spectrum of options which we can match to the age and interest of pupils. Young children may enjoy simulating the action of constructive and destructive waves in a line dance using gentle gathering and pushing movements for the constructive waves and bolder plunging movements for destructive breakers. I have found James Reeves' wonderfully evocative poem *The Sea* (in Farjeon *et al.*, 1970) invaluable for drawing attention to contrasting weather conditions and wave types with middle and upper juniors. Although quite a sophisticated poem, the imagery of the moody dog and the abundance of sound effects in this poem make it accessible and particularly appealing to junior children.

On page 157 is another poem inspired by coastal fieldwork and by reading Reeves' poem. This young poet's vision of people and the sea both eating rock at the seaside is a remarkable feat of lateral thinking of the kind that might identify her as a future primary teacher!

You can sample beach materials using metre (or smaller) sampling squares. It is best if these have an internal grid of string or wire so that items can be taken system-atically, for example at each

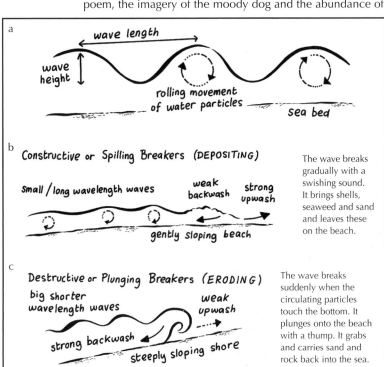

Figure 2 | The basic differences between wave type help to explain deposition and erosion.

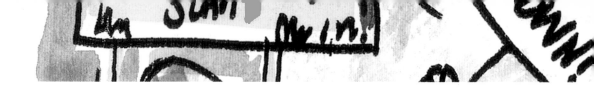

intersection. This introduces the idea of scientific sampling and helps pupils to understand why it is unsatisfactory just to pick up attractive pebbles. If sampling grids are not available use plastic hoops instead. Being fairly robust these can be thrown across the beach to achieve something approaching a random sample. In examining pebbles and other items questions to consider include:

- Is the material natural or manufactured?
- Are all the pebbles from the same kind of rock?
- What are the similarities and differences?
- How do shapes, sizes, weights and colours compare?
- What might be the reasons for the similarities and differences?
- Where may the pebbles have come from?

Photo | John Halocha.

The nature of these investigations and the level to which they can proceed will relate not only to the age and experience of the pupils but also to their mathematical ability, and there are fertile opportunities here for developing differentiated follow-up activities. It is not necessary to be an expert geologist to understand this kind of work. Developing observation and enquiry skills are the chief aim. However, it is helpful to use a good pictorial earth science guide such as How the Earth Works (Farndon, 1999). There is also potential for developing ICT skills using 'branch and sort' computer software.

Learning about rivers

Key stage 1

The wet sand tray is a good place to begin early work on rivers. Hilly landscapes can easily be shaped by hand, and valleys identified (Figure 3). It is quite difficult to demonstrate the flow of water simply by sprinkling 'rain' from a watering can, because the loose sand is so porous. But after trying this to show that water sinks into some surfaces, the' landscape' can be waterproofed with thin polythene film and the experiment repeated. A sprinkling of dry sand over the polythene should show that when streams develop they are capable of carrying loose material away and depositing it somewhere else. Ponds or lakes can be engineered too. Teachers of older infant classes, who might not have a sand tray on hand, will find a length of plastic guttering and a selection of sands and gravel of different grain sizes useful for creating investigations into water flow and its capacity to move things (Figure 4).

River work might equally well begin from a study of water and how it moves around the school. Whether your school has a flat or sloping roof there will be some system for transferring rainwater to the ground. Leaky flat roofs, a scourge of many modern school buildings, can even become a temporary asset, giving a good focal point for discussion of how surface water finds its way down. Models of buildings with flat, sloping and arched roofs could be undertaken as a

Figure 3 | *Creating a river landscape in a wet sand tray. Photo: John Halocha.*

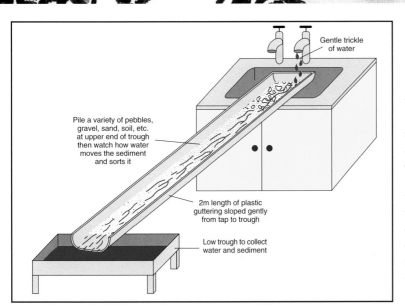

Figure 4 | *Guttering is used to show how running water can move things.*

technology project and evaluated for their effectiveness in shedding water. A sloping roof amply demonstrates the idea of a watershed, in this case the roof ridge, which separates the flow on one side of the roof from that on the other. In exactly the same way, a ridge of high ground in a real landscape, known as a 'watershed', directs water down into one river system or another.

There are many ways of illustrating the idea of a river system for young children. You could start by getting them to draw a simple pictorial diagram to show the progress of water via gutters and drainpipes to the ground in your own school. If you can direct a flow of fresh water from a drainpipe (or any other source) onto the surface of the playground, pupils can watch what happens if there is no organised drainage system (Figure 5). Finding an unobstructed path, the stream of water will create a temporary river flowing down to a low point and will possibly meander on the way. It is also interesting to overfill existing puddles and watch where the water goes. Pupils might make detailed observations of puddles forming and disappearing and make maps of puddles and the likely courses of excess water in the playground – a place with which they are already very familiar (Figure 6).

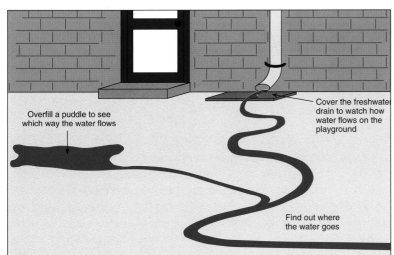

Figure 5 | *Where water goes in the playground.*

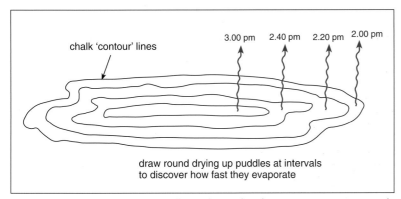

Figure 6 | *How a disappearing puddle can be used to demonstrate evaporation and simple contour patterns.*

Photo | John Halocha.

If there is a small stream or river in your immediate area an opportunity to observe this will be part of the physical component of your locality study. While young pupils should obviously not be encouraged to throw rubbish into the water they could watch how the current carries a biodegradable float, such as a cauliflower floret, downstream. If the river is liable to flood, the absence of houses near the river or the presence of flood protection barriers might form a good discussion point which illustrates the interaction of people and their environment. The study of an accessible riverside environment, such as a local small park, nature reserve or public footpath, might also form part of the work and provide opportunities for simple visual investigations into how clean the water is, whether people throw rubbish into the river, who uses the riverside and how this environment might be cared for or improved. Some schools have 'adopted' a riverside area and make a real contribution to improving it with local authority help. Ruth Brown's strikingly illustrated parody, *The World that Jack Built* (Brown, 1991) would be an excellent support for such work.

Key stage 2

Running water within the school grounds – in guttering and drains, overflowing from puddles and trickling down slopes, both natural and constructed – provides an initial focus for the study of rivers at key stage 2. Probably the most important single concept in this work is that water flows downhill. This may seem to be too obvious to mention but is critical to any understanding of river systems. Many pupils enter secondary school believing that rivers 'begin' at the coast and flow inland. Perhaps this is because they have seen a tidal river, or have been asked where a river 'rises', or have not fully understood the maps they have come across. The word 'mouth' adds to this confusion; after all, in the limited and personalised experience of primary pupils, things usually do go in via mouths rather than come out!

As well as shaping the physical landscape, rivers also exert powerful influences on human decisions about land use such as where to site a capital city or a chemical plant, where to graze cattle rather than grow crops, which is the best route for a new motorway, or even just where to spend next Sunday afternoon. Rivers are part of the natural water cycle, vital to industry, trade, factories and farming; needed for public water supply and transport; havens for wildlife; attractive for sport and tourism; beautiful, calming and mysteriously appealing, but also dangerously powerful and terrifyingly destructive in times of flood. So, as well as providing bridges across the geography curriculum, river studies also offer convenient routes into other subjects. From this point of view, a collection of picture books, story extracts, video-clips, news bulletins, newspaper extracts, photographs and artists' work in a number of styles will be valuable resources to support your work.

Another key idea is to think of a river as a system with distinctive 'inputs' and 'outputs'. The inputs are rain, snow and hail (collectively called precipitation); the output mechanisms of evaporation, percolation (seepage) and surface run-off ensure that water is returned as

efficiently as possible to the atmosphere or oceans. As a naturally harmonious system, a river constantly readjusts itself so that it can operate efficiently. Provided its capacity is not exceeded (in which case, of course, there will be a flood) the river basin achieves a carefully adjusted water transfer operation from the hills to the sea. Many of the physical features that are of interest in a river basin can be seen as by-products of the system's effort to adjust to current circumstances and maximise its efficiency. For example, if it encounters an obstacle, in the form of either a natural feature or something dumped in the water, the river will attempt to remove it by 'erosion'. Conversely, if something slows the current so that it begins to carry more sediment than its volume and speed will allow, it will simply drop any excess as 'deposition'. This approach places the river in the context of the water cycle.

Pupils' understanding of river systems can be built up gradually over a period of time, assisted by good classroom resources and the use of the school buildings and grounds. A large sand and gravel tray or outdoor 'pit' which can be raked up to form a mountain area and 'rained on' with a spray hose, will reveal some of the features of that developing system. Pouring water onto a selection of soil, sand, gravels and small pebbles banked up in a length of guttering will illustrate how water moves and sorts its sediment load (see Figure 4). The introduction of a waterproof layer (plastic or polystyrene sheet) will demonstrate the principle of springs. Those who live near beaches can also study river systems in miniature (Figure 7). Large picture resources (Y3/4) or a good video sequence (Y5/6) such as 'River Landscapes' in *Resources Unit 11-13 Geography* (BBC, 1985) will do much to draw these practical experiments into a real context. It is not so straightforward, however, to communicate the timescale over which the system develops and operates or to consider the formation of river features inside a classroom. Ideally, then, where fieldwork is possible it should emphasise the different forms that rivers can take.

A large number of geographical terms have been used in this section, some of which will be more familiar than others. Nowhere is technical and commonplace vocabulary richer than in the description of water features – stream, burn, beck, falls, force, ghyll, puddle, pool, pond, mere and tarn are just a few examples. Most regions have a rich variety of colloquial names. There are also many problems with definition. For instance when does a stream become a river and when does a river become an estuary? You are unlikely to find a precise answer as there are no

Figure 7 | *Not a river delta from the air – coal dust deposits being transported in tiny streams of water on Hartley Beach! Many other river features may be seen in miniature on the beach. Photo: Liz Lewis.*

'official' sizes at which these terms apply. Indeed, if there were a precise definition, say in terms of a flow rate of 'so many cubic metres per second', would it mean anything to the average person anyway? What these terminological uncertainties do imply, however, is that teaching pupils to name features is really an issue of inducting them at their own level into the whole nomenclature debate. 'What do you think we might call a feature like this?' is therefore always an appropriate question.

Many pupils, quite apart from not being able to employ a vocabulary of comparative sizes, may not even be familiar with the idea that physical features have 'proper' names like people do. We may need to be very conscious of saying 'the River Trent' rather than just 'the Trent' in the early stages of any work. Adding the three longest rivers named in the national curriculum, your own local rivers, and any others that pupils know about, to an enlarged UK map, is a simple way of raising this awareness and a good way of initiating discussion about sources, tributaries and mouths. It will help pupils if you emphasise the direction of flow by starting at the source and ending at the mouth when you draw a river on the map.

New terminology becomes easier to grapple with if it is encountered regularly. It is a case of being alert to the opportunity to 'drip feed' additional ideas whenever the chance arises. The use of appropriate stories or newspaper cuttings can greatly assist this work so the literacy hour is a good time to support geographical learning. You could, for example, ask the class to make up a name for a river that is unnamed in the story, and to relate the features of this river

to what they have been learning by drawing an imaginary map to show where they think it begins and ends and where along its banks the story is set. Freed by the imaginary setting from the constraints of achieving locational accuracy, pupils will be able to show you what they know about rivers and maps – a useful interim assessment opportunity. It is essential, however, that we are well prepared to support their understanding of real rivers too, and this involves developing our own subject knowledge.

Rivers fieldwork

Undertaking independent fieldwork on rivers may seem daunting to anyone who does not regard themselves as a 'geographer' but it doesn't need to be. By asking key questions such as What/where is this place? How did it get like this and how and why is it changing? you can generate ideas which you can incorporate into your initial planning. Visiting the site will also help to stimulate your thoughts. If possible you

Photo | *Liz Lewis.*

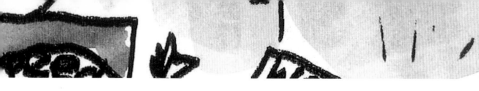

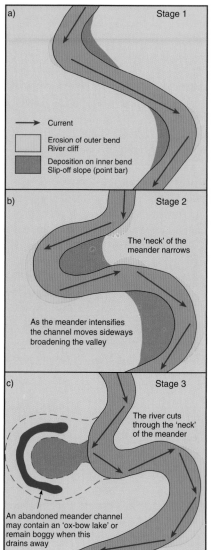

a)   Stage 1

→ Current

Erosion of outer bend
River cliff

Deposition on inner bend
Slip-off slope (point bar)

b)   Stage 2

The 'neck' of the
meander narrows

As the meander intensifies
the channel moves sideways
broadening the valley

c)   Stage 3

The river cuts
through the 'neck'
of the meander

An abandoned meander channel
may contain an 'ox-bow lake' or
remain boggy when this
drains away

Figure 8 | *How erosion and deposition can lead to the formation of an ox-bow lake.*

should get pupils to devise their own questions, such as those listed below. This is important as it gives them a sense of ownership and increases their commitment to the work. These questions can be audited to demonstrate the integration of physical, human and environmental elements in a typical set of pupils' questions.

- Will it be safe to paddle?
- Is it very deep?
- Who does it belong to?
- Where does the river come from?
- Why does the river get dirty?
- How fast does the river go?
- How long is the river?
- What would happen if you fell in?
- Are there any fish in the river?
- Will we see otters and things?
- Why do rivers bend and twist?
- Where does the water come from?
- Are there any boats on it?
- Does it flood over the edges?
- How is the river bank changing?

A good place to see the physical processes at work in a river is at a bend in the channel known as a 'meander'. As the current swings round the outer side of the bend the full force of the water is thrown against the bank, loosening the soil and scouring away rock so that it becomes eroded (Figure 8). On its own water usually has little effect, so the importance of the bedload as a scouring tool needs to be highlighted once pupils begin to recognise the erosion process. The vertical bank or 'river cliff' which the river cuts on the outside of a bend can be anything from a few centimetres in height to a spectacular feature. Over time, further erosion undercuts the bank making it unstable, with the result that it collapses into the water.

When choosing a fieldwork location bear in mind that it should afford as many of the following opportunities as possible:

- A guided walk along a stretch of riverbank, ideally incorporating a photograph trail
- Opportunities to identify, describe, sketch and label landscape features
- A chance to locate features and follow the route on large-scale maps
- A stopping point which gives a vista over the whole area
- Safe locations at which to investigate the river at first hand, e.g. measuring
- A chance for groups of pupils to do their own investigation under adult supervision
- A place to have lunch, visit the lavatory and play safely

Before you go, see that the pupils are equipped with a field notebook in which they can record their observations. Over a period of time this will build up into a valuable record of

Photo | Paula Richardson.

work undertaken. You might also decide to take some more specific equipment such as measuring tapes, stop watches, laminated maps of the area and biodegradable floats (e.g. cauliflower florets). Cameras of all sorts will be invaluable.

Follow-up work needs to be completed quickly, while the experience is fresh. One way of presenting the work is a large interactive working display which will spawn comparative work, foster discussion and provide factual material which pupils can present to other classes. You might also use the display as a way of reporting back to parents and raising enthusiasm for more fieldwork in your school. The promise of a real audience will help to sustain interest but be wary of prolonging the follow-up beyond its natural limits. There may, however, be opportunities for developing the work by focusing on issues and changes. For example, you might challenge pupils to plan a new riverside park. This will involve them in mapwork exercises and will show how human activities relate to the physical landscape. Topics of this kind are brimming with potential not only across the geography curriculum but in other subject areas as well.

Doing geography through enquiries

I have tried, with reference to both river and coastal studies, to combine an introduction to some essential understandings in physical geography with a vision of the enquiry approach in action, some advice about fieldwork, and some suggested pupils' activities.

I have avoided matching individual activities to particular age groups because I don't believe it is possible to be prescriptive about this with any real credibility, especially in the context of someone else's class. As teachers we are all best-placed to judge the capabilities of individuals in our own teaching groups and to select well-matched activities to extend their learning. Some schools may intend to develop a single programme of work on rivers, others to integrate the work with locality studies or other curriculum areas, and yet others to visit and revisit the theme throughout the junior years in differing contexts. All these modes of organisation are possible given the flexibility of the national curriculum and the intention that we should adapt any QCA scheme of work to fit our local opportunities and the needs of our pupils.

I have concentrated on the broad vision of field-based enquiry in the knowledge that there are already numerous published resources to assist the design of classroom-based work and individual activities, many of which can also be easily adapted to support the outdoor opportunities encountered in your own particular locality or region. For me as a teacher the beauty and strength of the enquiry approach is that it is open-ended. Once the broad context of the investigation is set, and the issues of interest defined by pupils and teachers working together, then individuals can enter that enquiry at their own level and proceed within it as far as they are able.

Any enquiry that is worthy of the title 'geographical' must incorporate both physical and human elements. It can achieve this by examining, in the context of real places, those relationships which are at the very heart of geography – the effects of the world's varied physical conditions

on human activities and of humans on their sensitive physical home, the Earth. By raising awareness of and fostering thoughtful responses to the many important and often politically sensitive environmental issues with which our world is currently struggling, we shall be providing the kind of learning that an informed and sensitive next generation will desperately need.

Developing understanding of the physical landscape is in some ways the most difficult aspect of geography but very worthwhile because that understanding can enrich appreciation of the natural environment and so feed the desire to protect and conserve it thoughtfully.

Sir Halford Mackinder, the prominent educationalist and pioneer of academic geography, who seems to have harboured a healthy scepticism about the sterile rote learning of the 'capes and bays' tradition, had this to say about school geography in the 1920s:

> *You cannot have taught them much by way of facts, but if you send them out curious, seeking to learn, and with an idea that there is order in the world and that their experience must fit into a corresponding order in their minds, you have done all you can hope to do.*

This liberating thought provides an ample justification for enquiry learning.

Our role in primary physical geography, then, is not to present our pupils with a watered-down version of our own half-remembered or recently-resurrected, senior school experience. It is rather that, inspired by our own local landscapes and supported by well-chosen secondary resources, we should invite and nurture the natural curiosity children have about their own world. All primary teachers, whatever their subject orientation and expertise, can surely embrace this role with confidence.

References and further reading

BBC (1985) *Resources Unit 11-13 Geography*. London: BBC.

Blathwayt, B. (1989) *Be Quiet, Bramble*. London: Walker Books.

Brown, R. (1991) *The World that Jack Built*. London: Red Fox.

DfEE/QCA (2000) *Geography: A scheme of work for key stages 1 and 2 (Update)*. London: DfEE/QCA.

Dow, J. (2001) *Webster's Walk*. London: Frances Lincoln.

Farndon, J. (1999) *How the Earth Works*. London: Dorling Kindersley.

Farjeon, E., Reeves, J., Rieu, E.V., Serraillier, I. and Graham, E. (1970) *A Puffin Quartet of Poems*. Harmondsworth: Penguin.

Hughes, S. (1998) *Out and About*. London: Walker Books.

Jackson, E. (2001) *Barnaby Bear at the Seaside*. Sheffield: Geographical Association.

Lewis, L. (2002) *Where is Barnaby's castle?* Sheffield: Geographical Association.

Lewis, R. (1988) *Come Back Hercules*. London: Macdonald.

Stott-Thornton, J. (1997) *Waves*. London: Kingscourt.

Useful websites

Meteorological Office: http://www.met office.gov.uk/education/curriculum/index.html

Maritime and Coastguard Agency: http://www.mcga.gov.uk/c4mca/mcga-home

Environment Agency: http://www.environment-agency.gov.uk

Mountain Rescue Council: http://www.mountain.rescue.org.uk

Wildgoose: http://www.wgoose.co.uk

Glodwish House

Marrowdow

Settlement

This chapter explains the key concepts used to help children understand what settlements are, what happens in them and how they change. Sections on the early years and key stages 1 and 2 offer practical ideas for planning and teaching the settlements theme within the primary geography curriculum.

Whether they live in an isolated farmhouse, commuter village, market town or a large city, all children come to school knowing something about settlements. For example, the pupil in your class who lives on a farm may have heard how it has belonged to the family for many years and know why his/her parents are determined to continue making a living from it. Another pupil, from an inner-city area, may have a strong sense of community and identity based on their experiences within a particular ethnic culture. These two examples are particularly meaningful because they also illustrate how your pupils' experience relates closely to current issues and debates. Although farming communities may appear isolated, they are actually connected to European agricultural policy and consumer demands. Similarly, in inner city areas, issues of social identity, integration and global migration can be found within the school catchment area.

Settlements are fascinating places and often provide really vivid examples of geographical processes and patterns at work in the wider world. They are also rich in opportunities for pupils to investigate real issues and geographical questions (Baldwin and Opie, 1996). They can be quite complex too! So, the first section of this chapter explains the geographical concepts which help to explain, as well as raise questions about, any settlement across the world. As you and your pupils plan geographical enquiries in your chosen settlements, you can decide how appropriate each of these concepts might be when writing your learning objectives.

Key ideas

1. Basic human needs

Studying any settlement can help pupils to understand the importance of basic needs. In the past, people tried to find safe places in which to build settlements. Living on high ground made it easier for them to spot enemies from some way off and to look out for possible food supplies. On another level, feeling 'safe' may involve being close to people similar to yourself who understand your values and beliefs, hence the growth of particular cultural communities. But geography also helps us to understand that the world is not that simple. For example, many farmers around the world have chosen to farm on the slopes of active volcanoes. It's not at all safe but they have decided that it is worth the risk as the fertile volcanic soil helps crops to grow well and ensures a regular supply of food. Human beings also need water, fuel and dry buildings in which to live. As pupils study a variety of settlements across the globe they will begin to understand that meeting basic needs can lead to all manner of outcomes. None of these responses is 'right' or 'wrong', they are simply different responses and ways of interpreting the physical environment.

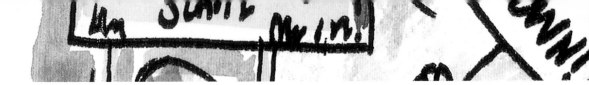

Water is an interesting example. As part of their topic or project work pupils may see images of people in dry areas carrying water over long distances. These people will be very careful how they use that water. In other cases pupils will discover that water is supplied to settlements through taps and study how much of it is wasted. Thus, by looking at contrasting settlements it is possible to introduce the issue of sustainability into your enquiries. Learning about migration can also help pupils to understand how people who are forced to leave their settlements struggle to survive, as well as raising the question of how people in other places react to newcomers (Asquith, 1999).

2. Similarities and differences

As they compare settlements pupils will begin to understand that while every village, town, city or suburb is unique, it is nevertheless possible to see patterns that are repeated in settlements further afield. Take Sheffield for example. It is a unique city that houses the headquarters of the Geographical Association. Sheffield also has a unique location, history and heritage. However, it has many similarities with other large cities both in Britain and across the world. Its industries have had to adapt to changes in global demands; some parts of the city require urban regeneration; attempts have been made to improve the transportation system by introducing a tram service. You may teach in a village where most adults drive away each day to work in other places. While your village is unique, a similar commuter pattern can be seen in hundreds of other villages. By allowing pupils to investigate a range of settlements, you can help them understand how places can have special characteristics as well as illustrating more general patterns and processes.

The way that settlements are changing provides another slant for examining similarities and differences. As multinational corporations spread across the globe, it may be argued that settlements are becoming increasingly similar. In Britain, our town and city centres are now dominated by a predictable range of chain stores, while small, individual shops disappear or are driven out to areas with cheaper rents. What type of town centre would pupils want to be in when they have grown up? Do they want to identify with somewhere special, somewhere different, somewhere unique? Or do they find it reassuring to know that services of a similar type and quality can be found in many places? Exploring these kinds of questions involves using imagination, prediction and other thinking skills. It also moves geography away from factual and descriptive accounts of places to consideration of how places change and what settlements may be like in the future.

3. Context

Many schools are now using commercially produced packs that focus on particular settlements, both in Britain and countries around the world. Although many of these packs are well-produced, they often lead to settlements being studied as isolated units. As a result pupils fail to understand how the place they have studied is linked to other areas. For example, if you are using a pack on an Indian village, it is important to know how that village is linked with nearby villages and the local town, what links it has with places further afield, etc. It can be difficult to find the answers to these questions, but as pupils' ICT skills improve we may be able to move away from commercially produced packs and investigate other settlements and their wider links by communicating directly with the pupils and teachers who live there. Understanding such links can give pupils a much more realistic appreciation of how the

modern world is connected. Physically, people travel much more widely nowadays and pollution is transmitted through the air and water. Cultural links occur through travel and overseas visits. Electronic communication is also rapidly linking settlements so that it is less important for people to move around. If geographical enquiries can help pupils to understand how people and settlements are linked in new and sometimes exciting ways, they will be better equipped to understand the complex world in which they live.

4. Terminology

As pupils learn about settlements they can easily develop misconceptions about the terms they are using (Ferris, 1998). If we tell them that a village is rural and towns are urban there is a risk that we might be preventing them from developing a full understanding of the patterns and processes happening today in the landscape. A village may be in a 'rural' location but rely heavily on people visiting it as tourists from nearby towns, and have very few inhabitants who actually work in agriculture. An attractive 'rural' village may in fact have a population which mainly works in nearby towns and cities and has very few connections with the surrounding countryside. It might be interesting to investigate how a town or city can spread its 'urban' effects. For example, pupils might investigate local supermarkets to find out how far their home delivery services extend through e-based shopping facilities. Ask pupils to question DIY and furniture stores to see how far out of town they will deliver goods. If you are in a village, find out where the working population actually does work. Why do they choose to travel a long way and spend money on transport when they could live closer to the town?

5. Settlement hierarchies

Settlements vary in size from individual dwellings, hamlets and villages to towns and cities (Figure 1). When you are planning settlement enquiries try to ensure that pupils are able to investigate a range of places while at primary school. This will help them make links and identify patterns. There are many ways of developing investigations. Here are some suggestions:

- How many people live in this settlement?
- What do they do?
- What types of homes can we find?
- What is the land used for?
- How is this place linked to other places?
- What facilities and services can be found in settlements of different sizes?
- Why do larger places have more facilities such as shops and places to eat?

Another way of investigating settlement hierarchy is to find out how some places specialise in certain activities. For example, seaside towns rely mainly on holidaymakers while other towns specialise in industries based on information technology. This may be taken a step further by looking at land use in settlements. Two towns may have similar numbers of people but cover very different amounts of land. So, the position of settlements within a hierarchy may not tell us all we need to know about them. Another key concept here is that there are reasons why settlements develop at specific sites. As these reasons change over time settlements need constantly to evolve. Those that fail to find new functions may decline and may even disappear.

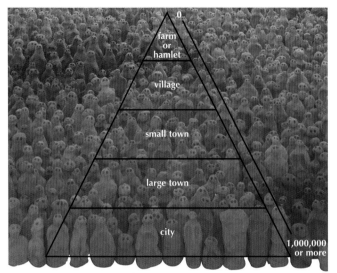

Figure 1 | Expanding settlements. Photo: Field for the British Isles, 1993, Antony Gormley, Terracotta. Approx. 40,000 figures, each 8-26cm tall. Courtesy of the artist and Jay Jopling/White Cube.

Settlement Size

People live in settlements of different sizes. Places are called hamlets, villages, towns or cities depending on the number of people who live there. However some places are called cities because they are given special permission by a king or queen.

6. The physical setting

When we are planning investigations in settlements, it is all too easy to ignore the landscape. However, the physical setting has a crucial impact. You might think about where water flows through a town or city. Are streams and/or a river actually covered over? A topical issue would be to investigate new building on land which may flood, thus linking aspects of physical geography with demands for housing. Investigations on air quality in parts of a village, town or city could reveal interesting results. You could also bring a new dimension to your geography lessons by involving people working in environmental management. The Environment Agency has specialists who are able to support work in primary schools.

Another angle is the impact of hills and valleys on land use. Misconceptions can be addressed here by making comparisons with settlements around the world. If pupils believe people prefer not to build on very steep slopes, access data on some fast-growing cities in South America. Again on a global scale consider why people build in dangerous places. Cities such as San Francisco and Istanbul are built in earthquake zones. Why are people prepared to risk living there? To broaden pupils' understanding, the concept of lack of choice could be introduced by studying why people continue to live in villages, towns and cities of Bangladesh that are not only prone to severe flooding but, according to some predictions, could be submerged if sea level rises in future.

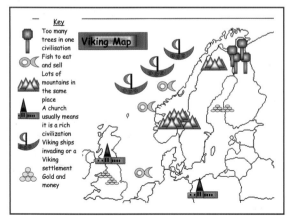

7. Issues

All the previous paragraphs contain examples of issues which pupils can investigate. They are highlighted here to demonstrate how the problems affecting settlements are constantly changing. To retain both your own and pupils' enthusiasm for geography, keep a look-out for new issues and problems. Your pupils may have good local knowledge and be able to identify suitable questions. Read local newspapers and talk to parents and local business and community people about

Pupils' work | Shelley Primary School, Horsham.

new developments. Watch regional TV news and topical interest programmes. Regularly walk your school's locality with your geographical eyes open. Can action groups, for example the local wildlife trust, highlight new issues or support pupils' investigations? Can you draw on national/international initiatives? For example could your class put together a proposal based on their own investigations that would put forward a case for their town or city to be chosen as European City of Culture? Could they suggest reasons for holding the next Commonwealth Games in their town or city, or even more controversially, in a village location?

8. Futures education and sustainable development

As we have seen, settlements are constantly changing, sometimes as a result of things which are happening locally or because of events in more distant parts of the world. As pupils begin to understand these forces they could consider what their settlement might be like some time in the future. They could also ask what they would like to see happen. Hicks and Holden (1995) have developed a wide range of interactive strategies for developing these concepts with primary children.

Settlements of all types and size have an effect on the environment. The amount of land needed to support a place is known as its 'environmental footprint' (Figure 2). Sustainable communities and people living in the developing world tend to have a relatively small environmental footprint. However, in Western Europe we can only support our current lifestyles by drawing resources from many other areas. It is estimated that the Isle of Wight has an environmental footprint around two and half times its own size. The equivalent figure for London is well over a hundred. How then could settlements become more sustainable? Is it actually possible? What happens to all the rubbish created in our town and where does it end up? Where does all the waste water and sewage go? Should towns and villages have their own wind farms? Energy is lost when generated electricity is moved a long way, but would wind turbines affect the appearance of the landscape? What do people in other European countries and around the world both think and actually do about some of these issues? Such investigations will show that many innovative ideas are being developed and that maybe we can learn from people in other places (Chambers and Garner, 2001, pp. 4-6).

9. Looking afresh at our locality

Familiarity can dull our geographical awareness! All our pupils live in a settlement of some sort and primary geography encourages them to look closely and actually question what they may well take for

Figure 2 | *Each settlement has an individual environmental footprint.*

Photo | Stephen Scoffham.

granted. Why has the village shop closed even though new homes are being built in our village? Why do our parents drive us to school each day even though we live quite close to school, and how does this affect our environment? Our streets have some unusual names: what do they tell us about what happened in our part of town in the past and why has it changed? We all know the river that flows through our part of the city, but where does it start and where does the water end up? Another valuable enquiry here might be 'and what effect do we have on that water?'.

Familiarity can also lead to pupils thinking that because their settlement operates in a certain way then all other places ought to be the same. Geography teaching in primary schools can help to dispel this misconception, largely by studying localities in a wide range of locations around the world which helps pupils to witness and value a variety of ways of life. We can also help by encouraging pupils to think about alternative ways of living in settlements. Do we really need to use so much electricity for lighting our streets? How might we make our town centre much easier and more enjoyable to use for people with various special needs? Geographical enquiries of this kind allow issues of inclusion to be developed in a practical and meaningful way within the curriculum.

Introducing settlements in the Foundation Stage

All six areas of learning in the Foundation Stage curriculum provide opportunities for helping young children to observe and think about the world. One of the best ways of introducing them to the concepts associated with settlements is to use their own experience. You might begin by talking about their own home and people important to them. Ask pupils to think carefully about their room at home; whether they share it with a brother or sister or have a space of their own. Get them to make a model using a box (a shoe box is ideal) as a frame. They can make a model of the bed and other items or use ready-made objects such as dolls' house furniture. Ask them what they like and dislike about their room. Why are things put in particular places? Did they decide or was it done for them? How would they like to make their room better? For example, would they change the colour of the walls or make more room to play? These activities begin to get pupils thinking about different areas and spaces, or as geographers express it, land use patterns. Other questions you might consider asking pupils are:

■ What can you see out of the window?

■ What sounds can you hear when you are in your room?

■ Where are the sounds coming from and what might be making them?

■ Do you like where your room is in your home?

■ Would you like a room which is sunnier, or away from a noisy brother or sister?

If you have a classroom pet such as a hamster, more key settlement concepts can be introduced by getting pupils to observe and ask questions such as:

■ Why do we keep the cage in a particular part of the classroom? (It needs to be away from draughts and bright sunlight) (Concept: shelter)

- Why does the hamster move food, bedding and toys around its cage? (They are creating different areas and spaces) (Concept: land use)
- Why do we need to keep the cage clean, and provide water and food as well as interesting things for the hamster to play with? (Living creatures have a range of daily needs) (Concept: services)

If you are able to arrange walks in the school grounds and beyond, encourage people to think about why birds and animals live in particular places. For example birds build nests well away from where cats can find them and rabbits make burrows in soil that is easy to dig. Such walks are also an excellent opportunity to use geography vocabulary linked with settlements. These words might describe different buildings – house, shop, factory – or items of street furniture – letter box, road sign, traffic lights. If you provide pupils with an aerial photograph of your school it will give them further information about your area. Research by Plester *et al.* (2003) suggests that even very young children can successfully use such resources.

Pupils at this age can also be introduced to settlements further afield through the use of stories. For example, *The Village in the Forest by the Sea* (Birch, 1995) introduces readers to life in a seaside village in Kenya. New words are introduced and pupils can be encouraged to talk about the similarities between their home and the one in the story. Also, how would they describe their settlement if they met the two girls? Activities like these can introduce young pupils to the rich diversity of people, lifestyles, climates and places we find in settlements all over the world.

Settlements in key stage 1

At key stage 1 geographical skills can be planned into investigations on settlements as pupils begin to ask geographical questions. For example, why is there so much litter in the street outside our school? When they investigate this problem pupils may find that the litter comes from take-aways in the nearby shopping centre or that people in the car park leave litter because there are no bins. Where should new bins be placed and who could do this? This small-scale study encourages pupils to think about land use in their locality, public services and the quality of the environment. It also helps to build a sense of community and forge wider links.

Pupils of this age are capable of making quite detailed and carefully observed models. They could be given the task, in groups, of making a model of a place they visit. Arrange for them to take plenty of photographs and encourage them to look at the buildings from different angles; the front will not be enough. Can you get up high to take some oblique photographs to help add details to roofs and so on?

A variation on this activity is to divide pupils into 'expert' groups. Each group is responsible for a different task: collecting information on buildings needed for the model, on people, street furniture, vehicles, natural features such as trees, bushes and streams. As pupils are building the model, adults can ask geographical questions about how the settlement works. Why are the parent and toddler parking spaces closest to the supermarket entrance, for instance? Why have no houses been built near the river? Why are the streetlights in particular places? What is made in that factory and why does it take up so much land? One way of consolidating this work is to arrange a visit to a nearby model village. As well as enabling pupils to study a whole settlement at a manageable scale, this will also help develop geographical skills of observation (Halocha, 2003).

Pupils like to work with visitors to their school. Be on the look-out for people you know who might be visiting settlements in other parts of the world: it may be parents travelling on business, a governor visiting relations abroad, or the school's police officer going on a course in another European country. Speak to these people before they travel and explain that you want to study settlements in different parts of the world. Your class could prepare some simple questions to give to the traveller before they leave. When the traveller returns they could be invited back and become the focus of a special report. This activity naturally lends itself to, and will benefit from, the use of ICT, but the extent of this will depend on how much time your traveller can give to the project. A CD-Rom of digital photographs could be left after their visit and used to prepare a simple illustrated report written by your class. Such an ICT activity will be much more meaningful than one devised simply to achieve objectives in the ICT Orders.

Pupils' knowledge and understanding of settlements can also be developed through the use of atlas and mapping skills. Here are some questions that will focus attention on your local area:

- Working with a large-scale map, who can find all the shops?
- Who lives the furthest away from school?
- Where are the best play areas to visit?
- What's the safest place to cross the high street?

Catling (2002) suggests many more activities using atlases and globes to develop pupils' knowledge and understanding of settlements and other geographical themes.

Settlement investigations for key stage 2

You can help to ensure that pupils investigate a range of settlements while in primary school by careful whole-school planning of fieldwork experiences. For example, if your school is in a city suburb, can you arrange a day's fieldwork to a nearby village? At some other appropriate time could a residential visit ensure that pupils experience a contrasting settlement such as a coastal town? By taking these opportunities you will provide pupils with direct experiences which will help them develop some of the key concepts discussed earlier such as settlement hierarchies and functions.

More complex geographical activities can be developed at key stage 2. You can use oblique aerial photographs to ask questions about settlements, develop geographical vocabulary and assess pupils' understanding. Collect these photographs (laminated if possible) from books, calendars and settlement study packs if your school uses these. You should then prepare a supporting set of small cards containing settlement words. Ask small groups or individual pupils carefully to place the cards on their photograph. Some of the words you write down might be 'red herrings' designed to prompt discussion. You could also have some blank cards for pupils to add their own vocabulary if not provided by you. Choose your bank of words carefully. If you include 'settlement' along with 'village', for example, you can see which term pupils use and assess their understanding. Including both 'swimming pool' and 'leisure facility' allows you to show how important collective words are in describing settlements (Figure 3).

Land use investigations often make an effective base for fieldwork at key stage 2. These involve a wide range of geographical skills such as careful observation, discussion and classification, reading and making maps, using keys, and interpreting findings using

| River | Supermarket | High Street | Park |

| Flats | Office block | Car Park | Football Ground |

mathematics (e.g. calculating the percentage of land used for different purposes). You will need an up-to-date large-scale Ordnance Survey map (1:1250 or 1:2500 scale) of your locality for this work. The OS website lists the range of maps. Your LEA geography inspector may also be able to help.

One way of organising the work is to divide your locality into small areas so that groups of 'surveyors' can map a manageable area under close adult supervision. Decide on a range of categories and associated key colours, such as 'transport', 'housing' and 'leisure areas', before going out. This initial discussion can draw on pupils' existing knowledge and provides a real

Photos | John Halocha.

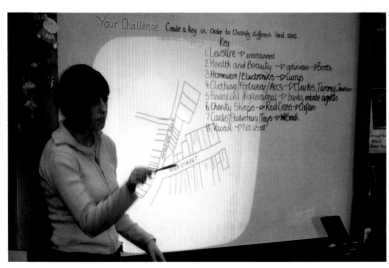

insight into their understanding of what happens in their home area. Recently, I heard a ten year old explain that a corner site was now derelict because it belonged to a bank which had merged and now they only needed one site in the high street. Clearly a young geographer!

When you have completed your land use survey you can go on to make further investigations. For example: How much land is given over to transport? Why are all the estate agents near each other? What might a land use survey in ten or twenty years' time show? ICT can be used to collect photographic evidence to include with your final land use map. These resources will make an excellent interactive display in your class and around the school. Finally, do ask your geography subject leader to keep these completed materials. They will provide your future colleagues in the school with an excellent historical resource to use as the basis for their own enquiries. For many more ideas on doing land use surveys see *Primary Geographer* (Geographical Association, 1997).

Currently, some academic geographers are researching how people interpret where they live (Holloway and Hubbard, 2001). Older people often have fascinating insights into what happens in settlements and how they have

changed over time. If you can plan a geographical enquiry that includes a study of how your school locality has changed, try to arrange a visit from a local resident who has known your area for many years. They may have old photographs, but if not, arrange for your pupils to find old photographs and, ideally, maps to have ready to discuss with your visitor. Ensure that questions and discussion ideas are carefully planned. If you are lucky enough to have two visitors who can come at different times, your class may find that they present two rather different versions of what happened in the past.

Activities of this kind will introduce young people to the concept that people see settlements in different ways depending on who they are, their background, family history and interests. In this way you will be helping children to understand that settlements are so much more than just the physical features we find in them; they include all the experiences, hopes, fears, pleasures, arguments, memories, beliefs and values of the people who live in them – and this is true for settlements all over the world.

References and further reading

Asquith, S. (1999) 'Questioning Kosovo', *Primary Geographer*, 39, pp. 4-7.

Baldwin, H. and Opie, M. (1996) 'Child's eye views of cities', *Primary Geographer*, 26, pp. 16-17, 20.

Birch, B. (1995) *The Village in the Forest by the Sea.* London: Bodley Head.

Catling, S. (2002) *Placing Places.* Sheffield: Geographical Association.

Chambers, B. and Garner, W. (2001) 'Sustainable cities', *Primary Geographer*, 43, pp. 4-6.

Durbin, C. (2001) 'Urban myths', *Primary Geographer*, 45, pp. 14-16.

Environment Agency website: http://www. environment-agency.gov.uk

Ferris, S. (1998) *Children's Contrasting Perceptions of City and Countryside in a Rural and an Urban School at Key Stage 1.* Unpublished dissertation, Liverpool Hope University College.

Geographical Association (1997) *Primary Geographer, 30: Focus on land use.* Sheffield: Geographical Association.

Halocha, J. (2003) 'It's a small world', *Primary Geographer*, 51, pp. 18-19.

Hicks, D. and Holden, C. (1995) *Visions of the Future: Why we need to teach for tomorrow.* Stoke on Trent: Trentham Books.

Holloway, L. and Hubbard, P. (2001) *People and Place: The extraordinary geographies of everyday life.* London: Prentice Hall.

Massey, D. (1999) 'The social place', *Primary Geographer*, 37, pp. 4-6.

Ordnance Survey website: http://www.ordnance survey.co.uk

Plester, B., Richards, J., Shevelan, C., Blades, M. and Spencer, C. (2003) 'Hunt from above', *Primary Geographer*, 51, pp. 20-21.

we use water for..

growing things

drinking

having a bath

cleaning things

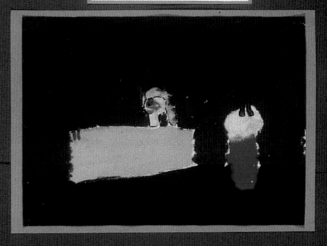

flushing the toilet

having a shower

cooking, drinking, washing up

Environmental geography

What is our knowledge worth if we know nothing about the world that sustains us, nothing about natural systems and climate, nothing about other people and cultures? This question by Jonathan Porritt, the environmental campaigner (see Foreword), appears on the first page of the geography national curriculum (DfEE/QCA, 1999). It draws attention to the fundamental importance of an environmental perspective in the modern world. Whether we admit it or not, we live in times of unprecedented change. For the first time ever people have the power to change the course of nature. Nobody can accurately predict the impact of our actions on the future of the Earth. What we know for certain is that we are having a major impact on the atmosphere, land and oceans and are affecting the balance of life on a planetary scale. It will fall to future generations to grapple with the problems that we are only dimly recognising today.

This chapter offers practical help on how environmental geography may be addressed within the context of good geography teaching. It lays emphasis on practical approaches and offers a number of case studies as templates for teachers to adapt to their own circumstances/issues/localities, including one of an eco-school. Environmental geography provides many exciting opportunities for engaging pupils' curiosity and interest in the world about them. However, to capitalise on this potential the national curriculum and Foundation Stage requirements should be regarded more as starting points than as ends in themselves. The following considerations should underpin work in environmental geography for all ages:

1. Fieldwork is fundamental to all geographical education and involving pupils in activities outside the classroom is especially relevant to the effective delivery of environmental geography.

2. Environmental geography is an extremely useful vehicle for appreciating other people's attitudes and values and for encouraging pupils to express their own opinions, develop tolerance of alternative opinions and their own ability to analyse and reconsider their own stance on a particular issue.

3. Unique opportunities are offered within environmental geography to cover issues in the context of other countries, thereby raising pupils' awareness of other cultures and values.

4. Environmental geography also offers a means by which the current citizenship and creativity agendas may be effectively addressed within a geographical context, thereby raising the profile of geography (see Chapters 2 and 21).

What is environmental geography?

Environmental geography specifically examines the interactions between people and their environments, studying both how people adapt to, and how they alter, specific environments. Studying environmental geography involves using geographical skills, knowledge and understanding to explore issues affecting the quality of life on Earth. It also involves considering attitudes, values and opinions and thereby provides an ideal opportunity for pupils of all ages and abilities to contribute to real issues that are of concern to them.

The flavour of environmental geography can be seen from the level descriptions for geography:

■ At level one pupils 'express their views on features of the environment of a locality'
■ At level two pupils 'recognise how people affect the environment'
■ At level three pupils 'recognise how people seek to improve and sustain environments'
■ At level four pupils 'explain their own views and the views that other people hold about an environmental change' (DfEE/QCA, 1999, p. 31).

Environmental geography should also be seen as one of the components of education for sustainable development. Since its inception the geography national curriculum has recognised the strong links between environmental education and environmental geography. In the Foundation Stage young children may be introduced to environmental geography in a number of ways, e.g. via play activities. Examples of this approach may be found in *The Early Years Handbook* (De Bóo, 2004)

There is clear potential for using environmental education topics as a vehicle for delivering all national curriculum subjects and particular aspects of literacy and numeracy. However, it is essential that teachers devise clear schemes identifying the geography within any environmental education taught, otherwise, it is all too easy to lose the focus and drift into other subject areas. Teachers (and indeed pupils) must be clear in their own minds where the geography appears in any activity they undertake.

Delivering environmental geography

Some general principles

One way of studying environmental geography is to start with the pupils' own environment and expand out to the wider world. All pupils need to study their local area, both as part of the Foundation Stage and key stages 1 and 2. The study of more distant environments as demanded in key stage 2 does not preclude their inclusion earlier in the curriculum.

The local environment is a primary resource and has the unique quality that pupils can interact with it. It has the advantages of being:

■ free
■ accessible
■ familiar

Local people are usually willing to be involved in pupils' activities. Indeed, developing links with local residents is an important and often overlooked aspect of developing environmental awareness, in addition to being a significant way in which geography can contribute to the citizenship agenda.

Working in the local area, pupils can collect their own information on a particular environmental issue through questionnaires, tape recordings, surveys, field sketches, photographs and other recording techniques. This has the advantage of developing both their enquiry skills and additional transferable skills such as those relating to ICT. When they study issues in more distant localities pupils will, of necessity, need to use secondary sources such as CD-Roms, atlases, photographs, newspapers, magazines, links with other schools and the internet. There are many ways in which the information may be utilised, for example:

Which of these features do you have?

Ideally you need at least one from each category for local area work for children aged 5-11. A starter vocabulary is also essential for local area work.

Water features
Stream, pond or lake, river, estuary, coastal area

Landscape features
Hills, valleys, cliffs, mountains (showing evidence of erosion or deposition by water, wind or ice), woods, moorland

Physical features
Slopes, soil, rocks

Climate work sites
For weather surveys, micro-climate work (usually school grounds)

Local issues
Bypass, road-widening scheme, out-of-town shopping development, new housing estate, new reservoir, rubbish tip site, local improvement scheme

Sites showing the origins of settlement
Crossing point of a river, a route centre, a defensive site, site where water became available, old core of modern settlement, evidence of growth, development, decline

Buildings
House, cottage, rows of houses, housing estates, groups, rows of buildings with different functions

Transport
Safe place for traffic survey, bus station, bypass, airport, railway station

Industry
Farm, business, small manufacturing unit, warehouse, factory

Shops
Single shop, parade of shops, supermarket, hypermarket, shopping mall/centre

Leisure facilities
Library, museum, park, swimming pool, leisure centre, golf course

Settlements
House, hamlet, village, town, city (including suburbs)

Services
Fire, police, ambulance, hospital, doctor, dentist, refuse collection, recycling plant, post office

Figure 1 | Local features to study at key stages 1 and 2. Source: NCC, 1993.

- the creation of databases to manipulate quantitative data
- discussions based upon qualitative data with general trends identified
- analysis of visual material

Studying an environmental issue using information and data collected by the pupils themselves enables them to understand the issue better, how it affects both people and the environment, and how different people may hold different views from their own.

The enquiry process provides the most appropriate and most effective ways of delivering geography. This is recognised in the curriculum guidance for the Foundation Stage and in the geography national curriculum which contains sections on geographical enquiry and skills at all key stages. The strength of this methodology is that it can effectively deliver almost any aspect of geography, including environmental geography (Figure 1).

Environmental geography is not just about collecting data and information on an issue, it is also about getting pupils to communicate their ideas on that issue. This may be achieved through such activities as:

- the production of a newspaper
- role play within the class or for wider audiences, e.g. there is an opportunity here for a special assembly or a presentation at a parents' evening
- interviews with people involved in particular environmental issues
- drawings, photographs and other art work to produce exciting wall displays, which offers the opportunity to use ICT
- the production of a simulated television programme
- the creation of a web page/site
- using stories – the pupils could either be introduced to an issue via stories such as *Oi! Get off my Train* (Burningham, 1994) and *Peter's Place* (Grindley, 2001), or encouraged to write one of their own.

Possible issues/topics for younger pupils

There are innumerable opportunities for teaching environmental geography throughout the

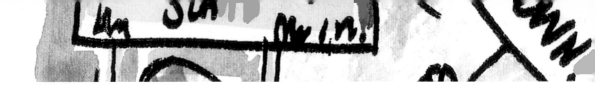

school. The suggestions given below could be adapted and developed for all age groups but are most suitable for the Foundation Stage and key stage one.

Looking after our school

Pupils could be actively encouraged to express opinions about what they feel is attractive and unattractive about their school and its grounds. Through writing and drawings they can devise plans for improvement. This theme could be part of an eco-schools development (see pp. 282).

Setting up a school council provides the ideal forum for pupils to share ideas of this kind, to plan their campaigns, and to give impetus to projects which involve pupils in making changes to their environment – a real lesson in citizenship.

Caring for animals

You could begin by asking pupils to consider what their pets require to survive. This might lead to a farm study/visit where pupils formulate their own opinions on the way farm animals are kept. Clearly this study might then involve more distant places, with pupils studying endangered wildlife habitats and species.

Looking after our water

Pupils could consider the various ways we use water and the absolute necessity for clean drinking water. This, of course, would naturally lead to a consideration of water supplies in other countries, encouraging children to empathise with those in distant lands and different cultures. Other activities for environmental geography include:

- Using stories to introduce or develop a theme, e.g. *The World that Jack Built* (Brown, 1991) and *Dinosaurs and all that Rubbish* (Foreman, 1972). The *Barnaby Bear* series (Jackson 2000-03; Lewis, 2001-03) from the Geographical Association also offers countless opportunities for very young children to explore environmental issues by following the adventures of this likeable bear
- Using poetry, e.g. *When Dad Cuts Down the Chestnut Tree* (Ayres, 1990)
- Creating a themed trail in a rural area (see the Ebchester School example on page 282)
- Creating a themed trail in an urban area
- Making a play area which pupils build themselves using resources such as bark chippings. (This has been successfully introduced into a South Tyneside school to avoid mud being brought into the school buildings.)
- Designing a butterfly garden with the help of older pupils
- Making nesting boxes for birds and siting them in appropriate areas
- Making attractive fences and boundaries. (A school fence bordering the Tyneside Metro was transformed by a series of murals, with help from the Passenger Transport Authority and the Groundwork Trust.)
- Improving the inside of the school with, for example, plants, signs, sculptures, paintings designed and produced by the pupils
- Transforming the playground by introducing street furniture (Tyneside Groundwork Trust has developed several playgrounds in this way. Schools have found that this simple approach has reduced bullying in the playground.)

The above ideas may be regarded as mini-projects and may provide a sound basis for further development in key stage 2.

Possible issues/topics for older pupils

The key stage 2 programme of study for geography requires pupils to learn about two localities and three themes. One of these themes has an environmental focus, with the precise requirements that pupils will study:

■ 'an environmental issue, caused by change in the environment (for example, increasing traffic congestion, hedgerow loss, drought),

■ attempts to manage the environment sustainability (for example, by improving public transport, creating new nature reserves, reducing water use)' (QCA/DfEE, 1999, p. 114).

Using stories to introduce and consolidate studies in environmental geography should continue at key stage 2. The *Storylink* packs (Dryden and Hare, 2000; 2001) provide an exciting range of ideas and approaches to integrating fiction with locality studies that could be adapted to suit different needs. Sometimes we forget that older children enjoy stories too!

The local area will often provide some excellent issues that pupils can investigate first-hand. Chambers (1995) describes a number of projects that explore children's environmental perceptions and it is always valuable to find out what children think about the place where they live. Some specific topics/ideas which could be developed and which involve a contrasting locality are suggested below, followed by two in-depth case studies. These ideas should not be regarded as being exclusively for key stage 2 – they can be readily modified for all age ranges.

Tourism

There are many examples of how tourism has had both a positive and a negative impact on the environment and these can be used to raise a variety of geographical issues. For example:

Intense visitor pressure on National Parks and mountain environments – People who visit such areas often contribute to the destruction of the very thing they have come to experience, i.e. unspoilt landscape, peace and tranquillity.

Pressures of tourism – This is well exemplified by, e.g. southern Spain, where small coastal fishing villages have been inundated by numerous foreign visitors who stay in the large tower-block apartments built to accommodate them. Torremolinos is an excellent example.

By-pass construction – In 1996 the Newbury by-pass highlighted the contentious nature of this issue and the opposing views of various factions in the locality. There are many examples of where local road building schemes have raised opposing feelings.

Rivers/coasts – The abuse of our rivers and coastal waters through misuse can, unfortunately, be all too readily studied, e.g. oil spills such as the disaster off the coast of north-west France in 2002, agricultural waste, sewage waste and chemical discharges.

Green belt land – Building on green belt land is a dilemma facing most urban areas and one that impinges directly on rural areas.

Quarrying – Depending on the school's locality, resources and access to sites, quarrying can be the focus of geography topics set in a variety of place contexts, for example:

■ an open-cast coal mine within a predominantly agricultural landscape (see page 280);

■ limestone quarrying within the Peak District National Park;

■ chalk quarrying in southern England;

■ precious-metal mining in an Equatorial rainforest, such as the Amazonian rainforest in Brazil;

■ iron-ore mining in Scandinavia.

Trees in our world

Trees provide a familiar but extremely useful focus for an exciting and wide-ranging environmental geography topic. Teachers could approach this in a number of ways but some possible aims could be:

- raising awareness of the value of trees to the environment, e.g. for landscaping, the production of oxygen and as a source of alternative medicines;
- allowing pupils to express and develop their own opinions on forest-related issues, e.g. coniferous monoculture, rainforest decimation;
- appreciating that there are different attitudes relating to the value of trees.

You might start the work with a local study. This could involve:

- mapping the school campus and/or local area to show where trees already exist, thereby raising pupils' awareness of trees and their local area;
- a visit to either a local park or woodland and/or, through a 'local expert', enabling pupils to identify suitable varieties of tree for their school, thereby increasing their knowledge and enabling them to make informed decisions;
- identifying, through discussion, the opportunities for enhancement of the environment by tree planting, thereby offering pupils an opportunity to express their opinions and ideas.

Armed with this knowledge and the opinions of a variety of people consulted by the pupils, the project could be advanced using questions such as:

Focus question	Teaching/learning outcomes
Do we want trees?	Enables the children to consider the value of trees
Why do we want them?	Offers the opportunity to express likes and dislikes
What trees will we plant?	Enables them to use their knowledge or highlights the need to find out
Which trees will grow in our environment?	This links to a consideration of weather and climate and so to other aspects of geography
How do we look after them?	Enables pupils to become conversant with how trees survive
How do we get the trees we want?	Issues of availability and scarcity
Who will plant them?	Possible involvement of the local community
How will we plant them?	Practical considerations of facilitating a change in the environment
Where will we plant them?	An excellent opportunity for pupils to produce alternative plans for the location and to present these so that a final decision can be made
Who will look after them?	Aspects of responsibility
How do we pay for them?	Considerations of costs and value

If these questions were to be formulated into a scheme to plant more trees in the school grounds, in order to achieve their aim pupils would need to develop their own ideas for raising the money, perhaps by writing letters to appropriate organisations seeking sponsorship, and organising money-raising events such as jumble sales, sponsored walks, 'swimathons' and quizzes. This could involve them in a series of community activities which would forge, develop and, it is hoped, sustain the community/school links. These activities could culminate in a high-profile event where a local dignitary takes part in the tree planting ceremony. The links between these activities and the citizenship agenda are very obvious, yet this project has clear geographical elements.

This tree project can be developed further to bring in studies of distant and contrasting localities. For example:

■ the destruction of rainforests – a case study of the Amazon or the Indonesian forests would be particularly useful here

■ the creation of a coniferous forest monoculture – as in the creation of coniferous forests in parts of Scotland.

Such an approach would require reference to secondary source material to resource the activities. This can be obtained from various organisations (for some suggestions see the list at the end of this chapter), many of whom have 'tailor-made' materials for just such a topic. Using such materials, pupils might be encouraged to ask and find answers to questions such as:

Focus question	Teaching/learning outcomes
What and where are the rainforests?	Raising awareness of geographic location
What is happening to them?	Understanding of the issues
Is this a problem? For whom?	Empathising with the local people
How can we influence what is happening there?	This raises major issues of empowerment.
What is it like to live here?	This would encourage pupils to empathise with a different culture

Once again, conducting this study within the context of a distant locality is best achieved by using an enquiry approach. Traditional tales such as *The People Who Hugged the Trees* (Rose, 1991) could also be used to amplify the work.

***Photo** | Anna Gunby.*

Case study: open-cast mining in Northumberland

Open-cast coal mining in Northumberland is a fascinating topic, highlighting many issues relating to environmental change in a UK locality. A template for making a study of this issue is outlined below. You could apply the same approach to a study of mining in a distant locality, such as the African goldfields or the Brazilian rainforest.

Possible aims:

- raising awareness of the impact of open-cast coal mining on the environment;
- allowing pupils to express and develop their own opinions on a contentious issue;
- appreciating that there are different attitudes relating to open-cast mining in a rural area;
- development of a variety of geographical skills and techniques.

You could focus the study with the following enquiry questions:

Focus question	Teaching/learning outcomes
What types of coal mining are there?	Increasing pupils' knowledge
Why is open-cast coal mining being carried out specifically in Northumberland?	Raises awareness of specific local conditions
What does an open cast coal mine look like?	Offers the opportunity for investigation/visit
How many jobs does the industry provide and what do these jobs entail?	Puts the study into a wider context
Is the quality of the environment more important than the provision of jobs for local people?	Asks pupils to consider the impact of closing the industry
To what extent do people living around the area of coal extraction benefit or suffer from this activity and in what ways?	Empathising with different groups
To what extent do the owners of the mine ensure minimum impact in the locality?	Understanding how environmental impact may be addressed
What representation have the local people made about the effects of this industry, and to whom?	Recognition of the role of local decision making/politics
How is the landscape reinstated after mining?	Understanding of techniques of reclamation
Is the reinstated landscape better than the original?	Asking pupils to express views
Has the land use changed?	Enables the impact of change to be assessed

Suggested activities to develop the work include:

- Whole-class discussions on the nature of coal and how it may be extracted. These discussions may be reinforced through pupils' own research on coal using books, CD-Roms and information from coal companies.
- The use of map skills involving the use of acetate overlays to delineate the mining area and to show the extent and nature of the development. Pupils can then draw their own conclusions about land use before mining took place and develop their own opinions as to the impact the mining has had on this landscape. They can also identify the dwellings and settlements in the immediate vicinity of the coal mines and can imagine what it would be like to live in that locality.

- A site visit. Before the site visit, pupils could work in groups to develop questions that they might choose to ask their guide. During the visit they could take photographs, make field sketches and make tape recordings (e.g. of their interviews or the noise of lorries).

- Part of the visit may also include a period of time spent in one of the settlements close to the mine. Pupils could collect a variety of additional information that might include, for example, interviews with local inhabitants, observations on the visual and aural impact of the mine, a dust survey (using dust plates) and a traffic count. The use of environmental recording devices could enhance this study and address ICT requirements.

- If a site visit is not possible, secondary sources such as photographs and videos could be used.

Follow-up activities might include:

- Making a display of photographs, maps, newspaper cuttings, field sketches and graphs of the results of the surveys. Pupils can use these as a resource and as a focus for the classroom activities.

- Having a class discussion on what pupils have discovered about people's attitudes towards open-cast mining.

- Creative writing activities in which pupils express their opinions and attitudes about what they have seen and interacted with. This could take the form of the front page of a newspaper and involve the use of a desk-top publishing package.

- Designing and mapping a landscape that would enhance the environment after the mine has closed. Pupils may choose to transform the hole into either a water park with wildlife habitats or a lake for recreational purposes; they may wish to return the area to an agricultural land use or develop a country park with woodland walks/mountain bike tracks with all the attendant facilities.

- A role play exercise centred on a proposal to start an open-cast coal mine in a new area, close to the school. The cast of characters in this scenario could include:

  - a local MP
  - a town planner
  - a conservation group representative
  - members of the parish council
  - a village shopkeeper

  - an unemployed parent of three children
  - an eight- or nine-year-old child
  - the mine owner
  - the haulage contractor
  - a farmer

This role play could take the form of a studio discussion for a television programme. It might be possible for a person from the local planning department to chair the meeting and draw together the threads of the debate. Such a person would be able to provide pupils with a considered decision as to whether or not the open-cast coal mine should go ahead. If you video the role play you could use it to stimulate discussion concerning the issues raised by open-cast mining.

Photo | *I.A. Recordings photographic archive.*

Photo | *Tina Horler.*

Rainforests – These provide a topical, though perhaps overused, context within which pupils can learn about our inability to sustain natural biomes. Mining and quarrying of minerals and metals in such an environment raise issues of sustainability and conflict. This opens up a wealth of views about the development of these distinctive regions.

As well as covering the environmental theme in the geography national curriculum, any work that you do on the environment will involve working at a range of scales, which is one of the other curriculum requirements. If you select your case studies carefully you may also be able to focus on a locality in a less developed country, thereby linking skills, places and themes in the way that was originally intended by the team that devised the curriculum.

The whole-school approach: a case study of an eco-school

Environmental geography can be taught using topics, themes and issues approaches but some of the most effective and integrated learning can be developed by taking a whole-school approach. Ebchester Church of England Primary School in County Durham provides an excellent example of how environmental geography may be delivered in holistic fashion. The head teacher, Richard Coombes, realised that the school environment, in particular the woodland, the stream and a substantial open grassed area, offered rich learning opportunities.

His enthusiasm encouraged parents, unemployed local people, local businesses and the pupils to co-operate as a cohesive body. Over a period of twelve years adults and children have worked together to conserve and develop wildlife habitats in the school grounds. This has provided an exciting, ongoing learning experience for all those involved, forging strong links within the community which continue to flourish and develop.

At the start each class was given an area to develop. It was their task to ensure the continued progress of the project, which they did through discussion and making drawings and plans. They also identified improvements that would enhance their area in the long term. This initial planning stage then led to decisions about what was wanted and final plans were produced. The plans provided the basis for pupils to communicate their ideas to those people who could help bring them to fruition, e.g. parents, local businesses, volunteers and experts. (Note that communication of this kind can involve the use of ICT in a variety of forms, e.g. PowerPoint presentations.) So far the projects undertaken at the school have included the following:

■ woodland area

■ weather station

■ pond

Photo | *Paula Richardson.*

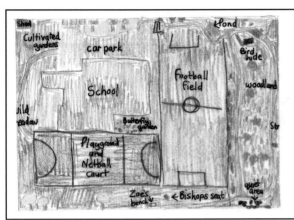

Figure 2 | *Trails devised by pupils at Ebchester CE Primary School.*

- amphitheatre
- sandpit area
- butterfly garden
- herb garden
- structured play area
- wild flower meadow
- memorial garden
- compost area
- organic vegetable garden (produce from which is prepared in the school dining room)

Each year new projects and initiatives are discussed and developed. One recent development has been the construction of an outdoor amphitheatre in which school productions can be staged. The theatre occupies what was formerly a piece of waste ground which presented a potential hazard to pupils playing near it due to its steepness. Now it offers a facility that is highly valued and utilised.

Making trails

Different areas of the school grounds, such as the butterfly garden and woodland area have been linked together by the creation of all-weather paths made from renewable materials. This has improved access to and between the areas throughout the year. The trails (e.g. a bird trail and a tree trail) and the various site developments were all planned by the pupils who used their own knowledge and imagination to draw the plans, with the help of local experts (see Figure 2).

Attracting visitors

Initially, the project was quite parochial but as the plans came to fruition Ebchester School began to involve other schools which resulted in considerable valuable and enjoyable interaction. Some pupils were appointed as wardens to escort visitors around the site and were given a warden's badge. This gave pupils:

- the opportunity to communicate information
- confidence
- the chance to express themselves

Worksheets were also developed by the pupils themselves for use by visitors from other schools (Figure 3). In devising the sheets the pupils used a great range of geographical skills including learning geographical terms, making maps and plans and using symbols and keys. One of the other benefits is that they developed a better sense of their own geographical situation which helped them when they studied contrasting localities.

Today, Ebchester School has a curriculum which, for the most part, can be covered within the boundaries of the pupils' own environment and, as a consequence, enables all subjects within the national curriculum to be addressed in a practical and meaningful way. In addition to subject learning, the work has enhanced pupils' self-esteem. As the head teacher puts it, 'All of the children benefit from their interactions with children from visiting schools and take great pride in sharing their school with others.'

FIND THREE COLOURS

What To Do

1) Take a little bit of the three colours you want.

2) Stick them on the three bits of sticky tape.

3) Tell me where you found them.

_____ _____ _____

4) Draw the whole object.

1	2	3

CREATURES

Go down to the woodland and see what insects are down there. Draw the insects that you have seen.

Pick out which one you like the best.

Next, draw which one you picked out.

Cut out the pictures of the insects.

Put them on a piece of paper.

Write down the names of the insects.

Figure 3 | Worksheet for visitors, produced by pupils at Ebchester CE Primary School.

In a project like this, involving the lower school a   e can ensure the continuity of all the activities. The secret of success at Ebchester has been the fact that this is a long-term, ongoing developmental process, which changes in emphasis from year to year because of variations in, for example:

- the pupils' interests and enthusiasm
- availability of resources
- expert help
- parental input
- curriculum developments
- contemporary environmental issues

The pupils have been central to the decision-making process and learnt a great deal from the interaction with adults and from planning the developments. This is an excellent example of the value of empowering pupils by giving them ownership of a project. This ownership also seems to stay with the pupils as the incidence of vandalism at the school is virtually zero and the older pupils who have moved on to secondary school still see the school as 'theirs'.

Ebchester School has now attained Eco-school status. This scheme promotes environmental education awareness in a way that links to many curriculum aspects including citizenship, personal, social and health education (PSHE) and, particularly significant for geography, sustainable development. Richard Coombes feels that gaining this status is recognition of the school's long-standing record of environmental work. The experiences at Ebchester School show how delivering environmental geography may be done in an imaginative and stimulating manner. It is possible to adapt this approach to your school situation and offer your pupils benefits that amply justify the time and effort involved.

Teaching environmental geography – some considerations

The very nature of environmental geography means that many of the topics, such as open-cast mining or the development of nuclear power, are likely to be contentious. Feelings about these issues can run very high and they need to be handled sensitively. Teachers therefore have a critical part to play in providing role models and it is essential that you are seen to be impartial and even-handed in dealing with the issue. It is also important that issues which create concern in young children, such as the possible extinction of animal species, are treated with particular care and that pupils are offered a final positive perspective.

Misconceptions

As in any subject, pupils are likely to suffer from a considerable number of misconceptions. In geography, these are especially significant when it comes to environmental issues as they may arise from, or lead to, prejudices. This was clearly illustrated by Palmer (1998) who was able to show that while children may be able to recognise and name unfamiliar creatures and landscapes this does not necessarily mean they understand their context or existence. Children's confusion about environmental issues can be considered as a mix of, among other things, factual errors, confusion over cause and effect and sweeping generalisations that suggest

Photo | *John Halocha.*

anything which is environmentally damaging contributes to every environmental problem.

The recent controversy over the fish stocks in the North Sea is a case in point and a neat way of introducing a European dimension. The issue revolves around the different perceptions of the viability and sustainability of the fish stock with the fishermen arguing for their right to fish in order to preserve their livelihood and way of life, while scientists argue that without severe limitation on catch size the future of the industry is in jeopardy. This is a complex issue and before it can be fully explored in the classroom some of the following initial misconceptions need to be addressed, for example:

1. Fish start and end their lives in a supermarket.
2. Commercial fishing is done with a rod and line.
3. Fish fingers are the fingers of fish.
4. That there will always be fish to eat (i.e. children take for granted the availability of a food supply).

The study of the North Sea fishing controversy offers pupils a real issue in which two opposing viewpoints are argued with both passion and evidence. The teacher's role in this study would be to facilitate the exploration of these opposing views and enable pupils to come to an understanding of the issue from which they could draw their own conclusions. The use of video clips, the internet and links with pupils in fishing ports are all ways in which this may be achieved.

A very positive and exciting way of overcoming misconceptions which may arise from (mis)interpretation of secondary sources, particularly when dealing with an issue in a distant place, would be to link up with a school in that country or locality. This can be achieved via the Windows on the World website (see useful contacts at end of chapter) which offers a means of 'pairing' schools. The exchange of on-the-spot information, opinions and perceptions would enrich pupils' learning in both participating schools and would make a considerable contribution to addressing the One World issue.

It is also very constructive to consider how other countries include environmental geography in their curriculum. In France, for example, environmental education is an important part of geography, science and civic education (citizenship) and there is a strong global perspective. This is perhaps something which needs to be further developed in the UK. One of the other central concepts is the idea of *patrimoine* or heritage. This addresses not only the protection of the natural environment, e.g. sustainability, but also the protection of relics of the past and the protection of the cultures of a group, a region or country. Older pupils in France are also taught about the role of the United Nations but, as in the UK, when studies of litter, waste, recycling and sorting are undertaken, the pupil tends to be the starting point.

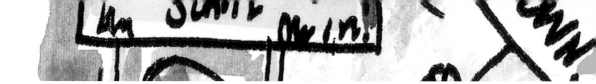

Conclusion

There is a wide range of resources available to enhance the teaching of environmental geography. To begin with, the local surroundings offers rich possibilities for fieldwork. Secondary sources include teaching packs and colourful and inviting information books. Many texts, however, focus on environmental education as opposed to environmental geography. It is therefore all too easy to be drawn into environmental education without fully appreciating the wealth of geographical opportunities that also lie along this route.

When it comes to finding out about issues, newspaper cuttings, local news footage, planning applications, information from pressure groups such as Greenpeace, and photographs collected over a period of time are all extremely valuable. The internet is another source to consider, especially if you can identify a few suitable sites in advance of lessons. One option is to get pupils to identify the issue they want to study for themselves and to organise appropriate resources. Flexibility and responsiveness on the part of the teacher is essential. It is important to remember that pupils need the opportunity to explore their own feelings, values and thoughts on environmental issues and to do so within a clear geographical framework. The best outcome of good teaching in environmental geography is perhaps that the pupils are able to feel that they can make a difference, no matter how small.

References and further reading

Alcock, K. (2001) *Our World: Early years activities to promote geographical understanding.* Dunstable: Belair.

Ayres, P. (1990) *When Dad Cuts Down the Chestnut Tree.* London: Walker Books.

Bowring-Carr, C. *et al.* (2000) *Transforming Schools Through Community Education.* Coventry: CEDC.

Brand, J. (1991) *The Green Umbrella.* Guildford: WWF/A&C Black.

Brown, R. (1991) *The World that Jack Built.* London: Red Fox.

Burningham, J. (1991) *Oi! Get off our Train.* London: Red Fox.

Chambers, B. (1995) *Awareness into Action: Environmental education in the primary school.* Sheffield Geographical Association.

Dean, K. and Jackson, E. (2003) 'A sense of place in play', *Primary Geographer,* 51, April 2003.

De Bóo, M. (ed) (2004) *The Early Years Handbook.* Sheffield: The Curriculum Partnership.

DfEE/QCA (1999) *The National Curriculum: Handbook for primary teachers in England.* London; DfEE/QCA.

Dryden, I. and Hare, R. (2001) *Storylink: Llandudno.* Sheffield: Geographical Association.

Dryden, I. and Hare, R. (2002) *Storylink: Kochi.* Sheffield: Geographical Association

Foreman, M. (1972) *Dinosaurs and all that Rubbish.* London: Hamish Hamilton.

Gadsden, A. (1991) *Geography and History Through Stories.* Sheffield: Geographical Association.

Grimwade, K. (ed) (2000) *Geography and the New Agenda: Citizenship, PSHE and sustainable development in the primary curriculum.* Sheffield: Geographical Association.

Grindley, S. (2001) *Peter's Place.* London: Andersen Press.

Hedderwick, M. (1994) *Katie Morag and the New Pier.* London: Random House.

Jackson, E. (2000-03) *Barnaby Bear* big books. Sheffield: Geographical Association.

Lewis, L. (2001-03) *Barnaby Bear* little books. Sheffield: Geographical Association.

NCC (1990) *Environmental Education: Curriculum guidance 7.* York: NCC.

NCC (1993) *Teaching Geography at Key Stages 1 and 2: An INSET guide.* York: NCC.

Palmer, J. (1998) 'Environmental cognition in young children' in Scoffham, S. (ed) *Primary Sources: Research findings in primary geography.* Sheffield: Geographical Association, pp. 32-4.

Rose, D.L.(1991) *The People Who Hugged the Trees.* Lanham, MD: Roberts Rinehart.

SCAA (1996) *Teaching Environmental Matters Through the National Curriculum.* London: SCAA.

Useful contacts

Centre for Alternative Technology, Llwyngwern Quarry, Machynlleth, Mid-Wales SY20 8DN.

Community Education Development Centre website: www.cedc.org.uk

Council for Environmental Education, School of Education, University of Reading, London Road, Reading RG1 5QA.

Council for the Protection of Rural England, Warwick House, 25 Buckingham Palace Road, London SW1 0PP.

Countryside Council for Wales, Maes-y-Ffynnon, Penrhosgarnedd, Bangor, Gwynedd LL57 2DW.

Eco-Schools website: http://www.eco-schools.org.uk

English Heritage Education Service, 23 Savile Row, London W1S 2ET.

English Nature, Environmental Education Section, Northminster House, Peterborough PE1 1UA.

Friends of the Earth, 26-28 Underwood Street, London N1 7TQ.

Geographical Association, 160 Solly Street, Sheffield S1 4BF.

The Green Teacher Co-operative Ltd, Swn Y Mor, Pen Yr Angor, Aberystwyth SY23 1BJ. Tel: 01970 626478. E-mail: djr28@tutor.open.ac.uk.

Learning Through Landscapes, 3rd Floor, Southside Offices, The Law Courts, Winchester SO23 9DL.

National Association for Environmental Education, Walsall Campus, University of Wolverhampton, Gorway, Walsall WS1 3BD.

Oxfam Education, 274 Banbury Road, Oxford OX2 7DZ.

Scottish Natural Heritage, 12 Hope Terrace, Edinburgh EH9 2AS.

Shell Better Britain Campaign, Red House, Hill Lane, Great Barr, Birmingham B43 6LZ.

Possible sources of sponsorship for grounds improvement

Groundwork Trust, 85/87 Cornwell Street, Birmingham B3 3BY.

The Tidy Britain Group, The Pier, Wigan, Lancashire WN3 4EX.

The Tree Council, 71 Newcomen Street, London SE1 1YT.

Watch Trust for Environmental Education, The Green, Witham Park, Lincoln LN5 7JR.

The Wildlife Trusts, Central Office, 47-51 Great Suffolk Street, London SE1 0BS.

WWF UK, Panda House, Weyside Park, Goldaming, Surrey GU7 1XR.

Acknowledgement

The authors wish to thank Richard Coombes and the staff and pupils at Ebchester Church of England Primary School, County Durham.

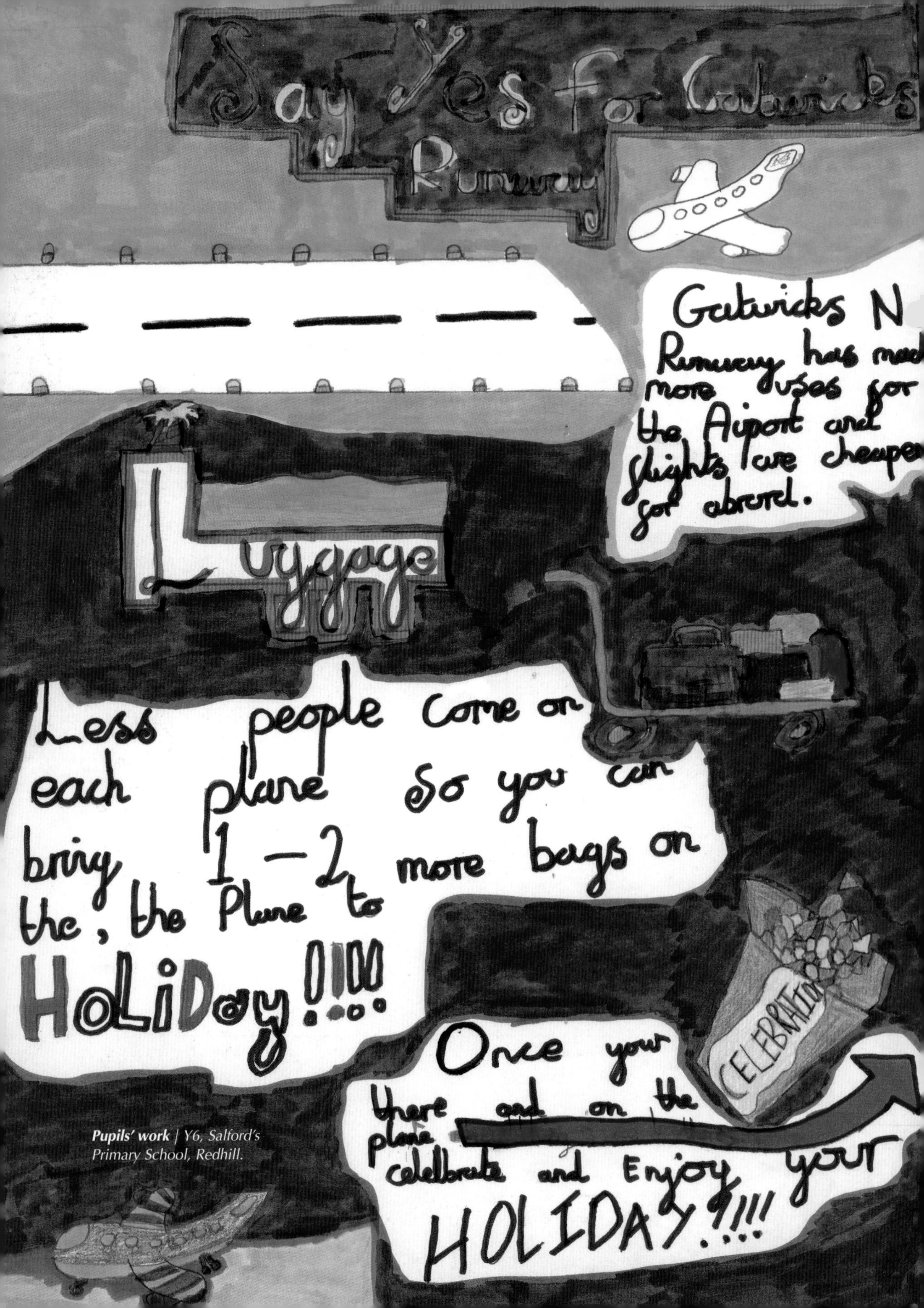

Say Yes for Gatwicks N Runway

Gatwicks N Runway has made more uses for the Airport and flights are cheaper for abroad.

Luggage

Less people come on each plane So you can bring 1 – 2 more bags on the , the Plane to HoLiDay !!!!

Once your there and on the plane celebrate and Enjoy your HOLIDAY !!!!

CELEBRATION

Pupils' work | Y6, Salford's Primary School, Redhill.

IN THIS CHAPTER YOU WILL FIND KEY IDEAS ON
CITIZENSHIP • DISTANT PLACES • ENQUIRIES • FIELDWORK • GOOD PRACTICE •
ISSUES • NATIONAL CURRICULUM • PROGRESSION

Citizenship, PSHE and primary geography

Many primary schools have citizenship and PSHE at the core of what they do. Preparing pupils for their lives now and in the future, developing their self-knowledge and their ability to understand and manage their feelings and relationship with others are all accorded high priority. Similarly, responsible behaviour, the importance of ethos, codes of conduct and the individual's role as an effective member of the school community are also emphasised.

There is a strong connection between primary geography, its content, teaching and learning strategies, and all aspects of children's lives and experiences. Not only is it difficult to separate them, it is also undesirable, as real life does not come in compartmentalised boxes. Our lives intertwine with those of others, both through space (locally, nationally, globally) and through time (past, present and future). As they plan their curriculum, schools need to take account of this link.

Primary geography investigates real contexts and real-life issues and situations. The primary geography curriculum encourages pupils to study places at a range of scales from local to global and to explore the interdependence of society, economy and the natural environment. Pupils study how people are influenced by and affect their environments and they are encouraged to develop a sense of responsibility for personal and group actions. As they explore 'live' issues pupils are encouraged to listen carefully to arguments from different viewpoints and to reflect on these points of view. Primary geography questions the values on which decisions are based and explores alternative ways of providing for human needs and wants.

Education for citizenship and PSHE also starts from the pupils and draws on their backgrounds, beliefs and concerns. It seeks to explore different values and attitudes, promote self-respect and tolerance of others, encourage respect for democracy, justice, law and human rights and develop social and moral responsibility. Pupils have the opportunity to compare values and beliefs, to examine evidence and opinions critically, to form conclusions, discuss differences and resolve conflicts. Real life and relevance helps to motivate the pupils, thereby creating more effective learning that involves communication, collaboration, self-directed learning, problem solving, researching and publishing findings. Pupils feel their work has purpose and a goal. Education for citizenship and PSHE involves not only pupils, but the whole school, their families and the communities in which they live, so strong links between home, school and community are essential.

Active learning

It is widely accepted that children learn in different ways (Gardner, 1993; 1999) so it makes sense to employ a wide range of methodologies in classroom teaching. Engaging pupils' interest through exciting, active learning is a very effective way of making the learning experience more memorable and effective. Government strategy emphasises this point:

Local Issues	■ Transport – development of by-pass, cycle routes, bridleways, pedestrian walks, safe routes to school ■ Development & redevelopment: Brown field/green field issues ■ Housing/industrial development ■ Siting of new supermarket ■ Waste – litter, vandalism etc., waste disposal, landfill sites, incineration, siting of collection points etc. ■ School grounds development ■ Children's playground issues
Global Issues	■ Fair trade (bananas, clothes, coffee etc.) ■ Global warming, destruction of rainforests ■ Energy resources ■ Poverty, hunger, homelessness, refugees ■ Human-rights, racism ■ Water resources ■ Waste

Figure 1 | *Exploring local and global issues.*

 *Primary education is a critical stage in children's development – it shapes them for life. As well as giving them the tools for learning, primary education is about children experiencing the joy of discovery, solving problems, being creative, developing their self-confidence as learners and maturing socially and emotionally* (DfES, 2003, p. 4).

If you want to promote active learning in geography, citizenship and PSHE there are a number of factors to take into account. In particular you need to:

■ Plan activities which help pupils to develop the knowledge, skills and understanding to enable them to act as responsible citizens

■ Investigate issues at a range of scales from the local to the global

■ Employ a wide variety of organisational strategies, from teacher-led, whole-class activities and discussion, to children working as individuals, in pairs and in groups

■ Use a number of different teaching and learning methods including role play, creative problem solving, critical thinking, communication, geographical enquiry including mapwork, fieldwork, analysis of photographs and links with the local community

Investigating issues

There are numerous issues which might be explored in a combined primary geography, citizenship and PSHE topic. Figure 1 indicates some of the possibilities.

Whether the issue is local, national or global, pupils will need to develop strategies to find out who the stakeholders are, who has the power to make the changes and who can influence the decision-makers (Figure 2). Pupils also need to think about who they feel they need to inform about the issue. The people will vary according to the issue.

Teaching activities

The following teaching activities illustrate how the principles outlined above can be applied in practice. They cover a range of topics, involve studies at different scales and are aimed at specific key stages (Figure 3). All the activities can be adapted and applied to other issues, not just the ones identified in this chapter.

Caring for the environment

Aim: To help the pupils investigate the impact of people's actions on the environment both locally and globally and on the lives of others as well as their own.

As a stimulus, read the story *The World that Jack Built* (Brown, 1991) which takes the format of the rhyme 'This is the house that Jack built' and focuses on industrial pollution.

Figure 2 | How investigating an issue can lead to change and action.

What is the issue?	Who has the power?	Who can influence the decision (stakeholders)?	What actions can the children take?
School grounds development	■ Governors of the school ■ Head teacher	■ Children, both teaching & non teaching staff, parents, local community	■ Research – collect evidence ■ Undertake an audit (e.g. green audit of school) ■ Collect information
Siting of factory, supermarket etc.	■ Board of Directors of a company ■ Managing director ■ Local councillors ■ Parish councillors ■ Planning Dept	■ Local people – individually, collectively ■ Voters, local constituents ■ Customers ■ Staff of a company ■ Shareholders of a company ■ Local action groups ■ Local media	■ Analysis and present data from evidence collected ■ Find out the opinions of others – questionnaires, debates ■ Inform interested parties (those whose will be affected) ■ Put forward the case in a rational, informed manner
Siting of new houses, new roads etc. – anything that impacts on the local community	■ Local Councillors ■ Parish councillors ■ Planning Dept ■ Highways Dept	■ Local people – individually, collectively ■ Voters, local constituents ■ Local action groups ■ Local media	■ Write letters in the persuasive or in the discussion genre to people who have the power to make changes ■ Draw up petitions ■ Make posters ■ Display their work in libraries/town halls/information centres etc.
Issues that impact on the whole country	■ Nationally – Government politicians	■ Voters ■ Action groups ■ Media – local/national	■ Inform people who need to know the opinions of those people who will be/are affected
Issues that impact on the whole of European life	■ Europe – European Parliament and European politicians	■ Voters ■ Action groups ■ Media – local/national/European	■ Contact media – local press ■ Invite an MP/local councillor into school ■ Hold sponsored events ■ Create and sell a fair trade directory for their area
Issues that impact on the world	■ Global/World leaders and politicians	■ Voters ■ Action groups ■ Media – local/national/world	

Activities:

■ Whole class: Select one of the pupils to take Jack's role. The rest of the class ask questions about his home, factory, attitude to the environment, the future, and so on. Is there anything Jack could do about the pollution? This activity helps to develop questioning skills and puts a pupil 'in role' in someone else's shoes to imagine the responses.

■ Use *Window* (Baker, 1991). This text-free picture book tells the story of the relentless encroachment of a town into the surrounding countryside. Through the window we see the changes in the boy's immediate environment, as a result of his own actions as well as the actions of his family and others.

KS1	Caring for the environment	**KS2**	Change in the local area
KS1	The world of work	**KS2**	New developments
KS1	Settlements	**KS2**	Waste management
KS1	An island development	**KS2**	Tourism and travel
KS1/2	Safe routes to school	**KS2**	A hotel in the rainforest
KS1/2	Fieldwork visits	**KS2**	Wants and needs
KS1&2	Visualisation and relaxation	**KS2**	Children's rights
KS1&2	Assemblies	**KS2**	Cultural diversity
		KS2	Going to secondary school

Figure 3 | Topics for key stages 1 and 2.

The world of work

Aim: To introduce pupils to the world of work

Activities:

■ Using the class role-play area, ask pupils to act out the different jobs done by adults (e.g. travel agent, post office counter clerk, shopkeeper, lorry driver). As they play encourage them to talk about their work, the clothes they have to wear and what they like and dislike about their job. Provide a dressing-up box, or simply hats, which the pupils wear when they are in role, to help bring the activity to life.

■ Give pupils a job with a geographical context to perform (e.g. booking a holiday, packing the suitcases and travelling to a certain destination). Ask them to choose a destination and to find out something about the place they want to visit, the type of clothes they will need and how they are going to get there.

■ Instruct each pupil in the group to take on a role and to behave in a certain way all the time (e.g. sulky, bossy, lazy, bad tempered, nasty, and so on). The pupils then explore, through role play, what would happen if they set about their given task as a group, but behaved in their allotted manner.

Settlements

Aim: To introduce pupils to the concepts of settlements and routeways.

Activities:

■ Introduce pupils to road playmats, construction kits and other similar resources.

■ Build a model village or town with certain buildings and features using a variety of materials (junk modelling as well as commercially produced equipment). How do you get from A to B? What is this building? What is it used for? Who works there?

An island development

Aim: To help pupils to investigate the impact of change on people's lives using *Katie Morag and the New Pier* (Hedderwick, 1994). As the new pier is built on the Isle of Struay, Katie Morag feels both excited at the prospect of seeing Granny Mainland more often and sad when she realises that her friend the ferryman will lose his job.

Activities:

■ After reading the book, ask pupils to discuss (in groups) who lives on the island and what work they do.

■ Compare the pictures on the first and the last two pages of the story. What things are changing on the island? Why? How do you feel about the changes? How do you think the islanders feel about the changes? Encourage pupils to express views about these changes.

Photo | Greg Walker.

- Using a writing frame put forward the case for and against building the new pier.
- In role (e.g. Struay parish councillor, islanders, ferryman, travel agent, Katie Morag, fisherman, conservationist) debate the issue of the new pier and the changes it will bring. Who is in favour of the changes and who is against?
- Discuss how the pier will change the lives of the islanders. Compare the jobs people did before and after its construction. Who is now better or worse off?
- Make a tourist map and brochures advertising a visit to the island covering a range of topics such as weather, landscape, buildings, jobs, wildlife and environmental issues.
- List the types of foods that are produced by the farmers and fishermen on the island. The pupils then write a menu for the ferryman's café using local produce from the farms and the sea, giving the dishes appropriate names. Discuss how you think visitors will respond to them.

Safe routes to school

Aims: To consider the impact of traffic on our lives, whether at a local, regional, national or global scale; to investigate issues of safety, pollution, health, management of scarce resources and sustainability from a survey on traffic; to find out how places are connected and how people move between those places.

Activities:

- Group work: Devise a questionnaire to find out how most pupils travel to school. Ask them to carry out a survey and then analyse the data (use ICT if appropriate).
- Individual work: Using a local map (use Local Studies software or pictorial maps with younger children), ask each pupil to mark his/her route to and from school showing the places where they cross a road with an 'X'. Now get them to conduct a safety audit (risk assessment) of the route they take and means of travel that they use. Can they identify danger spots (e.g. junctions, corners, parked cars), or hazards (cracked pavements, potholes, bad parking, speeding)? Are there any hazards that affect a particular group of people (disabled, elderly, people with pushchairs)?
- Group work: Ask pupils to look at each other's routes to school and the danger points. Discuss how the danger could be reduced.
- Group work: Ask the pupils to plan a safe walking route with safer crossing places (pelican and zebra crossings, bridges, islands, school crossing patrol points) and measures to reduce traffic speed (speed humps, speed restrictions). Ask your local Road Safety Officer to come to school to work with pupils on the project.
- Involve local councillors and show them the safe routes, explaining the danger spots. If pupils feel it is appropriate, they could ask the councillors to act on their findings.

Fieldwork visits

Aims: To focus on specific geographical themes or topics (fieldwork, contrasting locality, rivers, settlements); to develop investigative skills; to build co-operative and team-work skills and to encourage greater independence and self-confidence.

Activities:

- Whole class: Before the visit, ask pupils to reflect on the environment in which the visit is taking place, using geographical enquiry questions (Where is this place? What is this place like?). Exchange ideas, discuss and anticipate areas of danger.

Categories	Possible issues and problems
■ Pollution	■ litter and dog mess – health hazards ■ exhaust fumes – health problems ■ CFCs (chlorofluorocarbons) – destroying ozone layer – cancer
■ Conservation of energy	■ use of exhaustible fossil fuels (coal, oil, gas) ■ excessive burning of fossil fuels – acid rain, green house effect, global warming ■ use of nuclear fuels
■ Waste	■ wasting energy in the home ■ disposable society ■ use of plastics and material which is not biodegradable ■ unhealthy and unsightly disposal of plastics
■ Protection of wildlife and nature	■ destruction of woodlands and rainforests ■ destruction of animal habitats

Figure 4 | *Local environmental issues fall mainly into one of four categories.*

- Group work: Ask pupils to design their own rules and own codes of practice for the visit.
- Whole class: Discuss emergency first aid procedures with personnel from St John's Ambulance.
- Group work: List the contents of a first aid box. Get pupils to design one of their own.
- Group work: Find out about the weather data for the place to be visited, given the time of the year, as well as the terrain, and likely fieldwork activities. Either develop a kit list for the visit and pack a bag, or describe what will be needed.

Change in the local area

Aim: To make pupils aware that they can make important contributions to environmental issues and to remedying problems.
Activities:

- Whole class: Exchange ideas about a range of environmental problems affecting your locality. The issues pupils raise mainly fall into four categories, all of which will create a link between geography, citizenship and PSHE (Figure 4).
- Whole class: Discuss what is likely to happen in five, ten and fifteen years' time. What would the pupils like to happen? If the future is to be different, what needs to change?
- Group work: Ask pupils to discuss and list the actions they could take to help ensure the future quality of the Earth and the health and safety of the people who live there.

New developments

Aims: To explore different ways to improve the area (e.g. derelict site), by looking at different views of the issue and its resolution; to investigate how developments affect local people, for example, the impact on existing shopkeepers of building a new shopping mall; to encourage pupils actively to engage in the improvement of their local environment.
Activities:

- Introduce pupils to the Development Compass Rose (see p. 210). This framework helps pupils to discuss any issue, resource, photograph or artefact in a balanced manner at either a local, regional, national or international scale.
- Get pupils to debate the development by putting them in role (e.g. developers, parents, children, supermarket company, local wildlife trust, local councillor, house-building company). Give the pupils role cards with the different opinions, ensuring there is a strong emphasis on health and safety issues (e.g. dangerous unlit site, stranger danger, traffic,

Photo | *Anna Gunby.*

Enquiry questions	Learning Activities
■ Domestic waste: What sort of waste do we produce at home and school? ■ What sort of waste is found in a modern dustbin? ■ Do we produce less/more/ the same/different waste than in 1900? ■ Why?	**In groups:** Pupils design a questionnaire to use at home and at school Research work to find out the percentage by weight and volume of rubbish found in a modern bin. Comparison of above information with the same information from 1900.
■ Is all the packaging used on goods really necessary? ■ How do supermarkets and markets dispose of their waste? ■ How much waste and packaging is there in a packed lunch?	**Whole class:** Visit local supermarkets and markets to find out about how they dispose of their waste. Look at the compacters. **In groups:** Undertake a waste audit. Write ten 'top tips' for reducing waste in the classroom. Design a poster to encourage children who bring packed lunch to school to use reusable containers and packaging.
■ Industrial Waste: Why do we produce more hazardous industrial waste today than we did in 1900? ■ What are the problems of disposing of industrial waste?	**Whole class:** Arrange for a spokesperson from a local place of employment to talk about their efforts to minimise the effect of their waste on the atmosphere, rivers and so on.
■ Where is our nearest landfill site? ■ How do we get from school to the landfill site? ■ How can I ensure I am safe when I visit the site?	**In groups:** Locate the site on maps and on aerial photographs. Plan a route to the site. Whole class: Discuss the safety aspects of a visit to the site. **In groups:** Design a set of safety rules for the visit.
■ What happens at the landfill site? ■ How are the cells on the landfill site made? ■ Does the site affect the local community? ■ What will happen to the site when it is full?	**Whole class:** Make a site visit if possible. Listen to a talk by the site manager, relating to site management, how the cells are made and filled in, any affects the site has on the local community and how the land will be rehabilitated in the future.
■ How can we all help to reduce the amount of waste we throw away? ■ How does reduce, reusing, recycling, composting and minimising help the environment?	**Whole school:** Arrange a visit to school by 'Cycler' Robot. Contact UK Waste or your local Groundwork Trust to make your request. **In groups:** Sort cleaned waste into four categories (reuse, recycle, compost, minimise) and discuss how this will help the environment
■ Make a sculpture out of waste products. ■ Can you make a fruit bowl, to hold apples, bananas, oranges & grapes, out of paper?	**In groups, pairs or individually:** Make a sculpture or model out of waste. **In groups:** pairs or individually Make a fruit bowl out of newspaper. (Use any technique, paper mache, weaving and so on.)
■ Can we design and make clothes out of waste products? ■ Can we put on a fashion show of our clothes?	**In groups or pairs:** Design and make clothes out of waste products. **Whole class:** Produce a clothes show presentation.

Figure 5 | *Enquiry questions and activities for a unit of work on waste management.*

pollution). Try to reach a consensus. Discuss who the most powerful stakeholders were and why they were powerful.

■ If possible, debate a real 'live' issue within your locality such as the re-development of a brown field site and involve personnel from different groups with interests in the development.

Waste management
Aims: To learn about waste and its management (RRR – reduce, reuse, recycle); to increase awareness of the role of minimisation within school, the wider community and the industrial and business community; to help pupils to understand that everyone (including themselves) can make a difference to global environmental problems.
Activities: See Figure 5.

Tourism and travel
Aims: To develop sensitivity towards other cultures, challenge bias and make pupils more aware of stereotyping and the differing viewpoints; to investigate the environmental impact of tourism (litter, traffic congestion, footpath erosion) and review its impact on the local economy (jobs, income, new building work).
Activities:
Travel and tourism links to the geography national curriculum (investigating real people in

real places and the impact they have on the environment) and provides a context for the following enquiry questions:

■ Who are tourists?

■ Why do tourists travel?

■ Where do tourists travel?

■ When do tourists travel?

■ How do tourists travel?

■ Why do tourists travel to particular places?

■ What effects do tourists have on environments, economies, societies, cultures and host populations?

■ What differing attitudes are held about tourism and its effects?

■ What are the possible futures for tourism? (after Mason, 1992)

Extension activities:

■ Extend the above activities by using holiday images found in tourist brochures.

■ Give each group of pupils a collection of images showing (a) well-behaved tourists (b) tourists who show no respect for the host population. Ask pupils to discuss and list behaviours under 'good holidaymaker' and 'bad holidaymaker'. Compare lists. Compile a cartoon strip which shows examples of good and bad behaviour. Write a code of behaviour for tourists.

A hotel in the rainforest

Aims: To explore the impact of tourism on a community and to recognise that change can have both positive and negative effects on different sectors of the community; to explore concepts of fair play, power relationships and invasion of territory; to extend the children's knowledge and understanding of people and places.

Activities:

■ Read *Where the Forest meets the Sea* (Baker, 1987), a story about when a boy and his father take a boat to reach a remote tropical beach and an ancient rainforest in North Queensland, Australia.

■ Locate this rainforest on the globe. Now look for other rainforest areas. Discuss the vegetation and animal habitats found in the rainforest.

■ Create a rainforest environment in the role-play area.

■ In groups of three or four, give pupils a photograph of an unspoilt tropical beach scene. Ask them the following questions: What is this place like? How is this place changing? Why is it like it is? How do you feel about this place? Who might use the area shown in the photograph? Share responses with the whole class.

■ Discuss arguments for and against the development of tourism in this area of tropical beach/rainforest. Which groups of people support and oppose the development and why? How do you think the native Australians (Aborigines) feel about the changes?

Photo | *Diane Wright.*

Figure 6 | *How to explore an issue in role.*

1. Ask pupils to study a photograph of an unspoilt tropical beach and to discuss who might use it.
2. Divide the pupils into six groups each assuming one of the following roles: local resident, local person who earns living from the sea, local child, local government official, travel company representative, holiday maker.
3. To enable the pupils to get into role, ask them to discuss in their small groups: Who they are? What job they do? How they might use the beach in the photograph? How do they spend most of their time? Are they rich or poor? What parts of the beach might they need for work? What parts might they use for pleasure?
4. Ask each group to imagine that a hotel is going to be built by the travel company. How does the hotel change things? Will their way of life be affected? Will some changes be for the better? Will some changes be for the worse? Do they want the hotel to be built?
5. Ask each group to list three statements which express their views (in role) about this proposed development.
6. Ask the pupils to form six new groups, each comprising representatives from the original groupings.
7. In new mixed groups and still in role, children take it in turns to give views and discuss their statements, to listen to the views of others and to consider whether their view has changed.
8. Pupils return to their original groups, discuss what they have heard and then and elect a spokesperson.
9. Organise a mock public enquiry and allow each spokesperson to speak for two minutes.
10. Ask pupils to vote 'for' or 'against' the hotel and discuss the idea of 'rated' votes: that the local government and travel company groups votes are rated at two compared to the votes of the local residents, local child, local person who earns living from the sea and holiday maker, who are rated at one.
11. The rating provides an opportunity for discussion of who holds the power and why this might be so. In some countries groups, such as women and youths, have virtually no say in their country's development. Discuss whether this is fair.
12. Out of role, as a whole class discuss the following: Do they feel their group had a fair say? How much did their own group's interests affect the decision to build or not to build the hotel? What issues did the groups raise? Who held the power? Why do they think this was?
13. Write an account of the role-play activity in the role of a journalist attending the public meeting.

Activity adapted from Mason, 1992.

- Ask pupils to study the photograph of the unspoilt tropical beach and to discuss who might use it. Put them in role, e.g. Melanie Forest (conservationist), Darren Plant (botanist), Lesley Fortune (Director of Grandiose Hotel Developments Ltd), Peter Campaign (town councillor), Jeremy Beach (barrier reef coastal ranger), Miranda Holidayinn (Midsummer Hotel), Danny (the local boy), Polly (windsurfing club), Melanie Law (local resident) and debate how building a new holiday complex on the tropical beach will affect the area. Who is for the changes and who is against?
- Use a writing frame to put forward the cases for and against the building of the new holiday complex (note taking, discussion and persuasion genres).
- Make a tourist map and brochures advertising a visit to the rainforest hotel. Include all the information and pictures of what you can see and do.

Extension activity:
- Who has the power? The teaching activities can easily be extended to look at who are the most powerful stakeholders (Figure 6) or adapted for other development situations: e.g. new supermarket, by-pass, building or extending a regional airport, building a theme park at the coast.

Wants and needs
Aim: To develop pupils' understanding of the differences between 'wants' and 'needs', and 'quality of life' and 'standard of living'.
Activities:
- Whole class: Explore with pupils what they think they need to lead a happy, healthy life.
- Work in pairs: Ask pupils to select nine items from the previous activity and rank them according to which they feel are most important. Use diamond ranking with the most important at the top, the least important at the bottom and the others in between.

- Compare the results from different groups and discuss the diamond ranks. Try to compile an agreed class list if possible.
- Group work: Ask pupils to reflect on what they really need to be happy and healthy and to have a good quality of life. Ask them to revise their original lists of wants and needs. Do they wish to move some items up or down the list?

Extension activities:

- Whole class: Have a class discussion about the meaning of 'quality of life' (open spaces, pure air, friendship, leisure time etc.) and 'standard of living' (bigger house, better car etc.).
- Group work: Ask pupils, in their groups, to produce a poster illustrating 'quality of life'.
- Whole class: Explore with pupils who is responsible for our quality of life and reflect on how increasing our standard of living could have an impact on the environment and other people's lives, both now and in the future.

Children's rights

Aims: To identify basic human rights such as food, shelter, clothing and love; to consider how rights need to be balanced with responsibilities.

Activity:

- Group work: Look at a simplified version of the UN Convention on the Rights of the Child and think about a contrasting locality or place you are studying. Do the children there have the same rights as you do? Discuss the reasons for any differences.

Cultural diversity

Aim: To explore similarities and differences between the two cultures.

Activities:

- Read *Masai and I* (Kroll, 1993), a story about a young American girl who begins to imagine what life would be like in a Masai village when she learns about their culture at school.
- Compare and contrast the daily routine of the two girls. Look for similarities and differences.
- Ask the pupils to write a daily diary for each girl.
- Ask the pupils to write a letter to the Masai girl, telling her about their life. They can then write a reply as if they are a Masai girl.
- Use video material, websites and books to compare and contrast the lives of children in other parts of the world.

Going to secondary school

Aim: To ensure pupils are confident when starting at the secondary school.

Activity:

- Arrange for year 6 pupils to work with year 7 pupils from the secondary school to produce a new induction booklet for the secondary school. This will give pupils a stronger sense of ownership and team membership and increase their confidence. The new induction book should include maps and plans of the secondary school and a route showing how to get there. It could also include bus timetables, information about sports facilities and drawings of the different blocks or buildings.

Visualisation and relaxation

Aim: To celebrate the beauty of the earth and encourage a sense of awe, wonder and mystery.

Activities:

■ Whole class: Ask pupils to close their eyes and relax and imagine they are on the sea shore (or any other environment being studied in geography). Look carefully at the sea and the way it moves. What colour is the sea? How is the water moving?

■ Look carefully at the shore. Is it made of sand, stones, shingle or a mixture? Pick up a stone or a shell. Feel its shape, size, texture and weight. Touch two other things you find on the beach. Look at their shape ... and colour ... and size ... and feel their ... then put them back on the beach. Sit on the beach and listen. What two sounds can you hear? Who or what is making the sounds? Feel the elements of the weather on your body – is it sunny/windy/rainy?

■ In pairs: Ask pupils to open their eyes and then with a partner discuss parts of their visualisation, make notes about their feelings and share ideas with the class.

Assemblies

Many assemblies address elements of citizenship education and PSHE linked to geography. They also celebrate the community of the school and its ethos, and draw on the backgrounds, traditions and cultures of pupils and the communities in which they live. Common themes are the differences and uniqueness of people, preparation for adult life, giving something back in return for everything one has, and caring for others. Assemblies may be times when outside speakers such as council officers, charity workers and environmental campaigners visit the school to discuss issues with the pupils. By taking a global as well as a local perspective and by thinking about the future of the planet in its largest sense it is possible to reinforce and revisit the topics and activities outlined above.

References and further reading

Baker, J. (1987) *Where the Forest meets the Sea*. London: Walker Books.

Baker, J. (1991) *Window*. London: Red Fox.

Brown, R. (1991) *The World that Jack Built*. London: Red Fox.

DfES (2003) *Excellence and Enjoyment: A strategy for primary schools*. London: DfES.

Gardner, H. (1999) *Intelligence Reframed: Multiple intelligences for the 21st Century*. New York, NY: Basic Books.

Gardner, H. (1993) *Multiple Intelligences: The theory in practice*. New York, NY: Basic Books.

Hedderwick, M. (1994) *Katie Morag and the New Pier*. London: Red Fox.

Kroll, V. (1993) *Masai and I*. London: Puffin.

Mason, P. (1992) *Learn to Travel: Activities for travel and tourism in primary schools*. Godalming: WWF UK.

Section 5

Managing the curriculum

Places Near and Far

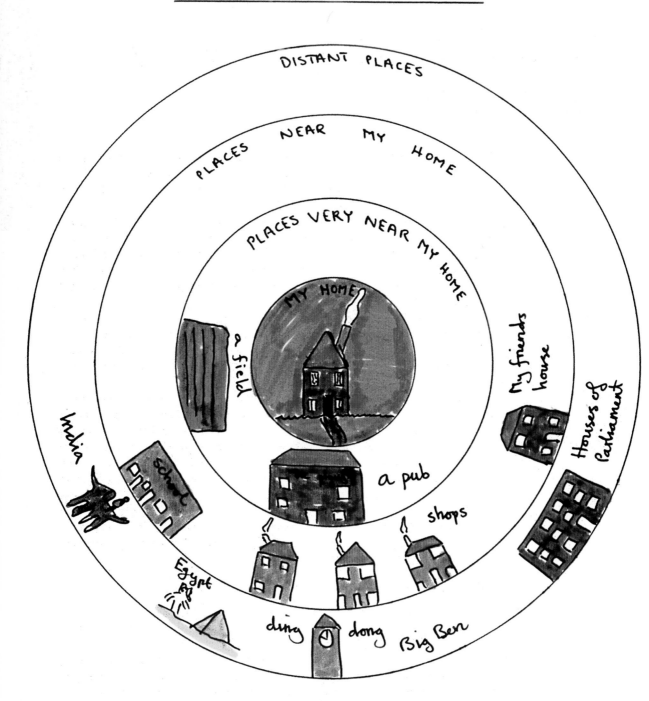

DISTANT PLACES

PLACES NEAR MY HOME

PLACES VERY NEAR MY HOME

MY HOME

a field

My friends house

India

School

Houses of Parliament

a pub

shops

Egypt

ding dong Big Ben

Photo | Stephen Scoffham.

IN THIS CHAPTER YOU WILL FIND KEY IDEAS ON
CREATIVITY • GOOD PRACTICE • NATIONAL CURRICULUM • PLANNING • PROGRESSION • QCA UNITS

Planning the geography curriculum

Geography teaching and learning should be an enjoyable, creative, stimulating and magical experience for pupils and teachers alike. Geography is about places, people and issues both locally and globally, all of which are of interest and concern to everyone. After all, we all live in the same world!

First-hand experience and practical work should underpin much of what we do in the name of geography and the wide boundaries of the subject should allow us to provide pupils with a wonderful learning opportunity. Over the past few years, however, for many and varied reasons geography has not had the time or the space in the primary curriculum to allow it to develop to its fullest extent. Fortunately there is now a new air of optimism nationally about what is important in the primary curriculum.

The national curriculum programmes of study provide the basic material with which to formulate a geography programme, but this still leaves us with significant scope for creativity. Government advice now encourages teachers to be flexible in their planning and to use their own professional knowledge and judgement to create the curriculum which is right for their own particular school and circumstances (Ofsted, 2002).

Teachers have also been urged in the DfES *Excellence and Enjoyment* strategy for primary schools to 'take a fresh look at their curriculum, their timetable and the organisation of the school day and week, and think actively about how they would like to develop and enrich the experience they offer children' (DfES, 2003, p. 9). However, with this new-found freedom comes the responsibility to ensure that the geography curriculum is planned so as to provide the best experience for pupils in our primary schools. Successful planning enables teachers to devise a coherent and exciting curriculum that promotes continuity and progress in children's learning. Planning takes many forms and schools often have a standard format for all subjects which they have devised themselves. The suggestions and advice offered in this chapter try to give stimulus and leadership to teachers who wish to create an exciting geography curriculum within their own school.

The planning process

The planning process is best seen as an ongoing cycle in which all staff take part (Figure 1). Naturally the subject manager has a key role in developing plans and units of work, updating resources and helping colleagues to devise a policy for the subject. Nevertheless, it is important that all teachers in the school have a sense of ownership and shared understanding of this process. It is equally important that once any plans and policies have been developed they are kept under review, are evaluated from time to time and revised as necessary. As Figure 1 indicates, many people are involved in curriculum planning and important decisions will be taken at each stage about how to deliver geography in your school.

In devising and modifying the curriculum you would be wise to take note of recent government advice as the boundaries are always changing in education. The annual subject

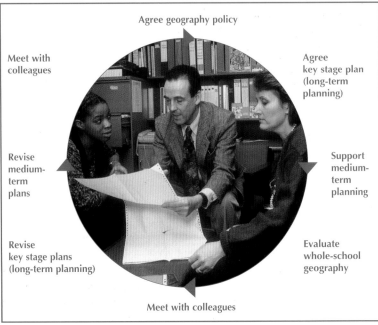

Agree geography policy

Meet with colleagues

Agree key stage plan (long-term planning)

Revise medium-term plans

Support medium-term planning

Revise key stage plans (long-term planning)

Evaluate whole-school geography

Meet with colleagues

Figure 1 | *The planning process.*
Photo: www.JohnBirdsall.co.uk

reports produced by Ofsted are a useful source of inform-ation and ideas. Over the last ten years these have repeatedly stressed the importance of fieldwork, enquiries and local investigations. They also stress the value of a wider world perspective. However, recent reports (Ofsted, 2003, 2004) also identify a number of con-cerns regarding the way schools are teaching geography. These include poor use of ICT, over-reliance on activity sheets, tasks which fail to challenge pupils adequately and lack of progression in learning. We need to see these comments not merely as negative statements but as useful signposts to aid our planning to produce a creative and exciting geography curriculum for each of our schools. So what other factors should we take into consideration as we plan our geography work?

1. The needs of the children

One of the best starting points is to think about what pupils actually need. What previous experiences have they had both within school and in their pre-school time? How do we build on these experiences and extend them? As well as exploring their own locality, pupils should experience a different environment from the one in which they live, be it city, suburb, village or rural area. We may also need to consider the smaller things in life. For example, have the pupils had the opportunity to travel on a train or bus?

2. The local area

The local area is an important resource for geography teaching and we need to use it fully. Remember that it includes the school building, school grounds and local streets and buildings. Ask yourself a few questions. How can we define our local area and what features does it contain? How best can we use it? What is missing? It may be helpful to talk to other schools working with children of different ages about how they use the same area, so that a range of different and progressive activities can be developed at each key stage to engage and challenge pupils. Many children have their first experience of fieldwork in the local area, and the experience is essential in helping them to learn about the real world. Practical experiences outdoors underpin classroom knowledge and skills and provide a context for learning about the real issues facing every community, wherever it may be.

3. Skills and resources in the school

Many teachers feel very insecure about their own personal knowledge in geography and about their skill and ability to teach it. Continuous professional development (CPD) is one way of strengthening subject knowledge and can help to introduce new teaching ideas. Your

own personal enthusiasm, e.g. for map reading, rock collecting or travel is also likely to be infectious. Resources collected from distant places or contrasting environments can provide real ingredients for teaching, bringing a little of the magic of the place into the classroom.

4. Time and pressure on the subject

The school day always seems packed with activities, and the core subjects, together with literacy and numeracy, dominate the curriculum. It is essential to be creative when planning geography and to find 'spots' or opportunities to enhance the time that has been officially allocated during the year. You might like to think about teaching geography in the following ways:

- As a smaller element within another curriculum topic
- In a talk or session with a visiting speaker
- In assembly
- In class discussion – Who saw the news on TV last night?
- During story time or when reading books
- In displays both in the classroom and around the school
- In a concentrated block of teaching time such as an 'Environment Week'
- As a homework task or holiday activity

Continuous teaching units, such as Unit 16: What's in the News? are also well worth considering as they allow you to spend a few minutes discussing topics on a regular basis – perhaps a local issue in the paper, a volcanic eruption, or school visitors from a distant place.

5. Linking geography with other areas of the curriculum

Geography has a unique role as a cross-curricular subject. It contributes to many aspects of the curriculum such as literacy, numeracy, citizenship, ESD, PSHE and thinking skills. It is important to think about how to exploit these links when planning geography and to make the most of any opportunities which may arise as a result. Other curriculum areas where overlaps are likely to occur include science, art, music and history.

Using the QCA schemes of work in planning

The QCA schemes of work were written to help teachers turn the requirements of the national curriculum into realistic and interesting classroom activities. They provide excellent examples of topics which might be taught but are offered only as guidance and to act as a stimulus for schools to adapt or draw up their own material. For example, Unit 13: A contrasting UK locality – Llandudno shows how to use fieldwork to study an area and raises a range of geographical questions. However, there is no need to study Llandudno itself. You can apply the format of this unit to any contrasting locality which your school decides to study.

Photo | Paula Richardson.

The other important aspect of the schemes of work is the lead they give in terms of the varying time allocations and possible curriculum links. Here is an illustration: geography Unit 4: Going to the seaside can be combined with history Unit 3: What were seaside holidays like in the past? to form a cross-curricular topic that might include a field visit and provide extended links to art and science.

What are the features of a good geography curriculum?

While the national curriculum identifies the range of material to be covered at different levels, good practice in geography teaching requires a far wider reach than this. Some of the features of creative planning include:

- Recognising that pupils have their own experiences and knowledge of the world
- Giving the pupils vital concrete experiences outside the classroom, and in a variety of locations
- Giving pupils access to a wide range of visual materials, including pictures, artefacts, maps and data, together with accurate and up-to-date resources
- Using maps in the context of studies of real places, and helping pupils to develop an understanding of scale
- Enabling pupils to learn to use a range of ways to describe the world, e.g. specialist vocabulary, maps, diagrams, models
- Integrating skills, themes and real places
- Providing activities which are enquiry based and encouraging pupils to look for issues, ask questions, and analyse their findings
- Encouraging pupils to link together different aspects of geography using key ideas such as pattern, change, process and interaction
- Including studies of a range of places on a global scale

Developing a curriculum plan

It is generally helpful when planning the geography curriculum to think in terms of long-, medium- and short-term plans (Figure 2). These should stem naturally from the school geography policy which in turn identifies the features of the subject that you feel are important. More advice about creating a geography policy will be found in Chapter 25. Some helpful examples of long-, medium- and short-term plans can also be found on the DfES Standards website (see end of chapter).

Photo | Paula Richardson.

| Whole-school planning | **Produce a policy statement for geography which addresses:** |

- the aims of geographical education and the distinctive contribution of the subject at key stage 2
- geography's role in reinforcing literacy, numeracy and ICT (basic skills)
- geography's contribution to personal and social education and to elements of the whole curriculum, such as environmental education
- opportunities for a wide range of teaching and learning experiences, including fieldwork, in geography
- how much time will be devoted to geography in the key stage

| Long-term planning (key stage) | **Produce a key stage plan for geography which outlines:** |

- how often and in what depth geographical work will feature during the key stage
- whether geography will be taught separately or linked with other subjects
- progression from geographical work undertaken in key stage 1
- the enquiry focus and the broad sequence of content for each geography unit
- how places, themes and skills will be integrated in each geography unit
- development of the skills of geographical enquiry

| Medium-term planning (e.g. half a term) | **Produce a plan for a unit of work which includes:** |

- a more detailed list of specific enquiry questions
- a sequence of teaching and learning activities
- learning objectives (knowledge/understanding/skills) and assessment opportunities
- resource needs and fieldwork arrangements
- the amount of time needed

| Short-term planning (e.g. lesson) | **Produce a lesson plan which clarifies:** |

- the lesson focus or question and the learning objectives
- the way in which skills are integrated with place studies and thematic work
- learning activities and, if appropriate, assessment opportunities
- how the pupils should be grouped, how resources are to be used, how other adults can be involved
- additional strategies for teaching the most and least able pupils
- opportunities for feedback to pupils

Figure 2 | *Planning geographical work at key stage 2. Source: SCAA, 1997.*

Long-term planning

Traditionally a school develops a long-term plan for each year group to show how the requirements of the national curriculum will be covered through a range of topics and discrete subjects. This long-term plan is vital as it gives an overview of what is to be included and helps to check that a balanced curriculum is being delivered. It is also useful in identifying overlap or where gaps in provision occur and as a way of checking continuity and progression in topics through different key stages.

Medium-term planning

The medium-term plan sets out the body of the topic to be taught and is concerned with the details within each unit of work. Broad learning objectives are central to medium-term plans. These specify the concepts, knowledge, skills and attitudes we expect pupils to acquire during the work done on a unit. They also provide a focus for assessment. The medium-term plan should:

- Focus on the unit of work and suggest timings
- Ensure equal access to the curriculum for all pupils in the group and identify ways of differentiating the material
- Relate to the programmes of study
- Specify broad learning objectives and suggest some assessment activities

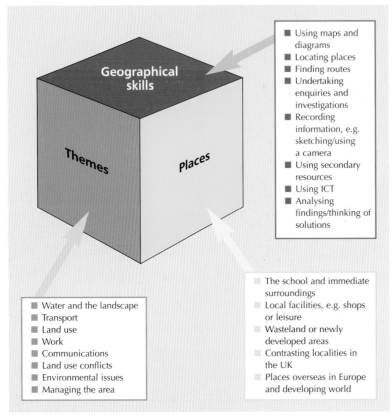

Figure 3 | *The integration of geographical skills, themes and places.*

Geographical skills

- Using maps and diagrams
- Locating places
- Finding routes
- Undertaking enquiries and investigations
- Recording information, e.g. sketching/using a camera
- Using secondary resources
- Using ICT
- Analysing findings/thinking of solutions

Themes

- Water and the landscape
- Transport
- Land use
- Work
- Communications
- Land use conflicts
- Environmental issues
- Managing the area

Places

- The school and immediate surroundings
- Local facilities, e.g. shops or leisure
- Wasteland or newly developed areas
- Contrasting localities in the UK
- Places overseas in Europe and developing world

- Pose key questions and encourage an enquiry approach to learning
- Identify activities to deliver the objectives
- List appropriate teaching resources
- Identify how ICT can be incorporated into a unit
- Identify possible links to other areas of the curriculum

You need to ensure that the work you do relates to real places and that wherever possible a global dimension is included. When pupils learn specific mapwork skills, for example, they should be encouraged to apply these to an actual context so that they have meaning, rather developing them as a free-standing skill. One of the principles underpinning the national curriculum is the notion that skills, places and themes should be integrated. This idea can be expressed diagrammatically using a cube to represent the geography curriculum (Figure 3).

Some of the questions you might want to consider as you draw up your medium-term plans are:

- Which aspects of the work will be enhanced by the use of ICT, including digital cameras?
- What geographical vocabulary can be developed?
- Where are the best opportunities for fieldwork and practical work?
- How can the use of maps, atlases and globes be developed most effectively?
- Which key questions will develop enquiry skills and provide a direction for the work?

Photo | *Stephen Scoffham.*

Short-term planning

Short-term planning or lesson planning is very detailed and personal to the teacher. It should reflect and maximise the individuality and personal flair and enthusiasm of the person concerned. Having said this the lesson plan should indicate:

■ The learning objectives and content for the lesson – what the pupils will be able to do, know and understand by the end of the lesson

■ What sort of assessment will take place

■ Differentiated learning experiences linked to individual children

■ Classroom organisation/grouping

■ Links with the classroom assistants/support teacher if appropriate

■ Resources needed

Progression and continuity

Progression is a key part of any curriculum. In geography, pupils should be introduced to greater breadth and depth as they move through their studies. They will also need to be provided with opportunities to examine increasingly complex issues and have the chance to develop an understanding of the attitudes and values people hold about a range of issues and places. For instance, how do people in their area feel about the proposed location of an incinerator? How varied will the views be that are expressed?

You also need to consider whether the places you decide to study in geography form a balanced sample and whether pupils will develop an increasing knowledge of different parts of the world.

Photo | Kate Russell.

Summary

The points below provide a summary to help you create a well-constructed curriculum plan for geography across the key stages.

1. Make geography real by studying actual places and the issues and problems that affect them.
2. Combine work on places, skills and themes rather than teaching them in isolation.
3. Think about the order in which pupils study topics. Allow for pupils to return to key ideas and aspects several times during their primary years.
4. Build progression into your planning by introducing more demanding skills and widening the range, complexity and methods of study.
5. See that you use assessment and findings from any research you have read to inform your planning.
6. Adopt an enquiry approach by structuring your teaching units around a series of key questions. Share these with your pupils.
7. Exploit natural links between geography and other subjects, especially ICT.
8. Try to provide a mix of local and global material in all years. For younger pupils 'local' can mean in their known world, while 'global' means any new experience which extends beyond it.
9. Increase the proportion of distant place work as children progress through the years.
10. Ensure that the pupils have regular exposure to geography throughout the primary years. They should have some, even if only brief, contact every term.

The Geographical Association declares in a statement circulated to all primary schools that 'geography makes a major contribution to our physical, intellectual, social and emotional development' (Geographical Association, 2004). The statement also identifies a number of aspects of learning which help to place geographical learning at the heart of the curriculum. Good geography teaching offers many opportunities to ensure this contribution by stimulating pupils' interest in both their surroundings and in the variety of physical and human conditions on the Earth. In turn, good planning will help to achieve a geography curriculum which will help pupils to understand more about their world and their place in it.

References and further reading

BBC (2003) *BBC Watch: Barnaby Bear goes to Brittany* (video). London: BBC Educational Publishing.

Cuncliffe, J. (1983) *Postman Pat Takes a Message.* Leamington Spa: Scholastic.

DfEE/QCA (1998) 2000/ *Geography: A scheme of work for key stages 1 and 2 Update.* London: DfEE/QCA.

DfES (2003) *Excellence and Enjoyment – A strategy for primary schools.* London: DfES.

Geographical Association (2003) *Finding Time for Things that Matter: Geography in primary schools.* Sheffield: Geographical Association.

Jackson, E. (2000) *Barnaby Bear goes to Brittany.* Sheffield: Geographical Association.

Ofsted (2002) *The Curriculum in Successful Primary Schools.* London: Ofsted.

Ofsted (2003) *Annual Report of HMCI of Schools: Standards and quality in education 2001-2.* London: Ofsted.

Ofsted (2004) *Annual Report of HMCI of Schools: Standards and quality in education 2002-3.* London: Ofsted.

SCAA (1977) *Expectations in Geography at Key Stages 1 and 2.* London: SCAA.

Websites

The DfES Standards website has useful examples of medium- and short-term plans, as well as helpful exemplifications of work at both key stages: www.dfes.gov.uk/teachinggeographyatkeystages1and2

Through My Window

by see wan

I can see cars	I can see a shop
I can see house	I can see a road
I can see people	I can see birds

Assessment of geography

A ssessing geography in the school curriculum can be made into a very complex operation. The purpose of this chapter is to provide advice to teachers and subject co-ordinators about realistic approaches to assessing pupil progress in geography that will enhance pupil learning in ways which will not place unrealistic demands on teachers.

> ❝ *... you cannot claim to be teaching without undertaking forms of assessment, and by implication, this assessment activity helps ensure the quality of what is taught (and how)* (Lambert, 2000).

Teachers face demands to raise standards at both pupil and whole-school level in all national curriculum subjects. There is also an expectation that procedures for monitoring teaching standards will be in place. While it is very positive always to strive for improved performance, the measurement of performance should not become such a priority that it detracts from the vital processes of planning and teaching. It needs to be recognised that when so much emphasis is placed on the core curriculum it is often only possible for co-ordinators to negotiate minimal time for geography. Assessment procedures need to be manageable within the teaching situation. Also, teachers will be more prepared to promote and enhance assessment initiatives if they actively contribute to the process of learning. One of the key messages in this chapter is that most useful assessments take place during normal teaching periods and assessment procedures need to be relevant to both pupils and teachers alike.

What is assessment for?

Assessment can fulfil many purposes within the geography curriculum. Co-ordinators and teachers need to be aware of these diverse purposes in order to determine which assessment procedures can enhance individual and whole-school standards.

It is important for a school teaching staff to have a common understanding about how assessment procedures can have diverse functions. They need to agree which assessment procedures will enable their school to achieve its objectives. The list in Figure 1 is not exhaustive and co-ordinators and teachers may find it useful to explore ways in which assessment procedures can be used to support the geography curriculum. Once the underlying purposes of assessment are agreed, the next stage will be to develop appropriate procedures.

As far as geography is concerned, recent Ofsted reports have repeatedly drawn attention to weaknesses in assessment, which they regard as unsatisfactory in many primary schools. As one of the latest reports puts it, there is a need for 'greater challenge' and 'better assessment' in geography teaching (Ofsted, 2004).

Assessment can be used to:

1. help motivate pupils to learn
2. inform planning
3. raise standards
4. find out what pupils know and understand
5. inform general standards within a school
6. fulfil statutory requirements
7. provide evidence of progress to external agencies
8. evaluate teacher performance

Figure 1 | Functions of assessment.

Formative assessment	Summative teacher assessment	National curriculum summative tests
Assessment for learning happens all the time in the classroom. It is rooted in self-referencing and involves both the teacher and the pupil in a process of continual reflection and review about progress. When teachers and peers provide quality feedback, pupils are empowered to take the appropriate action. Teachers adjust their plans in response to formative assessment.	Assessment of learning is carried out at the end of a unit or year or key stage or when a pupil is leaving the school. The aim is to make judgements about pupils' performance in relation to national standards. Teachers find standardisation and moderation meetings important quality assurance opportunities. Teachers often use information about pupils' performance in summative tests formatively.	The national curriculum tests and tasks provide a standard 'snapshot' of attainment at the end of key stages. There are also optional tests for years 3, 4 and 5. A pupil's performance is described in relation to the national standards or levels.

Figure 2 | *Different types of assessment.*

The assessment process

Assessment can take many forms – from formal testing to discussion with individual pupils. The system of national testing within the core curriculum (i.e. English, mathematics and science) has tended to equate assessment with testing and grading pupils according to their performance within a standardised framework. Not only does this approach help to formulate the level of performance of a cohort of pupils, it also provides information to teachers, parents and pupils about individual pupil success or lack of it. This formal approach to assessment can significantly influence the attitudes pupils develop towards learning.

An alternative approach is to view assessment not as a labelling and sorting procedure but rather as a procedure that can aid teachers to enable pupils to achieve their potential (Figure 2). It cannot be emphasised too much that teacher/pupil-centred assessment can enhance both individual pupil and whole-school standards. As the Assessment Reform Group (1999) notes, it has been 'proved without a shadow of doubt' that informal assessment with constructive feedback to the pupil will raise attainment.

Primary school geography is not subject to the pressures of external assessment procedures and national league tables and thus there is opportunity to use assessment as a vehicle to promote learning and understanding. Assessment can also be used to encourage positive attitudes to learning geography among pupils of all abilities whether they are slower learners or gifted and talented. In short, pupils' assessment needs to:

- promote learning
- promote success
- provide active involvement of the pupils in their own learning
- provide positive feedback to the pupils
- aid the development of positive attitudes among the pupils
- inform future planning

Photo | *Paula Richardson.*

The learning process

If assessment is specifically designed to facilitate pupil learning it is necessary for the teacher to know what needs to be learnt. The national curriculum defines the content which has to be covered in geography. For notions of progression it is best to turn to the level descriptions (see page 318). However, as all practising teachers will be aware the sequence of learning adopted by individual pupils fails to fit into neat categories. The art of successful teaching is to promote an environment that facilitates diverse approaches to learning and successful classroom dynamics. It is also imperative to engender positive attitudes and provide pupils with the potential for success. Assessment procedures can provide teachers with the information they need to create stimulating and challenging lessons that match the needs of the pupils with whom they are working. Generally, it is best to pitch activities at a level where the majority of the pupils will achieve the learning targets. The minority who will find this work challenging can be identified and their learning targets tactfully modified, while at the opposite extreme those who are exceeding the targets can have planned extension activities.

There is a further problem. The subject-driven structure of the national curriculum often does not sit happily with the way primary school pupils actually learn. Subject areas can create artificial barriers. Primary teachers have to be aware of the interplay of different areas of knowledge. For instance, pupils who undertake a geographical activity often need skills, knowledge and understanding in a number of curriculum areas (e.g. a river study will draw on mathematical and scientific ideas). It is important to consider if pupils will need these prior to developing new learning in geography. This is achieved by assessing the pupils and continuing the assessments while they are working. In order to create a positive learning environment teachers needs to:

■ identify clear learning objectives – i.e. know what is to be taught
■ provide pupils with information about what they are aiming to learn
■ have a strong notion about what pupils already know and understand
■ be aware of any misconceptions pupils may already hold. about the subject
■ have a strong notion of progression
■ be able to assess the learning that has taken place in order to organise future planning

Successful learning takes place when pupils have ownership of their learning, understand the goals they are aiming for, possess the skills they need, and are motivated. Thus for any successful learning to take place it is essential that the teacher is aware of the learning needs of their pupils.

Different forms of assessment

Any assessment process needs to be efficient, informative and require the minimum amount of time to administer. Primary teachers have difficulty simply meeting all the subject requirements of the national curriculum within the school day. In addition, the assessment requirements for the core curriculum can be quite onerous for many schools. It would be extremely difficult for a geography co-ordinator to advocate any assessment and recording procedures that are time-consuming or complex. It is therefore important that both geography co-ordinators and class teachers have a clear view about the purposes of assessment within the context of teaching geography. There are many simple assessment procedures. These include:

■ Observing pupils, not only watching but also listening to them recount how they carried out their work and what they found out.
■ Asking pupils open-ended questions that give them the opportunity to develop their own ideas.

■ Providing pupils with tasks that demonstrate that they have acquired skills or new learning.

■ Providing pupils with opportunities to present their work in different ways, e.g. maps, diagrams, writing, drawing, speaking, role-play.

Assessment will both be formal and informal, formative and summative. We often underestimate the amount of informal assessment that goes on in the classroom through observation and discussion with pupils during day-to-day teaching and learning activities. This professional judgement is an important component in, and starting point for, more formal assessments which might include written work, mapwork, oral presentations, group work, posters, role play and fieldwork. A range of assessment techniques is necessary to assess a pupil's full range of ability; some pupils are better at mapwork, others may find oral presentations easier. Also, some types of assessment lend themselves to a particular topic; role play and simulation are often good ways to explore the range of views connected to an environmental issue such as a local by-pass; posters are effective in allowing pupils to express their views of a locality and recognise how people affect the environment. Similarly, completing a table will allow pupils to compare two localities; using a large-scale map of the local area can be an ideal way of illustrating geographical patterns. Whatever approach you adopt it is important to remember that assessment tasks should be an integral part of a unit of work. Assessment is not a bolt-on extra. As Black and William remind us:

> *Teachers need to know about their pupils' progress and difficulties with learning so that they can adapt to meet their needs* (Black and Wiliam, 1999).

Pupil self-assessment

Self-assessment can help to develop children's' understanding and encourage them to take a positive attitude towards their learning. It can also provide valuable evidence of their learning. However, teachers need to take due regard of the pupils' maturity and level of ability in setting self-assessment tasks. Pupils must also be encouraged to apply their learning to new situations.

The key element of self-assessment is to provide opportunities for children to talk critically about their work. They also need to understand the purpose of their work and appreciate how what they are doing is part of the 'big picture'. To be effective, self-assessment must be consistently applied throughout the curriculum and care must be taken by the teacher to ensure that pupils perceive critical reflection as a positive action and not a self-deprecating one.

Example

A group of Y6 pupils were carrying out a topic about Spain. One of the key targets was to be able to determine the location of Spain using atlases and maps. The pupils were told their targets and they discussed appropriate achievement indicators. They then had to define where different places were using their newly acquired skills. At each attempt the pupils were encouraged to reflect critically on how they could further improve their work. A similar self-assessment activity was carried out in a Y1 class where children were learning how to draw a picture showing the location of their home and different routes in the local area. Each time they carried out an activity they were asked to consider alternatives. As a result the pupils were better able to transfer their learning to other mapwork exercises involving features such

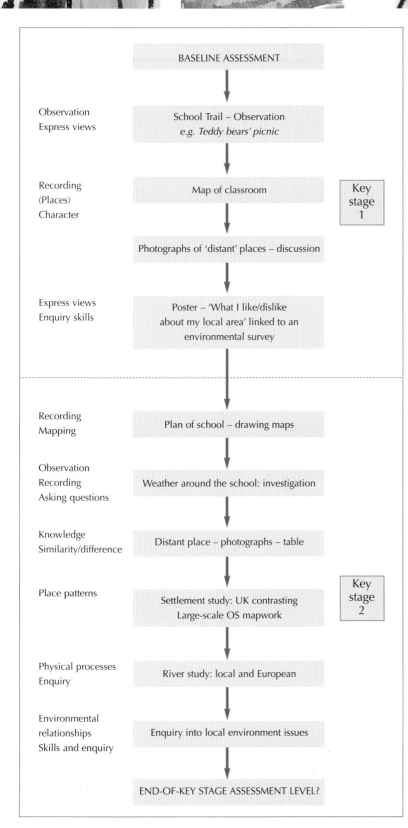

Figure 3 | *A model of possible assessment and recording opportunities.*

as shops and parks. Some even managed to develop this skill further by using simple symbols on their maps.

What do I need to record?

The opportunities for assessment will vary according to the school's own scheme of work. Similarly, the school's policy on assessment and recording may determine the amount of recording undertaken. It is neither possible nor necessary to keep or record every piece of work that a pupil produces. The records you keep will depend on the purpose of the school's recording system.

Records have a number of purposes. Among other things they can:

■ aid planning
■ inform parents
■ inform teachers
■ record progress
■ provide data for curriculum coverage

The key element in any recording system is that it is manageable and is relevant to the teaching and learning processes. For example, if a scheme like the one in Figure 3 was used it would be sufficient for you simply to indicate whether a pupil could or could not do certain things, like recognise the countries of the UK or be aware of places beyond their

Figure 4 | *Progression in geography.*

Pupils are expected to perform at the following levels in each aspect of geography:

- Levels 1-2 (typical key stage 1)
- Levels 3-4 (high level key stage 1/typical key stage 2)

Levels 1-2

At levels 1-2 pupils demonstrate that they have studied at local scale and are aware of places beyond their own locality. Typically their work will show the following attributes:

Geographical enquiry and skills

- drawing on limited experience and on resources provided, to ask and respond to simple geographical questions and to express their own views, using basic geographical vocabulary (geographical enquiry)
- using simple techniques and skills to undertake straightforward tasks, as demonstrated and supported by the teacher (use of skills)

Places

- recognising and describing 'where things are' in the simple contexts of the classroom, school grounds or local area, and being aware of some places in the wider world (location and context)
- identifying and beginning to offer descriptive observations about simple recognisable features of places (features and character)
- making simple comparisons between individual features of different places and recognising how places are linked to other places in the world (contrasts and relationships)

Patterns and processes

- responding to questions about 'where things are' by making observations about features in the environment and recognising simple patterns (patterns)
- responding to questions about 'why things are like that' by recognising and making appropriate observations about some physical and human processes (processes)

Environmental change and sustainable development

- identifying and describing easily recognisable examples of the ways people affect the environment and of attempts to manage these interactions (environmental change and management)
- recognising some ways in which change may damage or improve environments and affect their own lives (sustainable development)

Levels 3-4

At levels 3-4 pupils demonstrate that they have studied a range of places and environments (including the local area) at more than one scale and in different parts of the world. Typically their work will show the following attributes:

Geographical enquiry and skills

- asking and responding to geographical questions and offering their own ideas in the course of undertaking tasks set by the teacher, and being able to identify and give simple explanations for views held by others (geographical enquiry)
- using a range of simple pieces of equipment and secondary sources to carry out tasks supported by the teacher (use of skills)

Places

- knowing the location and contexts of places they study and some significant other places (location and context)
- describing a range of physical and human features of places studied, using appropriate geographical terms, and beginning to offer reasons for the distinctive character of places (features and character)
- making simple comparisons between individual features of different places and recognising how places are linked to other places in the world (contrasts and relationships)

Patterns and processes

- responding to questions about patterns in the landscape around them and making appropriate observations about the location of features relative to others (patterns)
- beginning to explain why things are like that and how things change by referring to physical and human features of the landscape (processes)

Environmental change and sustainable development

- identifying changes in the environment and beginning to understand how people both damage and improve the environment (environmental change and management)
- recognising how and why people may try to improve and sustain environments and identifying opportunities for their own involvement (sustainable development)

own locality by the end of year two. All the staff could decide on a few basic skills they agree to assess at some point during the year, e.g. map skills. In schools with more than one class per year group, class teachers might decide to carry out the same activity and compare performance. One of the most effective tools in helping to assess pupils' performance and progression is the subject portfolio.

Level descriptions

For both assessing and reporting pupil progress, the level descriptions in the geography national curriculum provide a useful reference point. Also useful is the QCA guidance for geography (www.qca.org.uk) which gives some examples of children's work. Assessing and

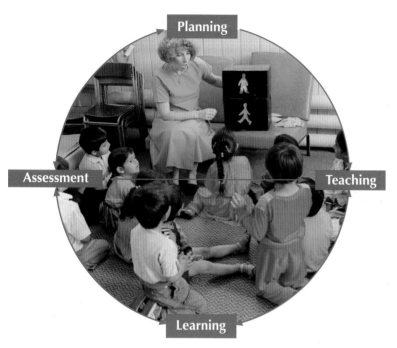

Figure 5 | *The interconnections between planning, teaching, learning and assessment. Photo: www.JohnBirdsall.co.uk*

recording pupils' attainment will help you plan progression in pupils' learning and for continuity in pupils' experience across the key stages. Remember though that, as in other subjects, level descriptions are not intended for the detailed assessment of individual pieces of work. Rather they provide end-of-key-stage summative judgements that serve as a 'best fit' description of a pupil's ability in geography.

The programme of study and level descriptions can be broken down into four main aspects:

1. Geographical enquiry and skills
2. Knowledge and understanding of place
3. Knowledge and understanding of pattern and process
4. Knowledge and understanding of environmental change and sustainable development

The QCA provides a notion of progression within each of these aspects (Figure 4). Assessment involves combining these different aspects. Pupils carrying out an enquiry into changes in their local area, for example, will be using a range of geographical skills and studying a range of geographical physical or human processes. Similarly, pupils who are learning about an overseas locality may be developing their mapwork and photograph interpretation skills as well as enhancing their ability to use the internet and ICT packages as sources of information.

How can assessment can help me in my planning?

Assessment of pupils' attainment can be used to inform any stage of the learning process (Figure 5). You might have information regarding the pupils' experience and ability that you can use in your initial planning. The teacher then builds up a hypothesis about what the pupils know and the activities that will lead to further progressive learning. The activities planned will relate to ability levels within the class. The differentiation will be by either outcome or by task. The assessments will be continuous as the teacher interacts with the pupils and there may also be a finished body of work that will also provide information about the success of the pupils in achieving learning targets.

Good assessment, recording and reporting systems will:

- allow you to assess the standard that pupils are reaching at any one point in time;
- help pupils to know how well they are doing and what they need to do better;
- provide information to help you plan more appropriate teaching and learning activities;
- provide information on pupils' progress for parents, colleagues in the school and for any transfer documentation to other schools.

Testing

The school will need to decide what to test and what criteria to use. It will be very hard to construct tests that could accurately indicate a level as defined in the level descriptions. There is a requirement for pupils to be formally assessed in the Foundation Stage as part of their Foundation Stage Profile.

Monitoring pupil work

This is a strategy where the co-ordinator reviews work carried out by pupils. The co-ordinator can ask to see all work or ask for a sample. A process for recording the results of such a process will need to be in place.

Monitoring planning

The co-ordinator receives teachers' medium-term and/or short-term planning. This will enable the co-ordinator to monitor implementation of the school geography policy and scheme of work. In addition, areas of development identified by individual teachers resulting in modification of planning can be monitored by the co-ordinator.

Observing teaching

This can be a very positive and beneficial approach. It is expensive to apply because some form of teaching cover is required as well as training for the 'observer' on how to observe positively yet critically.

Reporting on individual progress

Reporting to parents is a legal requirement at the end of the academic year. Report documents relate to individual pupils and it would be difficult to utilise them for whole-school assessment of standards unless, within the document, teachers are required to enter a level of attainment. It is not a legal requirement for a child to be allocated a level in the three primary key stages.

Teacher assessment of levels

This can be achieved by using the level descriptions and the 'best fit' approach (see chapter 00). It is a time-consuming process, especially if in the cause of consistency 'levelling' is required for all national curriculum subject areas.

Exemplification portfolios

Samples of pupil work are kept centrally. They indicate the standards of achievement and also the expectation of the standards within the school.

Figure 6 | Some possible strategies for monitoring standards.

Monitoring standards

The argument thus far has been that assessment needs to be continuous because the primary function of assessment is to facilitate planning and also to provide positive reinforcement to individual pupils. However, schools are also required to maintain and improve standards and it is the role of the geography co-ordinator to decide how to monitor these standards. Figure 6 indicates some of the options but schools may develop others. Whatever processes are used they need to have a logical cohesiveness that is contained within the school's geography and assessment policies.

Do I need to keep a portfolio of pupils' work?

You do not have to keep a portfolio of each pupil's work. However, developing a geography portfolio for the whole school is a good idea because it allows the co-ordinator/ subject leader to:

■ develop consistent judgements across year groups

■ come to a common understanding of level descriptions

■ help 'new' teachers in the school or year group to develop appropriate expectations

■ show aspects of progression and continuity across the key stages/school

■ monitor the quality and standards of geography across the school

A portfolio might include examples of pupils' work from different year groups to show progression, or it might have pieces of work from different abilities within the same year group to show differentiation. As well as pupils' work you could include photographs and teachers' notes, annotations to the work, and so forth. These annotations to pupils' work might include the following:

1. Details of the instructions given, resources used, amount of teacher support and whether the work was undertaken individually or in groups.

2. Brief commentary on what the work shows in terms of geographical knowledge, understanding and skills.

3. Other evidence of the pupils' work – discussion, comments to the teacher and so on.

4. Comments about how this work could be taken further and whether the work is in line with expectations.

Such a portfolio would make an invaluable tool for INSET activities related to geography and an excellent record to show to parents, visitors and visiting Ofsted inspectors. Subject leaders are playing an increasing role in monitoring their subject area and, for them, a good portfolio is an ideal alternative to visiting colleagues' classrooms to observe their teaching, particularly given the normal time constraints.

Teachers might also be encouraged to complete a proforma for their pupils' work using the headings in Figure 6. Staff could then discuss the relative merits of individual pupil's work and perhaps relate these to aspects of the level descriptions.

Conclusion

Assessment plays a crucial role in the learning process. It helps us to appreciate the different ways in which pupils learn, identify what they know already, and take account of any misconceptions they may possess. This information gathering process should take place within an environment where the pupils want to learn. Formative assessment has a particular value in supporting new learning. Ofsted inspectors are advised to look for the following clues:

> *Clues to the effectiveness of formative assessment are how well teachers listen and respond to pupils, encourage and, where appropriate, praise them, recognise and handle misconceptions, build on their responses and steer them towards clearer understanding* (Ofsted, 2000, p. 61).

Seen in the broadest terms assessment has two main functions. It can:

1. Provide a 'snapshot' of the knowledge and understanding achieved by pupils.

2. Be an aid to determine the next stage for planning areas of learning for pupils.

Within these two broad functions, assessment procedures can affect the learning processes that take place within the classroom. It is important to appreciate the crucial link that exists between assessment in the classroom and these learning processes. As an anonymous saying expresses it:

> *If you want a plant to grow it is no good just measuring it, you need to give it the things that will make it grow. Children are the same, measuring them is not enough! They need to be given learning that makes them grow.*

References and further reading

Assessment Reform Group (1999) *Assessment for Learning: Beyond the black box*. Cambridge: University School of Education.

Black, P. and Wiliam, D. *Inside the Black Box* http://www.pdkintl.org/kappan/kbla9810.htm (viewed 30 June 2004).

Lambert, D. (2000) 'Using assessment to support learning' in Kent, A. (ed) *Reflective Practice in Geography Teaching*. London: Paul Chapman.

Ofsted (2004) *Annual Report of HMCI of Schools: Standards and quality in Education 2002-3*. London: Ofsted.

Our front doors

4 | 0 are | blue

3 | 0 are | yellow

5 | 0 are | brown

5					
4	Ian	Tanya	Robert		
3	Phoebe	Claire	Natasha	Naomi	Jamie	Sarah
2	Robyn	Jason	Shannon	Abbie	Jenna	Gary
1	Rachel	Corinne	Rebecca	Daniel	Siobhan	Leigh

Geography, inclusion and special needs

The next few years are going to be exciting and interesting times for those who teach geography in primary schools – not that the last few years have been anything but exciting, given all the changes that have taken place. However, the government has now signalled that as part of its drive for inclusive education, primary schools are likely to have responsibility for many more pupils designated as having Special Educational Needs (SENs).

The Department for Education and Skills has also produced a strategy paper Excellence and Enjoyment (DfES, 2003) which outlines a new approach to the curriculum. In the foreword the Secretary of State for Education declares: 'Children learn better when they are excited and engaged … When there is joy in what they are doing, they learn to love learning'.

There is also a clear indication in the strategy paper that the government is going to extend and refine the coverage of value-added calculations, including the use of P scales. (For more information see DfES, 2003; performance descriptors below Level 1 are to be found at www.nc.co.uk.net/ld/Ge_perf.html and DfES, 2003, p. 24.) The aim of this 'new approach' is to enable teachers to demonstrate what a pupil has achieved against attainment levels that are relevant and valid for that pupil. This should have a powerful impact on recognising what individual pupils can achieve and is spelt out in Excellence and Enjoyment under the title 'Learning – a focus on individual children' (DfES, 2003, p. 3). Teachers will be able to set realistic targets for learning in geography that are related to the pupil rather than some 'outside norms'. This new view of assessment, coupled with a realisation that 'there will be different sparks that make learning vivid and real for different children' (DfES, 2003) means that practitioners can once again think creatively about the primary geography curriculum. Indeed, it is clear from the work of many primary teachers that geography (as well as other foundation subjects) might well be the spark that brings learning alive for many pupils, including those who experience a range of barriers to learning.

It is to be hoped that these initiatives will give SENCOs, geography co-ordinators and class teachers more scope to be creative in geography – particularly in relation to curriculum content, class and school organisation, as well as time spent on the subject. It is worth noting that the government strategy paper includes a case study from St Joan of Arc School, Bootle. Here the head teacher argues that: 'The curriculum needs to be exciting. In our case, this means building in quality not quantity so that children learn first hand' (DfES, 2003, p. 19).

The distinction between quality and quantity is a key point, particularly for those pupils who are unable to participate fully in school life, and will be returned to later. One of the implications is that SENCOs and geography co-ordinators are now empowered to devise an exciting geography curriculum based on the school locality and tailored to pupils' needs. This view is reinforced in the strategy paper where it states that 'schools will be encouraged to take a fresh look at their curriculum, their timetable and the organisation of the school day' (DfES, 2003, p. 12).

Photo | Paula Richardson.

Most primary school teachers know from experience how different teaching approaches can add quality and enjoyment to their pupils' learning. Inevitably, pupils have varying strengths and experience different learning difficulties. Good or effective teaching involves matching the creativity of the educator with personal knowledge of the capabilities of pupils. This pedagogical approach is likely to provide equality of opportunity and curriculum access. Subsequently, barriers to participation will be minimised and all pupils will be able to demonstrate their abilities and achievements.

When we are close to teaching it is easy to miss the progress a pupil is making, particularly if they are accessing the curriculum in smaller steps than their peers. Learning achievements are often clearer to regular visitors so it is important to adopt a team approach, involving other teachers, the SENCO, Teaching Assistants (TAs) and parents. Where a culture of team work prevails, teachers benefit from the support they obtain and can confirm that their particular approach is appropriate to the situation. This also helps to ensure consistency across the school and fosters the achievement and enjoyment of all. This approach is also vital to avoid the perception that only 'specialist adults' can cope with certain pupils – we need to develop the expertise of all staff for the benefit of all pupils.

Fieldwork

Fieldwork is crucial in geography and ensures first-hand memorable experiences for all pupils because it involves all the senses. The examples in *Excellence and Enjoyment* (DfES, 2003) can be used to argue the case for allowing more time for fieldwork activities. The head teacher of St Joan of Arc, Bootle mentioned earlier made the point that his school acquired two minibuses 'so that classes can make visits and build opportunities for enrichment'. Not every school can acquire a minibus, which is clearly a valuable asset – particularly if it is built to take wheelchairs – but one can often be borrowed from secondary schools, special schools or charities. A local university that has a geography department or trains teachers can also be a useful source of 'labour and transport' and help to support both residential trips and local fieldwork. An alternative is to establish links and relationships between a range of schools to share resources, teachers and pupils. Many special schools are delighted to join in partnerships with other schools as they all too often become isolated and the expertise and experience they have is not shared. The LEA advisory team should be a good starting place for developing such partnerships.

Fieldwork, though, does not need minibuses in order to be effective. A primary school in Islington planned regular fieldwork for its nursery class in the local area. This culminated in a trip on the underground to a local park for a picnic and gave pupils the chance to experience a different environment. The nursery teacher, nursery nurses, a member of staff trained in geography and senior staff at the school were all involved in the planning. The crucial aspect of this type of work, particularly for those pupils who experience barriers to learning, is careful preparation. You need to walk the route, identify suitable learning opportunities and carry out a risk assessment. If you take pictures, perhaps using a video, you will be able to use these as material to prepare the pupils before they go. When you return, the images will provide a valuable resource for pupils who either missed the trip or who need a visual reminder of what they saw. Another useful tool is the digital camera and the 'simple cassette recorder' as both can be used to record material for use back in school. Using video cameras and recorders also gives pupils a sense of ownership and pride in what they are doing.

When you are doing your planning you need to ensure that you are fulfilling your responsibilities under the Special Educational Needs and Disabilities Act (2001). This involves taking into consideration generic questions such as 'Who will describe the environment for blind pupils?' and 'Are there any problems regarding wheelchair access?'. You will also need to consider how to support pupils who are challenged by changes in routine or the freedom of relatively unstructured situations. Remember that each pupil is unique and that many pupils have undiscovered strengths which they can offer or contribute. Good planning involves assessing the needs of the class, groups and individuals.

Well-planned fieldwork enables pupils to build on previous skills and knowledge. Remember also that the work that develops as a result can be used as a school resource and put in the library on the geography shelves. This means that videos, ICT material, maps, plans and photographs will be available for all pupils and parents. This will be valuable because it:

1. Demonstrates to the pupils that their work is valued and is used again. (A key issue in the area of sustainable development.)
2. Enables pupils to revisit the work – to recap, remember, go back over and extend ideas. (This is important for pupils who learn more slowly or more quickly than their peers.)
3. Allows parents to see what their pupils have done. (Their children's work has a purpose.)
4. Means that others can use the work as a starting point in future years.
5. Enables the school to build its own quality resources for geography and special needs when there is a paucity of quality material in the market place.

Students on initial teacher training courses could be used, when placed in schools on teaching practice, to help develop local fieldwork material in conjunction with the SENCO and geography co-ordinator. For example, a school in the north-west of England recently used school experience students, in conjunction with the college tutor, to develop work related to mapwork and orienteering. Everyone benefited. The students involved gained valuable

experience, the pupils, particularly those considered to have special or additional needs, extended their learning, and the school added to its resources. One of the other outcomes was that the material provided evidence of how the pupils were progressing in relation to geographical skills and enquiry.

This approach means that fieldwork and the results of the fieldwork investigations can become part of the school's learning resource. It also means that the parent of a pupil who is a wheelchair user or is blind or deaf, for example, can see what pupils did last year and perhaps make suggestions about how to modify the visit. Making the results part of the school resource means that pupils can also explore the notion of change using their own archive. Studies which help pupils to recognise that local environmental issues are complex and contested could include the redevelopment of wasteground sites, new traffic management schemes and the loss of local play space.

In one school, for example, year 5 and 6 pupils examined the issue of safety outside their school. The pupils were convinced that, because of a recent accident, there ought to be a zebra crossing. They collected information in a number of ways including traffic surveys, interviews and questionnaires, and wrote letters to local councillors and MPs. Schrivener (2003) describes this type of work as action-orientated learning. In essence, the pupils examined what they saw as a real issue. This can be crucial in bringing geographical learning to life for pupils who have barriers to participation and who may not be 'switched on' by a more formal or academic approach.

Practical enquiries and investigations of this type provide meaningful activities over which pupils can exercise a sense of ownership. They also help pupils examine and clarify their values and attitudes which in turn helps to stimulate their involvement in the future of the environment. It is important that within the geography curriculum pupils are helped and enabled to recognise the personal role they have in the future of their planet. This sometimes begins with an investigation of a local issue that the pupils have identified and then moves out from the locality through the region to national and global questions. Relating the work to Local Agenda 21 is another way of helping the pupils to recognise that their voice is valued. If schools can build their own resources as a result of action-orientated fieldwork over a number of years they will enable pupils to feel that their work is valued. This surely means, as suggested by the head of St Joan of Arc School, that we need to go for a quality and not a quantity model of the curriculum and learning.

Basic principles

Following this discussion about geography fieldwork it is now possible to establish some fundamental principles regarding inclusion and pupils with special learning needs. These are:

1. All pupils are entitled to receive a geography curriculum that enhances their ability to be part of a democratic society now and in the future.
2. All pupils are entitled to receive a geography curriculum that increases their geographical knowledge in a manner that (a) increases their self-esteem (b) opens their minds to future geographical exploration.
3. All pupils have geographical knowledge. It is our task as teachers to:
   – recognise their existing knowledge
   – build/extend and develop this knowledge

– enable pupils to use their knowledge in present and future debates about the world we live in. (*Visions of the Future* (Hicks and Holden, 1995) is a valuable resource in this area.)
4. All pupils have a right to enjoyment which, as *Excellence and Enjoyment* (DfES, 2003) declares, is their 'birthright'.

These fundamental principles are clearly recognised in the national curriculum (DfEE/QCA, 1999): setting suitable learning challenges; responding to pupils' diverse learning needs; and overcoming potential barriers to learning and assessment. These basic principles and the ideas can also be matched directly to the concept of inclusion and the requirements of the Special Educational Needs and Disability Act 2001. *The Code of Practice* (DfES, 2001) makes it clear that:

> From September 2002 schools will be required not to treat disabled pupils less favourably for a reason relating to their disability and to take reasonable steps to ensure that they are not placed at a substantial disadvantage to those who are not disabled (DfES, 2001, p. v).

This places the emphasis firmly on adapting the environment to suit the pupil rather than trying to make the pupil fit an inappropriate context. It also provides a clear definition of SEN that can be used as a base line for planning and thinking. The definition makes it clear that pupils have a learning difficulty if they:

■ have a significantly greater difficulty in learning than the majority of pupils of the same age; or

■ have a disability which prevents or hinders them making use of educational facilities of a kind generally provided for pupils of the same age in schools within the area of the local education authority (DfES, 2001, p. 6).

The basic principles and the statements regarding disability and learning difficulty need to be placed within the context in which you are working – your primary school, its pupils and staff, its geographical location and its existing resources. To this must be added, I would suggest, the time available for geography and geography-related activities, especially fieldwork, given the importance attributed to the core subjects in the NLNS and Ofsted inspection regime. It is also important to consider the needs of those pupils who might be described as 'gifted and talented'. The key principle as always is responding to the pupils with regard to individual and collective learning needs and to build on strengths.

Research by the author in the last four years in the north-west of England suggests that the time currently available for geography in many primary schools amounts to less than one hour per week. This raises the very difficult issue of the balance between time for curriculum coverage and time for understanding subject matter. This problem is likely to be especially

Primary Geography Handbook

Photo | Paula Richardson.

acute for pupils with particular learning needs. It also raises the question of how to find time to extend a study of geographical issues both for those pupils who might be described as 'gifted and talented' and those who are enthusiastic about geography even if they are not high achievers.

Whatever subject is involved, education requires balancing curriculum requirements on the one hand with pupils' learning needs on the other. In order to keep self-esteem and motivation levels of both staff and pupils high, it is essential that this match is appropriate and that all involved achieve success. This means that in the primary school it will be crucial for the SENCO, the geography co-ordinator, the age-phase managers and senior mangers to work together to examine the following key questions:

1. What do we believe we can reasonably explore at either first or second hand with our pupils, within the environment in which we live, to develop their geographical understanding?
2. How can we ensure that all pupils are introduced to a fundamental core of geographical understanding – with clear possibilities for extension and breadth – in the time we are prepared to give it?
3. How can we minimise the barriers that some pupils experience in geography, including physical and sensory disabilities?
4. What resources, both human and physical (books, maps, videos, ICT, CD-Roms etc.), do we need in order to achieve the above?
5. Do we have an efficient and effective recording system that follows the pupils through from nursery to year 6 that will enable the secondary school to build on the skills and concepts the pupil has acquired and which identifies the fieldwork they have undertaken?
6. What are the possibilities for action research relating to geography and special needs that might involve a range of staff and relate to the school's CPD and staff development plan?
7. What links can we establish outside the school – special schools, primary and secondary schools, industry, colleges and universities that will enhance the quality experience in geography?

It is to be hoped that these guidelines will provide a framework within which pupils with special educational needs can undertake geography in a manner that opens doors, eyes and imagination. The alternative is to risk closing down their natural wonder about the world because we decide they are not capable of understanding aspects of the subject due to a low reading age or a behavioural, a physical or sensory difficulty.

An inclusive approach

There are many specialist organisations that can provide specific and detailed material for a range of learning needs. You can find out about these from the national curriculum website. However, it is also important to recognise that labels such as dyspraxic, dyslexic, BESD (behaviour, emotional and social development needs), blind and autistic do not delineate homogeneous groups but simply recognise a continuum of learning issues. So, for example, when a blind student made it clear to me that he would let me know if he needed help and asked me not to make generalised comments, he was in fact requesting not be categorised.

In other words, I was to look at him as an individual who happened to have particular learning needs, some of which might be related to his blindness, but others that might stem from his personal learning style, enthusiasm for the subject, self-confidence and so on.

It is therefore crucial in applying fundamental principles that we always focus on the individual learner and not the so-called 'disability with learning'. When examining the physical environment of the school, for example, we need to take two standpoints: general issues to do with education and teaching; and specific issues related to individual pupils.

This, I would suggest, is the hallmark of quality teaching. The principles that underpin quality teaching are the same for all pupils, not just those described as having learning difficulties. Recent educational thinking and research confirm the validity of this approach. Gardner's theory of multiple intelligence (1993); the notion of emotional intelligence (Goleman, 1996) and the idea of accelerated learning (Smith, 1998) all suggest that we need to adopt a broad teaching repertoire. This repertoire needs to based on a careful assessment of individual learning needs at any precise moment in time.

An example of this 'catholic' approach can be where pupils are put into groups of three. One pupil closes their eyes, another gives directions and the third acts as scribe noting the instructions word for word. The 'director' navigates the pupil from A to B in the classroom, the playground or the school field using simple instructions – right, left, forwards, backwards – using paces as the unit of measure. On completion of the journey the pupil who had their eyes closed draws the journey and the three compare notes. Each pupil gets a chance to play each part. It is fascinating for sighted pupils to see how accurate many pupils with sensory impairments can be in recounting their journey – whether via drawing or speech. This type of work can help in the development of graphicacy and spatial ability and illustrate the value of maps drawn to a uniform scale. One school in the north-west of England uses a large-scale street map of the area around the school projected and drawn on a wall to encourage pupils to tell and draw their journey to school. Pupils are encouraged to discuss a range of possible routes and give reasons for their choices. No route is necessarily wrong, but some may be better for specific purposes.

Clearly the work can be extended by the use of a compass and the use of games such as treasure hunts and orienteering. Activities like these also help to build trust and self-confidence. You might get pupils to draw and describe real and imagined journeys. If they have heard the story of *The Hobbit* they will be aware of a journey through different landscapes, or you might discuss the film version of *Lord of the Rings*. These stories could also be used as a stimulus for drawings (and maps) of an imaginary journey to a specific place or destination. Again this idea lends itself to pupils working in groups. Some pupils are particularly good at drawing real and imagined maps, some might provide the words and some could provide the illustrations. The final result might even be enacted as role play or drama.

In terms of problem solving it can be valuable to start at the local level. This allows pupils to bring some initial knowledge to the issue or problem. You might then extend the scope to include problems which, while they may not be evident in your locality, still involve pupils and will be perceived as relevant. For example, when you plan a day trip for infants to a castle think about how each group could be presented with different geographical problems. What is the best route? What is the quickest route? What is the cheapest route? What is the most interesting route? Other questions to consider are whether the pupils need a map, what they need to wear and where they could find shelter if it rains. By posing geographical questions you will be challenging pupils to use literacy, numeracy and other skills in seeking the answers or solutions.

The problems that you select can either be open-ended or set within parameters decided by staff. Environmental issues and Local Agenda 21 provide a particularly useful focus and have the advantage that they can be illustrated locally, e.g. graffiti. Again, Local Agenda 21 issues might be used as the platform for moving on to wider national and international issues. Try to encourage pupils to see that politicians and policy makers struggle to find answers as there are often no quick solutions. It can be difficult for pupils to recognise complexity and it can be salutary for them to recognise that teachers too may not have the answer and that our ideas change with 'new knowledge'. This is part of the essence of geographical thinking.

A variation on this theme is for pupils to collect used plastic bottles from the school and home over a specific period of time and store them in rubbish sacks. Speculate on how many sacks would fill the room. This can then lead to a debate on the impact of rubbish on the future of our planet and ways we might deal with it. Another example comes from a small village school in Wales. When the only newsagent in the village closed, pupils were asked 'How might we get newspapers?'. Working in groups they came up with a range of possible ideas. (For further ideas and suggestions look at the Eco-schools handbook or website (www.eco-schools.org.uk) which provide a range of resources and ideas which place pupils at the centre of the decision-making process.)

Practical activities of this kind which involve learning through the senses can be vital for some pupils and awaken their sense of awe. Again it is possible to move from local to national to global, particularly by using the internet and television material. Video news clips that pupils can go back to again and again are ideal for stimulating, reinforcing and extending learning, especially when combined with stories about the lives of people in other countries. Your school needs to reflect on how best to build a resource collection for geography, catalogue the material and place it in the library for future use. This material can also inform the debate with regard to the past and the future. Video clips and media articles collected over a period of time are an invaluable tool for helping pupils to recognise that ideas, values and feelings in geography change over time.

It is also important to consider the way the school and classroom are organised. We need to do this from a standpoint not of guilt or pressure from a government initiative but because we recognise how the learning environment we create has an impact on pupils, the subjects, the environment and our teaching in general. We also need to acknowledge that as teachers we have an 'emotional link' to our teaching and that some pupils will test this emotional link. Pupils who present challenges in behaviour and/or learning can put pressure on our personal credibility and resolve. It is easy to feel guilty and to say to yourself 'I don't think the pupil is learning. Help, what do I do next?'. An idea in one school has been to get a Teaching Assistant to take a pupil who presents a particular challenge within the spectrum of BESD out of the classroom with a digital camera to take some geographical pictures of the locality. This intervention works because some pupils need to engage in a task more physically than others and require the reassurance of one-to-one support. We need to recognise that giving pupils the occasional opportunity to work outside the classroom environment does not contravene

Photo | *Anna Gunby.*

the principles of inclusion. Inclusion is not so much about a pupil's physical location as accepting and valuing the unique contribution they can make. These tensions are part of everyday school life. It is our role as teachers to see that pupils are provided with a supportive learning environment.

A good example of this approach was developed in a year 5 and 6 class in an inner city school where a significant number of pupils had learning difficulties that resulted in challenging behaviour. Here the class teacher developed, over an extended period of time, the notion of peer support and collaborative learning. Pupils were encouraged and expected to help and support each other on both a cognitive and affective level. One particular boy who was prone at times to become upset, aggressive and throw objects was supported by a group of girls on his table. The year 5 boy might have several outbursts a day. The girls were helped to recognise the signs of distress – usually because he was becoming frustrated with his work – and would offer help, distraction or seek adult assistance. The crucial fact was that the behaviour was not seen as naughty but as something that was understood but not acceptable. It was also making the boy unhappy and he would often end up in tears and then become the subject of fun. The class teacher saw this as a class issue and enabled the pupils and the boy to take increasing control of his behaviour, which gradually improved over the period of a year. Two factors were crucial in addressing this problem: the trust between the pupils and all the adults involved; and the deliberate use of class discussion to help the class community recognise the role that they had in supporting each other in the management of learning and behaviour.

The same approach was also used during geography fieldwork where the boy would be in a group to provide support (and record data) but not do his cognitive work. Placing the emphasis on assistance meant that any questions posed to the group needed to be open ended to enable him to contribute ideas. In a different situation a pupil with learning difficulties might need very specific questions. Again, other pupils and support assistance could be used to create an affirmative environment. In literature this approach can come under various headings from circle time, to peer tutoring and mediation. (Mosley (1993) provides much information on these areas.) The main point is that by discussing the learning environment pupils become involved in a collaborative process. This is not dissimilar to the notion of action-orientated learning except that the centre of the action is not a particular subject or activity but the learning environment itself.

This approach has important ramifications for the notions of collaborative learning and fieldwork. I have observed a PGCE primary student who happened to be blind teaching in a city centre school where the pupils and the student managed the environment and interacted successfully, each using their strengths in a supportive manner. The key questions are:

1. How do we maximise the staff expertise in geography?
2. How do we combine staff expertise in geography and SEN to meet teaching and training needs?
3. How do we allocate time and space for geography?
4. How do we systematically build up a range of resource material, some based on pupils' work, so that it is available for all over the coming years?

5. How do we make maximum use of parents, outside speakers and visits to establish a notion of continuity and development in relation to geography?

6. How do we manage the physical environment to ensure that we comply with the Special Needs and Disability Act 2001?

Specific issues

All pupils will have specific issues or difficulties at certain times, in certain subjects and with individual teachers. We need to try different approaches, recognise we will make mistakes and discuss problems with pupils as they arise. Pupils can often give you valuable insights into why something has not worked but they need to trust you first and you need to listen to them. This will take time, so be patient. In order to examine specific learning problems a team approach works best, especially if you have an agreed common philosophy. Once you have examined the physical environment of the classroom and the school has begun to build resources, then you need to build on pupils' strengths and individual learning styles.

It can be difficult and time-consuming to ascertain what skills and knowledge pupils with SEN have if we are too focused on the notions of literacy and numeracy as discrete skills. It is important that we create time and space to discuss and listen to our pupils. The Cognitive Acceleration through Science Education (CASE) group developed an approach to science that might well prove valuable with geography (Simon, 2002). My simplistic interpretation is that the building blocks for learning depend on the respect and value which teachers accord to pupils' views. So, for example, the traffic crossing issue discussed earlier would be an ideal vehicle for this approach. The use of open-ended questions which the teacher builds on and extends is also crucial, for example 'I see what you are saying but what if… 'Suppose we did this what might be…'. The aim is to provide a scaffold which helps pupils explore their thoughts and ideas rather than pupils trying to provide the right answer that they think we want.

The notion of providing a scaffold, which is something as primary teachers you do already, can be extended with regard to the learning needs of specific pupils. The scaffold might involve:

■ changing the font, colour or font size of printed material;

■ providing a visual timetable for pupils on the autistic spectrum;

■ the use of a TA, parent or another adult/pupil;

■ the physical adaptation of the classroom or environment you wish to work in.

Some pupils require a very specific routine that provides the security in which they can work. A pupil who is described as having Autistic Spectrum Disorder (ASD), for example, may have problems in organising themselves and materials in both space and time. They can manage the familiar better than novelty and seem to prefer activities and routines that they have experienced before. Other pupils may require a range of strategies with regard to managing the behaviour which should be covered within the school's behavioural policy – but again a collaborative approach is crucial.

The increasing role of TAs should also have a positive impact on the organisation of learning. There is also much to be said for a learning log in which pupils record their work using a file, video clip, computer disc or some other electronic format. Not only does this provide feedback and a record for the teacher, it also helps to open a geographical dialogue which can stay with the pupil and take them through the junior years and into secondary school.

Corbett and Norwich (1999) have explored the links between general learning pedagogy and specific learning programmes for individual pupils. They recognise that the programmes are difficult to write and manage and argue that there is increasing evidence that individual programmes can isolate pupils. This is why teachers need to plan creatively for a range a different needs. Problem solving, a key part of geography, can be particularly hard to include. Some of the examples already provided – the plastic bottles, using the *Lord of the Rings* film, why we might need a zebra crossing, eco-schools – all provide opportunities for pupils to be creative within the geography curriculum. Finding ways of promoting geographical literacy is what matters most. The pupil whose reading age is not yet commensurate with their chronological age may have other geographical strengths, skills and talents on which you can build. As teachers it is our job to uncover and use these in order to build for success. The key is how we recognise this and build on it. Be patient because it can take time to unlock the key to a pupil's geographical learning, and always plan and aim for quality and not quantity.

References

Carpenter, B., Ashdown, R. and Bovair, K. (2001) *Enabling Access: Effective Teaching and Learning for Pupils with Learning Difficulties*. London: David Fulton.

Corbett, J. and Norwich, B. (1999) 'Learners with Special Educational Needs' in Mortimore, P. (ed) *Understanding pedagogy and its impact on learning*. London: Paul Chapman.

Cownes, E. (2003) *Developing Inclusive Practice*. London: David Fulton.

DfES (2003) *Excellence and Enjoyment: A strategy for primary schools*. London: DfES.

DfES (2001) *The Code of Practice*. London: DfES.

Gardner, H. (1993) *Multiple Intelligences: The theory in practice*. New York, NY: Basic Books.

Goleman, D. (1996) *Emotional Intelligence*. London: Bloomsbury.

Hicks, D. and Holden, C. (1995) *Visions of the Future*. Stoke on Trent: Trentham Books.

Mosley, J. (1993) *Turn Your School Around*. Wisbech: LDA.

Schrivener, C. (2003) 'Getting your voice heard and making a difference: Using local environmental issues in a primary school as a context for action-orientated learning', *Support for Learning*, 18, 3, pp. 100-6.

Simon, S. (2002) 'The CASE approach for pupils with learning difficulties', *School Science Review*, 83, p. 305.

Smith, A. (1998) *Accelerated Learning in Practice*. Stafford: Network Press.

Websites

National Curriculum Online: www.nc.uk.net/index.html

Quality Circle Time: www.circle-time.co.uk/

SEN Teacher Resources: www.senteacher.org

SEN Toolkit: http://publications.teachernet.gov.uk

Teachernet: www.teachernet.gov.uk/wholeschool/sen/

Photo | Stephen Scoffham.

IN THIS CHAPTER YOU WILL FIND KEY IDEAS ON
ASSESSMENT • GOOD PRACTICE • INSERVICE TRAINING • INSPECTION • NATIONAL CURRICULUM •
PLANNING • RECORD KEEPING • TRANSFER AND TRANSITION

The geography subject leader

Most primary teachers deliver the full range of national curriculum subjects so it is expected that they will have a working subject knowledge of both the core and foundation subjects. However, it is also expected that teachers will co-ordinate one or more curriculum areas and be responsible for developing it/them across the school. These can be very considerable demands.

As a teacher you will possess more extensive subject expertise in some curriculum areas than others. However, the subjects you are asked to co-ordinate are not always the ones in which you are strongest. This chapter provides guidance to all who co-ordinate geography whether or not they have specific subject expertise. It sets out the responsibilities of the subject leader/subject co-ordinator and will help you match your background and experience with your responsibilities, your school's agreed agenda for action and your own professional development needs. Acting as a subject co-ordinator is an integral part of the leadership and management structure of a primary school. It also provides you with the opportunity to develop your skills at middle-management level.

The role of the co-ordinator is crucial to the success of geography teaching in a primary school. As a co-ordinator you will need to champion the subject and demonstrate its importance to pupils and staff. In addition, geography lends itself to cross-curricular work and provides a context for exploring many other subject areas. Teachers are increasingly encouraged to be creative in how they plan and deliver all that is expected of them within specific time constraints. Therefore the more flexible and creative teachers become, the more effective the cross-curricular work will be.

Superior subject knowledge is not necessarily a criterion for the choice of a co-ordinator. In some cases, particularly in small schools, geography is co-ordinated by the head teacher or deputy head teacher in addition to their other responsibilities. In larger schools, teachers have responsibility for co-ordinating only one subject. However, geography is also frequently linked with history and sometimes with RE, with one co-ordinator responsible for two, or even all three subjects. Typically, while senior teachers co-ordinate the core subjects and receive responsibility allowances, geography is

Which of these descriptions best matches the position you find yourself in?

- You were appointed this year as an newly-qualified teacher and were asked whether you would co-ordinate geography.
- You are in a small school where everybody has to carry a wide range of curriculum responsibilities and there was no one else to take responsibility for geography.
- You are the head teacher and no one else felt able to do the co-ordinator job so it fell to you.

- You last studied geography when you were 14, but you are interested in outdoor education so it seemed natural to give the co-ordinator job to you.
- You are already the history co-ordinator and geography and history are seen as being very similar.
- You are a geography graduate and it was assumed that made you an ideal co-ordinator.
- You have attended INSET or further study linked to the work of geography co-ordinator and feel quite confident about the role.

Figure 1 | *Geography co-ordinators come from a range of different backgrounds.*

often co-ordinated by the least experienced newcomer who receives no responsibility allowance at all.

Geography subject co-ordinators come to their posts in a great variety of ways. The descriptions in Figure 1, derived from conversations with teachers in Cheshire, give a sense of the diversity of backgrounds.

So it is that the teaching experience and subject expertise of geography co-ordinators may vary considerably. You are likely to teach most of or the entire curriculum, although you may combine a generalist role with advice and support for geography; you may be a semi-specialist; you may teach geography full-time. Whatever your background, to be a successful curriculum co-ordinator you will need the following:

- the professional skills to work with colleagues collaboratively, sharing curriculum expertise
- a clearly defined role backed up by a job description which is linked to the school management structure
- a sound working knowledge of the geography curriculum and access to appropriate INSET for professional development
- a perspective of the geography curriculum in the context of the whole curriculum framework
- the ability to produce a key stage plan or plans which meet statutory and school requirements
- the ability to produce detailed topic or unit plans
- the support of the head teacher and governing body and full recognition in the school development plan
- clearly identified budget provision

The responsibilities of the geography co-ordinator

These are shown in Figure 2 and expanded in the paragraphs that follow. The order in which they appear is not intended to indicate importance. The first thing to do if you are appointed to this post is take stock of your situation: your priorities are likely to emerge from that process. Look at where your school is now, where you want to be in the future and how you are going to get there.

Developing your own expertise

Taking stock

Preparing for an inspection

Planning the geography programme

Making cross-key-stage links

Managing resources

Devising an action plan

Measuring pupils' progress

Developing school self-review

Providing support for colleagues

Figure 2 | The responsibilities of the geography co-ordinator.

Taking stock

Discover what is expected of the geography co-ordinator in your school. There should be a 'job description'; if not, ask the head teacher for one. Find out what geography is already being done, and what is going well. Talk to colleagues to find out what they expect of you, and what confidence/expertise in geography they already possess, so you can identify areas where they would welcome support. Look at coverage of a scheme of work – many schools adopt the QCA units of work, others devise their own.

You also need to check the place of geography in the school development plan; it should include costed development targets. Undertake a scrutiny of geographical work and look for links between geography and other areas of the curriculum. This will show you what the pupils actually do in their geographical studies. It will also indicate where geography is taught well. This evidence will support you when you examine the professional needs for yourself and your colleagues.

Bear in mind that some of your colleagues will require more support than others. For example, if you have an NQT in school, focus specific attention on supporting this colleague. An audit of resources will also be very valuable. Don't forget to look for resources such as information books and 'big books' which can be used to support geography in a cross-curricular way.

Examine the long-, medium- and short-term plans for geography. These will indicate progression and continuity of the subject. Remember that pupils should have a recurring experience of geography throughout the primary years, so if there are any gaps you will need to make plans to fill them. Make sure that colleagues are using the enquiry approach (see Chapter 7), and integrating the study of places, themes and skills. Think also about what geography you want the school to be doing next year – and in three years' time.

Planning the geography programme

The programme for geography should be clearly set out and each unit should be supported by a teaching scheme. You should also ensure that the programme sets out how continuity and progression in geography are to be achieved both across and within the key stages. Check that your colleagues are aware of the whole geography programme, not just the parts they teach; preferably, they should have been involved in its development. This will form the basis of your long-term plans.

Geography lends itself extremely well to the development of thinking skills such as information processing, enquiry and evaluation. As a school staff consider how you can develop other vital skills through your geography work. *Excellence and Enjoyment* (DfES, 2003), the government's strategy for primary schools, encourages teachers to be more flexible in their approach to teaching by making use of a wider range of teaching styles and strategies. This could form the basis for a valuable early evening INSET session.

The programme should also identify links with other curriculum areas and the opportunities to reinforce literacy, numeracy and ICT through work in geography. Pupils should be able to enjoy geography in a variety of ways (fieldwork, visits, visitors, ICT, videos) and these will have to be planned into appropriate points in the programme. It should also be possible to take opportunities for 'incidental geography' (e.g. current events, school assemblies, displays).

Managing resources

Your school should possess enough teaching resources to deliver the geography programme to pupils of different abilities through a range of varied learning experiences. The first thing to do is check which resources you already have and identify any gaps; consider the need for a good balance of books, maps, photographs, videos and software when you decide how they should be filled, prioritising expenditure sensibly by concentrating first on the main teaching units. The school library may contain resources which will support geographical work so include these in your survey; both fiction and non-fiction geography resources should support the literacy programme. Ensure that any purchases reflect priorities within the school strategic development plan (SSDP). Also be aware of local and national priorities. Make good use of the internet by compiling a list of useful websites both for teachers and pupils to use.

Ensure that your resources support active, investigative learning. In particular, ensure resources take account of different learning styles – this is required to support inclusion. Don't forget local resources – people, buildings, environments – which may support fieldwork. Make sure that your resources are easily accessible and that all your colleagues know what and where they are.

Keep your resources up-to-date and ensure that you receive a good range of product catalogues and resource lists. Joining the Geographical Association will give you access to up-to-date ideas via *Primary Geographer* as well as member's discount on a range of teaching resources. If you do nothing else, you should make subscribing your priority.

Measuring pupils' progress

It is the co-ordinator's job to ensure that colleagues are clear about how to assess geography, and that their judgements are in line with the whole-school policy and practice. Chapter 23 gives detailed information about how to assess pupils' work in geography.

Teaching plans should also identify clear learning targets; work across classes should be compared, and work recorded and reported to parents. Staff meeting time should be devoted to this assessment. Achievement should inform the planning of future work. Feedback to pupils is important; they should know their strengths and weaknesses and have the opportunity to assess their own work. Most important is that they are given time to reflect on their learning and how they can make improvements to it.

Providing support for colleagues

One of the most important roles of the co-ordinator is to facilitate INSET for colleagues. INSET sessions can range from 'one-to-one' advice and support; to the staff meeting after school; to a half day or whole day involving all your colleagues and sometimes including the support staff. Clusters of schools can be effective for networking. Make use of the Local Education Authority Advisers and County INSET courses. Finding informal time to chat to colleagues always pays off in terms of generating mutual support.

Here are some thoughts about planning activities for your colleagues. It is recognised that this offers quite a challenge and may make you feel quite uneasy as someone who will be expected to have 'all the answers'!

■ Good research and preparation are essential – you may not be the expert, but you should be able to refer colleagues to an appropriate source of help. The Geographical Association's publications, especially *Primary Geographer*, can be very useful here.

Photo | *John Halocha.*

- One-to-one support is time consuming, but can be well targeted to need. It is particularly successful when combined with 'team teaching'; for instance, when you visit an outdoor centre for the day or overnight.
- Staff meetings after school tend to be time-pressured so aim for a crisp start and try to make sure you get the business done before colleagues have to leave. The golden rule is don't try and cover too much in the time available.
- Half and whole days are undoubtedly best. They work particularly well when you can involve your colleagues in devising curriculum materials themselves. Try and show your colleagues that the geography curriculum is within their grasp; for instance, their lives have a major geographical component simply in their journey to work.
- Inviting an adviser or someone from a college to run a session can be very effective, but they need careful briefing – make sure they address your agenda!

Developing school self-review

Most schools have some system for reviewing whole-school issues, in particular the effectiveness of curriculum management and teaching. However, evidence from inspections suggests that school self-review is one of the weakest parts of a school's planning cycle. Ofsted reports indicate that monitoring is a very powerful tool for improving the learning environment and for raising standards of pupil achievement, but that most co-ordinators are not able to monitor directly or even indirectly the quality of geography in their schools.

It would be helpful if the senior management team could monitor geography teaching throughout the school. As a co-ordinator you should be involved in this process. Conducting peer reviews, where colleagues observe each other in a supportive manner, is also very valuable. This monitoring process can effectively tie in with performance management. Furthermore, if geography is a priority within the SSDP, then it would be completely justified to focus on geography for performance management objectives, either at whole-school or class level or for your own professional development. Some of the areas a self-review should investigate are:

- the quality of learning and teaching
- the pupils' achievements
- the pupils' response to particular activities such as fieldwork visits
- geography's contribution to pupils' spiritual, moral, social and cultural development
- leadership and management, staffing, accommodation and resources
- provision for gifted and talented/SEN pupils
- health and safety issues, particularly risk assessment of fieldwork sites
- action research in the classroom

It is best to identify specific areas for review to make the process manageable and to avoid superficial treatment of important issues. Evidence for the review will come from a variety of sources depending on the item being covered. It should be qualitative and professional, not

solely quantitative. You might include discussions with teaching and non-teaching staff including the head teacher, parents and pupils past and present; book scrutiny; an evaluation of resources, including the library; a review of industry visits and fieldwork; and the efficiency of financial planning.

Ultimately self-review is intended to help schools create a balance between consolidation and innovation. It needs to be built into curriculum management to ensure the effective delivery of a stimulating and challenging curriculum.

Devising an action plan

The outputs of review will help you devise an action plan for geography. The plan will need a three-year overview and a detailed one-year programme. You will need to consider the following:

- Objective: What are we going to do?
- Rationale: Why do we need to do it?
- Success criteria: What do we expect to achieve?
- Action plan: How are we going to do it?
- Personnel: Who will do the different parts of the action plan?
- Resources: How much will we need to spend?
- INSET: What are our training needs?
- Timescale: When will it be completed?
- Monitoring: Who will monitor progress, to whom will progress be reported?
- Evaluation: How will we measure success?

The completed action plan will identify priorities for geography and contribute to the SSDP. Given the right climate, drawing up the plan will provide an opportunity to celebrate success – something which can often be overlooked in these times of constant change.

Making cross-key-stage links

The geography co-ordinator has an important role in ensuring that pupils receive a coherent geographical education as they move through their school life. Prior learning must be acknowledged, celebrated and assessed at the start of pupils' formal education, and at all points of transfer throughout their school life. As Blyth and Krause argue: 'Young children become geographers before they become pupils, because they begin to think and feel about people and places they know, and also about unfamiliar and distant places, from an early age' (Blyth and Krause, 1995, p. 11).

Both pastoral and curriculum-related information needs to be exchanged, especially when pupils move school. This is most important at the key stage 2/3 boundary, which often marks a significant change in the style and organisation of the geography curriculum. The following suggestions will help ensure a smooth transition:

- sharing resources between your partner secondary school(s) to enable you to begin a curriculum dialogue;
- forwarding information about the localities, in addition to the local area, which your pupils have studied;
- acknowledging that secondary schools tend to use themes rather than places (localities) as the main curriculum building block. (Note: there are four key stage 2 themes which carry into key stage 3);

■ the high school acknowledging that your pupils will have a good working knowledge of their local area which can be shared with pupils from other schools in year 7.

Developing your own expertise

It is now recognised that continuous professional development (CPD) is essential to effective curriculum leadership. Your personal programme should be agreed in your appraisal and/or form part of the school's development plan. Most geography co-ordinators need two main things: externally provided INSET and the help of a curriculum adviser. The former will offer broad-brush advice, applicable to all schools, whereas the latter can be tailored to your individual needs. The problem is that both can be expensive. One way round this difficulty is to attend the Geographical Association Annual Conference which is held in a different location each Easter and offers both broad advice and specific support at minimal cost.

Your professional development can also be enhanced by action research. Finding out about the problems which pupils encounter in geography is both fascinating and rewarding. If you and your colleagues focus on a particular aspect of teaching you will be able to analyse your respective skills. This also gives you an opportunity to assess the work of your colleagues in a supportive and non-threatening manner.

Another approach to CPD is to use networks and cluster groups. Other schools will have the same problems as you have. By working with colleagues in similar schools you can develop an informative and supportive liaison with relatively no outlay. Cluster groups can often work well with feeder schools working with the local secondary school. This cluster work can also support transition between key stages 2 and 3. Work carried out by a co-ordinator can also be tied into performance management objectives.

The Teacher Training Agency has recognised that curriculum leadership is a demanding role and has been developing recognised and certificated qualifications. New standards for Qualified Teacher Status were put in place in June 1998 and larger requirements for all Initial Teacher Training (ITT) courses later that year. Further work is afoot not only to establish statutory requirements for ITT courses, but also to recognise and accredit the role of curriculum leaders.

Preparing for an inspection

In preparing for an inspection, the geography co-ordinator is best placed to make sure that the presence of geography in the school is clear (plenty of geographical work on display) and well integrated into the whole curriculum. Make sure that colleagues recognise the distinctive contribution that geography makes to the school curriculum (world knowledge, sense of place, pattern and process, environment, skills), and that they understand how it also contributes to basic skills and provides effective learning opportunities for all pupils.

You will need to find out if there was a reference to geography in the main findings in your last school inspection report and if your school prepared an action plan for this area. Reviewing the curriculum using the Ofsted framework will also help to identify those aspects of geography which could be better organised. The inspectors will certainly review the progress your school has made since their last visit. The inspectors will also be checking that

teachers are working to clear plans and have proper resources and that the pupils are challenged and motivated. This aspect of the work of a co-ordinator is discussed more extensively in Chapter 22.

How do I construct a geography policy?

The purpose of a geography policy is to inform parents, governors and other interested people about how geography is taught in your school and how it contributes to pupils' development. The policy should reflect a consensus of staff views on the place of geography within the whole-school ethos. Schools have a statutory obligation to provide a policy for each area of the curriculum.

Your policy should contain clear and simple statements about the nature and importance of geography. It should be brief and jargon-free, and supported by a scheme of work. The main elements of a geography policy are given in the outline below. It is included as an example of good practice, but your school will need to respond in its own particular way to match its own needs. It is based on the geography national curriculum programme of study, and although the specific requirement to follow the programme of study has been relaxed, it is important for any policy to offer viable schemes of work which deliver geography as part of a broad and balanced curriculum.

An example of a geography policy statement

Introduction

This policy outlines the purpose and management of the geography taught and learned in the school. The school policy for geography reflects the consensus of opinion of the entire teaching staff, brought about through discussion in staff meetings.

A structured framework has been designed, allowing for progression and continuity across the primary sector. The implementation of this policy is the responsibility of all the teaching staff.

The nature of geography

Geography is concerned with the study of places, the human and physical processes which shape them and the people who live in them. Pupils study their local area, and contrasting localities in the United Kingdom and other parts of the world.

Geography helps pupils to gain a greater understanding of the ways of life and cultures of people in other places. The study of the local area forms an important part of the geography taught at our school, particularly at key stage 1. Enjoyable geographical activities are planned to build upon the pupils' knowledge and understanding of the local area. Through our teaching of geography we aim to:

- stimulate the pupils' interest in and curiosity about their surroundings;
- create and foster a sense of wonder about the world;
- inspire a sense of responsibility for the environments and people of the world we live in;
- develop pupils' competence in specific geographical skills;
- increase the pupils' knowledge and awareness of the world;
- help pupils acquire and develop the skills and confidence to undertake investigation, problem solving and decision making.
- We hope that pupils will increase their knowledge and understanding of the changing world and will want to look after the Earth and its resources. We hope that they will begin to develop respect and concern for, and an interest in, people throughout the world regardless of culture, race and religion.

Roles and responsibilities

Each member of the teaching staff will have responsibility for the teaching of geography and they will need to ensure that their own knowledge is continually updated. The school has a geography co-ordinator to assist this process, and take specific responsibility for geography issues in the school. The geography co-ordinator should be a member of the teaching staff who has been trained to lead the teaching of geography. It is his or her responsibility to:

- support colleagues in teaching the subject content and to demonstrate good teaching where required;
- audit current practice;
- instigate and organise teaching programmes, planning documents and schemes of work where necessary;
- develop a school policy;
- resource the curriculum by:
  - keeping abreast of the latest teaching methods and materials by subscribing to the Geographical Association termly magazine, Primary Geographer, and storing up-to-date publishers' catalogues and information about new software and equipment in the geography curriculum file;
  - renewing, updating and complementing resources where necessary, and drawing up action plans accordingly;
  - building up a collection of video and audio recordings;
  - ensuring that colleagues are made aware of the resources available in school and their location;
- facilitate the standardised assessment of pupils' work at moderation meetings;

- be a consultant to colleagues, and plan the geography content of their teaching on a termly basis. Copies of the medium-term plan are located in the geography curriculum file;
- keep people informed of possible visits, exhibitions and courses.

Entitlement

Geography is one of the national curriculum foundation subjects with designated programmes of study linked to descriptions of attainment levels. Schools have been advised that as a starting point they might spend 33 hours per year on geography at key stage 1 and 36 hours at key stage 2.

At key stage 1 and 2 the programme of study is divided into a number of areas including:

- geographical enquiry and skills
- the study of key ideas such as change, patterns and processes
- the study of localities
- the study of themes

At key stage 1 pupils study their school locality and a contrasting locality either in the UK or overseas. At key stage 2 pupils study a locality in the UK and a locality in a country that is less economically developed. There are three themes at key stage 2 - water and the landscape, settlements and an environmental issue.

All pupils are entitled to access the geography programme at a level appropriate to their individual needs. Those pupils who work below the levels defined by the geography national curriculum for their key stage will be given additional assistance by the class teacher. Progress will be carefully monitored; if necessary, and with full parental understanding and support, internal special needs help will be made available. We aim to meet all the needs of individual pupils in the school, whether the pupils are designated as having special educational needs or are gifted and talented. The school fully supports the policy of inclusion.

Leadership

The geography co-ordinator will be given non-contact time in order to develop geography throughout the school. She/he will be expected to report back to the SMT in school of her/his findings. It will be expected that the co-ordinator will scrutinise planning and books, audit resources, examine CPD for the staff, observe lessons, model exemplar lessons, write and implement the geography action plan and deliver staff meetings and INSET.

The geography co-ordinator will be expected to develop the subject according to the school development plan and will work to develop learning and teaching according to the ethos of the school. The co-ordinator will work within any local and national priorities.

Implementation procedures

The national curriculum programmes of study define the content of the school curriculum for geography. We follow the advisory units for geography devised by the QCA to help schools deliver studies on particular topics. In our school geography is the main focus for teaching for one term each year. However, wherever possible, some geographical skills or place knowledge will be included in studies with other foci. Current events ranging from local issues to floods and earthquakes overseas are used sensitively and appropriately to promote learning.

Places

As well as undertaking specified place studies, pupils at key stage 2 also need to learn to locate significant places and environments in the UK, Europe and wider world. Whenever possible we develop this locational knowledge in the context of our work on places and themes. We also seek to give emphasis to the lives of real people in order to avoid stereotyping, give equal emphasis to the roles of men and women at all levels in society and create positive images of parts of the world that may be much poorer economically than ourselves.

Skills

The geographical skills outlined in the programme of study at both key stages are integrated with work associated with places and themes. Throughout the school pupils are encouraged to use appropriate geographical vocabulary to promote their thinking. We encourage practical activities and investigations within the classroom and further afield, together with the use of ICT equipment.

Themes

The study of places and themes are linked at both key stages. However, to allow in-depth study and ensure that the work is undertaken at a range of scales from local to national some studies are discrete. An investigative approach to geography involving pupils' active participation in enquiry, fieldwork, mapwork and the use of ICT is promoted throughout.

Where does geography occur in the curriculum?

- as a major focus for a topic
- as a smaller element within a topic
- as a discrete lesson/talk
- as part of an assembly

- in discussion (who saw the news last night?)
- during story time
- in displays both in the classroom and in other areas of the school
- as an integral part of residential visits. The school visits outdoor educational centres where geographical skills can be confirmed in a practical way. (All risk assessments are held with the head teacher, the teacher in charge, the governing body and the educational visits co-ordinator.)

Classroom organisation and teaching methods

Classroom organisation will depend on the needs and abilities of the pupils and also on the aims of the lesson. However, a variety of approaches such as whole-class lessons, group, paired and individual work are experienced by pupils in their geographical work. Access to resources is sometimes a determining factor in classroom organisation. Teachers are mindful of the ways in which pupils learn. The teaching of geography reflects different learning styles to ensure full inclusion.

We believe fieldwork helps to promote learning in all aspects of geography and we seek to incorporate it in all aspects of the curriculum plan. When they are engaged in fieldwork pupils are expected to behave in a considerate and responsible manner, showing respect for other people and their environment.

All pupils have opportunities to use ICT as part of their geographical work. They also have opportunities to use the following resources: globes, maps, atlases, pictures, aerial photographs, compasses, measuring equipment, cameras, books and games. Most of these items are kept in the resource area and are readily accessible to staff, who will select resources appropriate to the needs of the pupils. The interactive whiteboards are an excellent resource enhancing pupils' learning.

Assessment and recording

Pupils' progress is assessed and monitored during the year through normal teacher planning and observation. Individual records, and portfolios of pupils' work, will chart progress. A class record of areas/topics is kept to assist with future teaching and planning for continuity and progression. Pupils' attainment is also monitored against the expectations set out in the QCA units of work. Parents are informed of their child's progress at termly parents' evenings and via the annual reports which are sent home in the summer term.

Special needs

Differentiation in terms of learning objectives, tasks, teaching methods and resources will be planned for pupils with special educational needs. All pupils should have access to materials and opportunities suitable to their specific needs. Exceptionally able pupils need to be challenged with open-ended tasks which provide opportunities to tackle more complex issues and use a wider range of resources.

Review and revision

This policy was drawn up in 2004 by the geography co-ordinator and is based on the school's understanding of the geography national curriculum. It will be reviewed annually by the geography co-ordinator who will collect and collate the experiences and ideas of colleagues that arise during staff meetings and any designated INSET time.

Background documentation

This document was informed by reference to the National Curriculum (DfEE/QCA, 1999), QCA schemes of work (DfEE/QCA, 2000) and guidance from Cheshire LEA (Cheshire Advisory and Inspection Service, 1997).

Gorsey Bank Primary School, Wilmslow, Cheshire
Head teacher: Maggie Swindells, Geography co-ordinator: Fiona Pullé

References and further reading

Blyth, A. and Krause, J. (1995) *Primary Geography: A Developmental Approach.* London: Hodder and Stoughton.

Catling, S. (1993) 'Co-ordinating geography', *Primary Geographer,* 14, pp. 10-11.

Cheshire Advisory and Inspection Service (1996) *Planning the Curriculum in Cheshire Primary Schools.* Chester: Cheshire County Council.

Cheshire Advisory and Inspection Service (1997) *Putting it in the Right Place: Guidance for planning and organising the geography curriculum in your school.* Chester: Cheshire County Council.

DfEE/QCA (1999) *The National Curriculum: Handbook for primary teachers in England.* London: DfEE/QCA.

DfEE/QCA (2000) *Geography: A scheme of work for key stages 1 and 2. Update.* London: DfEE/QCA.

DfES (2003) *Excellence and Enjoyment: A strategy for primary schools.* London: DfES.

Krause, J. and Garner, W. (1997) *Geography Co-ordinator's Pack.* London: BBC Education.

Marsden, W. and Hughes, J. (eds) (1994) *Primary School Geography.* London: David Fulton.

Rainey, D. and Krause, J. (1994) 'The geography co-ordinator in the primary school' in Harrison, M. (ed) *Beyond the Core Curriculum.* Plymouth: Northcote Press.

Laura Murray   2nd May

What view would a bird
have of snawsdowns walk?

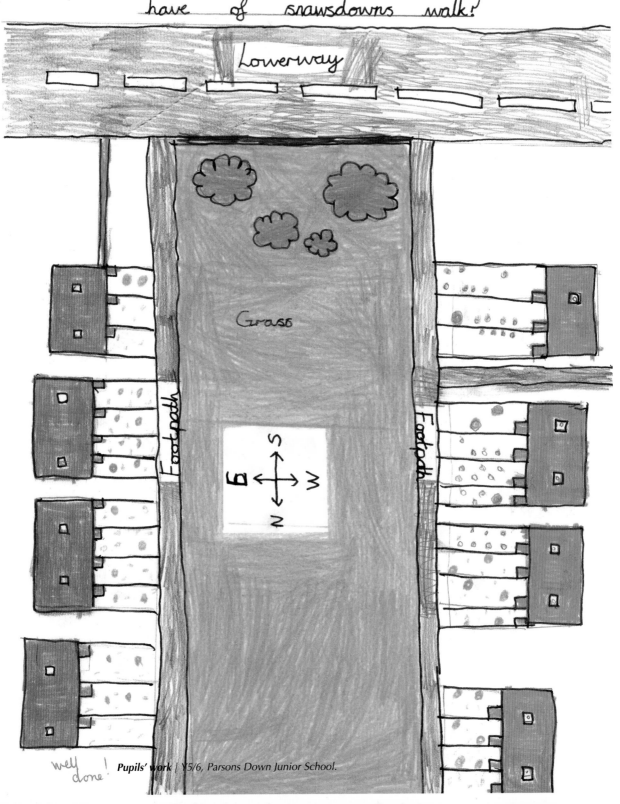

Lowerway

Grass

Footpath   Footpath

E ↔ W
S
N

well
done!

IN THIS CHAPTER YOU WILL FIND KEY IDEAS ON
DISPLAYS • GENDER DIFFERENCES • GOOD PRACTICE • INCLUSION • INSPECTION •
NATIONAL CURRICULUM • PLANNING • RECORD KEEPING

The inspection process

The quality of pupils' learning in geography and its contribution to the whole-school curriculum and management is at the heart of the inspection process. The 1990s saw the beginning of nationally organised school inspections, under the auspices of the Office for Standards in Education (Ofsted). Since then the system has been altered several times. Its purpose is now to: '… provide an independent, external evaluation of the quality and standards of the school'. The inspectors have to follow specific guidelines when they write their reports. They: '… must tell the school what it does well and what it needs to do to improve. They must explain how and why they have come to their conclusions' (Ofsted, 2003). They also look at whether or not the school has improved since its last inspection.

The statutory programmes of study for geography have also been altered several times since the inception of the national curriculum. The introduction of the national literacy and numeracy strategies in the late 1990s had a considerable impact on the subject from which it has still not recovered. Geography continues to be adversely affected even after the revisions to the curriculum which were implemented in 2000. However, government advice (QCA, 2002) reminds us that geography must be included in the primary school curriculum and recommends that, over a year, 30 hours or more should be allocated to geography in key stage 1 and 33 hours or more in key stage 2. Most schools still fall far short of this and the Ofsted annual subject reports (see Ofsted 2001; 2002a) make it clear that: '…insufficient time and resources are being allocated to promote adequate standards in geography' (p. 4). Furthermore schools are now being challenged to be more creative about how they organise their curriculum, to include foundation subjects such as geography more effectively in their lesson planning and to enthuse both staff and pupils in order to help raise standards and quality of learning. *The Curriculum in Successful Primary Schools* (Ofsted, 2002b) stresses the importance of effective humanities work and advocates the following:

■ Planning and organising the curriculum in separate subjects but making links between subjects

■ Blocking a few foundation subjects including geography over several weeks to reduce pressures on planning

■ Grouping no more than three or four subjects under a theme where there were strong links between them

■ Blocking practical work (e.g. geography fieldwork or mapwork) for several afternoons for a short period, allowing for intensive and sustained work

■ Devoting a whole day or sometimes a whole week to a subject or theme (e.g. Europe, news, our neighbours, near and far week)

Inspectors will bear these issues in mind, along with other ones highlighted in current annual Ofsted reports, so subject leaders would be wise to be familiar with recent documentation. The inspectors will also probably focus on issues arising from the school's self-evaluation and examine how geography contributes to the school's overall standards and achievement, leadership and management. In particular they are likely to consider:

- Time allocation
- Depth and breadth of the subject and its contribution to the whole-school curriculum
- Quality of teaching and learning, including enquiry, independent learning, fieldwork and other practical activities
- The extent to which geography teaching is contributing to pupils' standards of work in literacy, numeracy and ICT
- Your role as a leader and manager of the subject
- Whether geography is inclusive of all pupils

Inspection can help individual teachers by validating the geography they are doing with pupils or pointing out where weaknesses lie. Inspection is also useful for subject leaders; the identification of strengths and weaknesses by an experienced and objective outsider can enhance the good subject leader's judgements, confirm his or her hunches and validate or stimulate ideas about the way to progress. Inspection is intended to be supportive and developmental. It should be viewed with less fear or threat than has been the case in the past. Clearly an element of threat always remains, not least because the process is relatively infrequent and therefore unfamiliar. Good preparation always helps, and should make the experience more tolerable and positive. Working on the premise that the more you understand about the inspection process and its purpose, the more effectively you will cope with it, this chapter highlights what geography inspectors are seeking to find out. It is addressed particularly to the geography subject leader, and considers the three phases of an inspection – before, during and after.

Before the inspection

As subject leader, you should try to develop your own opinion of the strengths and weaknesses of your subject. It is important to prepare for the inspection properly through school self-evaluation, school improvement planning and your own action planning. You will then know where you stand as regards paperwork (the policy, schemes of work, resource listings and classroom practice) and the translation of this paperwork into real geography teaching and pupil achievement in geography. The scheme of work, which consists of the long-term/key stage and medium-term plans, is a key link as it helps interpret the policy in terms of practice.

Most schools now have a policy and a scheme of work, either of their own or developed from QCA's units or a combination of both. Sometimes these schemes will have been followed for a number of years but they may be going through a process of updating, experimentation or evaluation and alteration. The key is not the paperwork itself but actual classroom practice and the school's awareness of what it is doing in geography. Nevertheless, most subject leaders will have a file which contains all documentation relevant to the subject and includes some examples of levelled pupils' work as well. Indeed, many head teachers wisely expect this as part of the role of the geography subject leader.

Before the inspection you may find it useful to think through your answers to the questions below. They will help you to prepare for the inspection by focusing your attention on how you organise and deliver geography in your school and you may be asked about this during the inspection.

- Do you as subject leader have a vision for the subject in your school?
- What is the place of geography within the school improvement plan and the school self-evaluation process?
- Have you acquired a view of what standards pupils reach in different years or key stages? (Use www.ncaction.org.uk can help with this)
- What are your expectations about standards in geography in your school?
- Do you have a notion of how much continuity and progression pupils have in their geography learning, especially in map and atlas knowledge and skills?
- In what ways does the school monitor and evaluate geography standards?
- If you use the QCA units, to what extent do you modify and adapt them to match local needs and resources? (A key Ofsted issue)
- How is pupils' progress in geography assessed against standards, rather than just coverage? (Another key Ofsted issue)
- What do you do with the information you acquire from assessments?
- How do the staff and pupils feel about the subject? Do they enjoy it? Do you have evidence for this, e.g. feedback from pupils interviewed?
- Do you know how geography has improved since the last inspection?

Although it should not be necessary to make a special effort or put on a show for the inspector, most people like to smarten up and do their best to impress just as when they have guests at home. You can use the following checklist to help:

- Are the resources where they should be if they are not in use?
- Are they tidy?
- Do staff really know about the resources available? (High staff turnover can have a bad impact on resource awareness.)
- Are there geography displays around the school?
- Can you give examples of how you support colleagues in their geography teaching?
- Have you led any staff development sessions on geography, even sharing ideas while looking at pupils' work to assess coverage and standards?
- Has the school been involved in other continuing professional development (CPD) such as the best practice research process?

Before they come the inspectors will have analysed any relevant non-geography-specific school documents the school will have provided in advance. These will include:

- the school improvement plan
- the school brochure
- the school timetable
- school data concerning maths, English and science targets
- the school self-evaluation form (S4)
- the pre-inspection evidence commentary provided for the team

- Is there a policy?
- Is it clear in its intentions?
- Will it be easily understood by staff?
- If some expected aspects are not dealt with in this policy, then is the reader referred to other school policies?
- Is geography part of the school improvement plan?
- How much professional development for geography have the subject leader and other teachers received?
- What budget is allocated to geography?
- Is the co-ordinator responsible for managing it?
- Are the resources adequate?
- To what extent do pupils use ICT in geography?
- Is adequate time allocated to geography?
- Is there evidence of the enquiry approach in planning and policy?
- Is there an indication of how speaking, listening, reading, writing and data handling and ICT support geography?
- Is there any evidence of the subject's explicit contribution to social, moral, cultural and spiritual education?
- Is there a geography school portfolio or pointers to a methodology for staff becoming familiar with standards in the subject?
- Is there a balance of geography across years and, ideally, terms?
- Does the documentation provide information about the emphasis given to fieldwork, around and beyond the school?
- Are places, themes and skills integrated in planning?

Figure 1 | *A typical list of questions, which an inspector may seek answers to.*

An experienced specialist inspector will have a checklist of questions which he or she uses to analyse paperwork in order to be as informed as possible about geography and whole-school issues when the inspection begins. The inspector will only seek further evidence from you to check out ideas or follow up lines of enquiry pertinent to the whole inspection. Try out the questions in Figure 1 on yourself just to check that you are aware of current strengths and weaknesses.

During the inspection

So you have done your best to prepare yourself and your colleagues for the inspection. What happens now? Typically the inspectors will work as a team gathering evidence to follow up particular questions which they have identified that relate to the inspection framework and the pre-inspection evidence. This may include subject-specific details but it will often involve generic issues like leadership, management and curriculum range. So, rather than formal subject leader interviews, you can expect to be asked for information on, say, your leadership and management of geography, the time you spend teaching it or how you handle inclusion and challenge the most able pupils.

Even if there are many geography lessons happening, the inspectors may not observe all of them and some may be observed in part for a particular thread of evidence. Lesson observation is now taking less priority than in the past, although of course it is still the most reliable way of judging teaching. Teachers often ask: 'Should we decide to put off fieldwork until after the inspection?'. Remember, fieldwork is part of good practice in geographical education and is often particularly motivating for pupils. If fieldwork falls naturally within the lessons planned during the inspection period there is no need to change what you are doing. It is up to the inspector to decide whether or not to accompany off-site visits, and this will usually depend on how far away they are and the rest of his/her timetable.

Subject leaders also ask: 'What if no-one is teaching any geography during the inspection period?'. The smaller the school, the more likely this is to happen, especially if geography teaching is blocked into certain parts of the term or the curriculum is organised holistically. In the past, the absence of geography teaching has been seen as a possible indicator of unsatisfactory practice, but inspectors are now tracking all other types of evidence, including speaking with pupils, to reach a conclusion about the nature of the geography curriculum. It

is the inspector's job to find out how well geography operates as part of the whole curriculum and to assess if units of work are effective in creating high standards. It is not their role to check whether you are using some pre-conceived model of their own or anyone else's. However, inspectors are likely to do a better job and provide more useful feedback if they do see some geography teaching actually happening.

Finally, on the subject of timetabling issues, teachers should make it clear to inspectors which subjects are being addressed in sessions or lessons if it is not obvious from the timetable. Classroom teachers may also be asked informally about geography during lesson observations or as inspectors see them around the school. For example, the inspector might ask a teacher what support they have had from the subject leader, and whether the geography resources are sufficient. The inspection time is an intensive period and the inspectors will be trying to gather as much information as efficiently as possible.

Lesson observation

Lesson observation is a key source of evidence. When inspectors visit classes to observe all or part of a lesson, they will be collecting information which can be categorised and graded under these four headings:

- Teaching
- Learning
- Standards
- Achievement

They may also collect evidence and grade or note how the lesson contributes to:

- attitudes, values and personal qualities
- the whole school curriculum
- care, support and guidance
- partnerships with parents, other schools and the community
- leadership and inclusion

You will see the inspector making notes on a form. You can expect the inspector to sit alone, watch and listen. If it is possible to do so without interrupting the flow of the lesson, he or she may well sit next to a pupil or with a group and go through their work with them or ask them questions about what they are learning in geography to get an idea of their level of achievement. Inspectors often have the knack of sitting next to the pupil you would least like them to question on geography! A good inspector will soon sense this and seek to redress the balance.

Figure 2 | *The features of good geography teaching and learning.*

- a properly planned lesson with specific learning objectives related to geographical knowledge, understanding and/or skills
- pupils are challenged to use and transform geographical information in some way, not just to transfer it from one place to another
- lesson plans which can be related back to the scheme of work, whether school designed or QCA adapted
- objectives of the lesson are shared with pupils at the start of the lesson
- fun starters and/or finishers to lessons, including short place-knowledge quizzes, for example
- mapwork linked to real places
- an approach is adopted which promotes the skills of geographical enquiry through fieldwork and practical activities within the classroom, in addition to the use of secondary sources
- well-prepared, stimulating, up-to-date resources, ready at the beginning of the lesson, e.g. artefacts and/or music of distant or local places, photopacks, videos, maps, books, atlases, globes, CD-Roms, websites, digital cameras, e-mail and other ICT tools
- resources are matched to the age, ability and learning styles of pupils in the class, e.g. type and scale of map or photograph, text, worksheet, writing framework
- specialist geographical terms and vocabulary are used where appropriate, e.g. 'hill' in year 1, and 'landscape' in year 6
- topical examples and real situations are used whenever possible
- there is development of an understanding of patterns, processes and application of skills
- there is expectation that pupils will work through geographical questions for short or longer periods to collect, record, represent, communicate and analyse data, be it information collected during fieldwork or classwork
- the teacher has reasonable subject knowledge
- opportunities are planned for pupils to discuss and recognise and weigh up environmental, aesthetic and cultural attitudes and issues
- the incorporation of thinking-skills techniques into activities
- a creative approach to the subject
- there is a good pace to the lesson – with pupils aware of timings
- teaching assistants support learning well
- there is a clear summary to the lesson, involving pupils in explaining or showing what they have learned

Inspectors sometimes take the opportunity to speak with a teacher sensitively during a lesson, if they judge that doing so will not interrupt what is happening. It is good practice for them to give you brief feedback at the end of the lesson – possibly by evaluating the strengths and weaknesses of your teaching. Sometimes, because of a very intensive timetable, it can be very difficult for an inspector to achieve this but you may always request feedback at the nearest possible time later in the day.

Inspectors are interested in how pupils behave and respond during lessons. Evidence of good learning and pupil engagement in geography include the following:

- good behaviour whatever the learning style – individual, paired or grouped
- good co-operation with others when working in the classroom or in fieldwork
- an inquisitive attitude and an enthusiasm for other places, people and cultures
- reasonable confidence in tackling new or familiar tasks (e.g. atlas work)
- an awareness of the points of view which others may hold about local issues
- a willingness to comment sensibly on the social, moral, cultural and environmental issues their geography raises
- the importance of other relevant subjects and skills such as literacy, maths and ICT in geography work

What are the features of good geography teaching and learning? Good teaching is teaching that promotes effective progress and high standards. In the case of geography it has many of the features identified in Figure 2.

How do inspectors judge standards and achievement?

Inspectors will look for evidence which assesses the pupils' strengths and weaknesses relating to such aspects as:

- standards or levels for pupils at the end of key stage 1 and key stage 2
- inclusiveness for ethnic minorities, EAL, girls and boys, SEN and gifted and talented pupils
- knowledge of places, what they are like and how people adapt to living in them
- how pupils are developing their geographical skills, including enquiry, through the use of maps, fieldwork and other visual and written resources, including ICT
- pupils' understanding of the patterns and processes which affect people and landscapes and the environmental links between them
- how pupils develop their attitudes and values by weighing up different viewpoints
- pupils' ability to apply aspects of geography already learned
- pupils' ability to use suitable geographical vocabulary in oral and written form

What provides evidence about suitable progress?

Inspectors will be looking to see if pupils achieve as much as they can in geography. Do they:

- make at least sound progress within the activities provided?
- talk with reasonable confidence about current and previous work?
- use previous knowledge and skills to progress, where possible?
- use maps and atlases as ready tools to find places and talk about them?
- give explanations about features, patterns or processes based on their research or work to-date?
- demonstrate reasonable depth of knowledge about a range of real places, especially when they can refer to maps or their recorded work as a stimulus?
- talk clearly (at their ability level) about what they have as an individual learned in the lesson by reporting back to a group or the whole class – as the lesson organisation permits?
- demonstrate progress in their geography learning over time as well as in a lesson, through what is recorded in their books as well as what they are able to talk about?

Clearly, even if evidence relating to all the above was available, an inspector will not be able to collect it all in every lesson: inspectors, like teachers, are not superhuman! However, most of this evidence will be available for some of the time in every lesson and it is an inspector's job to make an objective judgement about the quality of pupils' geographical learning in a lesson by focusing on as many of the criteria listed above as possible.

What happens if no geography lessons can be seen?

If there are no geography lessons taking place at the time of the inspection, inspectors will use the other evidence that they normally refer to during an inspection (Figure 3) and which they collect in order to make judgements. In addition they will need to have discussions about their geography work with individuals or groups of pupils representative of the age range of pupils in the school. This usually involves talking to pupils, and checking their books, in relation to work they have done, also using, say, a map, a photo or some pertinent geographical resource to focus the pupils in order to explore their current understanding and to assess how easily they can transfer this to new situations.

Source	Inspector's focus
Samples of work from each year for an able, average and special needs pupils, with a major focus on years 2, 4 and 6.	What are standards like? Is there evidence of progression? Is there a range of work indicating varied teaching and learning styles?
Geography portfolio, if one exists	Does marking tell pupils what they need to do to improve?
Pupils' records and reports	Is pupils' progress assessed and recorded? How are records used? Do reports communicate the pupils' strengths and weaknesses? Are targets geographical?
Teachers' short-term planning and lesson notes (in addition to pre-inspection planning evidence)	Are there clear learning objectives? Is work differentiated?
Geography resources – equipment and paper resources, ICT such as computer files of pupils' work (e.g. PowerPoint presentations), favourite websites etc., stored on the network	Are there enough? Are they accessible? Are they sufficiently up-to-date? Is ICT an effective resource? Do resources offer equal opportunities?
The school library ICT files of pupils' work and geography-specific websites or software	Is there a geography section? How extensive is it? How up-to-date is it? Is there a variety of atlases?
Wall displays	Do they promote enquiry skills? Are they recent? Do they represent pupils' work?
Staff meeting minutes	Have all the staff discussed geography? Has the co-ordinator led in-service training in the school?

Figure 3 | *Sources of school geography information.*

After the inspection

Co-ordinators should receive oral feedback at the end of the inspection, and the feedback session should be scheduled for you. You may find you need to deal with it alone, but if you can be supported by another member of staff it is useful for them to take notes on the points the inspector makes which indicate strengths and weaknesses. Listening and taking key notes at the same time can be difficult in this situation. The inspector should indicate to you what his/her final judgements on the quality of the geography in the school are likely to be, and any key strengths or weaknesses. You will probably receive more feedback of direct use to you and the school at this session than you will in the written geography summary in the inspection report, the main purpose of which is to inform parents.

While you are not at liberty to disagree with the inspector, you may certainly have a professional dialogue and check facts which have been used to make judgements if you feel they are incorrect. A wise subject leader will also use the feedback session to pick up advice for the future which the inspector may be pleased to offer, by chance or design!

After the inspection it will be important to share your written notes of the inspector's oral feedback (and any responses you made to it) with senior management. As well as reading the published inspection report carefully you also need to see if geography is cross-referenced in other sections, such as leadership, management, social, moral, cultural and spiritual education. For example, lack of monitoring may be referred to in geography but it may also be weak across the whole school and will therefore have to be highlighted in the post-inspection action plan.

Currently Ofsted publishes a geography subject section in its report on the schools it has inspected. Occasionally geography and history may be reported on together if little evidence was available for one or both. However the inspection regime may change in the future, it is likely that the report will always be made available for an audience beyond the school.

The report will highlight issues for action which your school has to address. Geography may figure directly in the main action plan in which case the head teacher and governors will oversee this initially, although as co-ordinator you will need to extract the relevant parts for your own action. If geography does not appear on the main action plan, then it makes sense for you to look at your school and subject improvement plan, and to readjust your geography programme as necessary in the light of the inspection.

When you are drawing up a plan to develop geography, you might find it useful to use the following list of headings to help you to focus in on priorities:

- the target
- evidence that shows that you have achieved the target
- people responsible for the target
- the date by which the target is to be achieved
- any known financial or other resource requirements

Action planning should be seen as the final positive phase of the inspection process. You have prepared for the inspection, survived the experience and now should be able to use it to help you plan a systematic way forward in the light of any weaknesses identified. Even if all the judgements were positive, you will still need a maintenance action plan for geography to ensure that it receives the attention it deserves, and does not become sidelined or overlooked.

References

Ofsted (2001) *Geography in Primary Schools (2000-01)*. www.ofsted.gov.uk

Ofsted (2002a) *Geography in Primary Schools (2001-02)*. www.ofsted.gov.uk

Ofsted (2002b) *The Curriculum in Successful Primary Schools*. London: Ofsted.

Ofsted (2003) *Guidance on the Inspection of Nursery and Primary Schools*. www.ofsted.gov.uk

QCA (2002) *Designing and Timetabling the Primary Curriculum: A practical guide for key stages 1 and 2*. London: QCA.

Resources

Good geography teaching requires a wide range of supporting resources. In an ideal world teachers and pupils might collect their resources themselves. In practice the use of prepared teaching materials will considerably enhance the delivery of the subject. There are a huge range of resources available. Commercial publishers, charitable bodies and the media all produce structured materials designed to meet the latest curriculum requirements. The Geographical Association has an extensive catalogue for primary schools (www.geographyshop.org.uk).

While there are plenty of materials from which to choose, schools often find themselves severely restricted when it comes to funding. What is the best way of using limited amounts of money and what approaches will help appraise what is available? The information here is not intended to be comprehensive. However, it will serve as a prompt and help you make some informed choices. Over a period of time an enterprising teacher can assemble an impressive range of resources for little or no cost. Equally there are some resources which simply have to be purchased, such as:

maps at various scales	aerial photographs	atlases
globes	contrasting locality packs	textbooks
IT software	teachers' guides	magnetic compasses
thermometers	tape or digital recorders	cameras (digital/video)

Given the pace of change there is little point in listing specific products or titles here. For up-to-date information, consulting the educational press, visiting the Geographical Association website (www.geography.org.uk) and subscribing to *Primary Geographer* will prove invaluable. *Primary Geographer* is the only journal aimed at primary school geography teachers and it contains a wealth of ideas, articles, reviews and comment.

Maps and fieldwork equipment

Ideally any maps that you use regularly should be laminated to make them more durable.

Resource	Comment
Large scale OS maps (1:1,250 urban areas) (1:2,500 rural areas)	At this scale individual houses and gardens can be easily discerned. Invaluable for local area studies, fieldwork and enquiries
Medium scale OS maps (1:10,000)	A black and white map showing the pattern of roads, street names and main buildings which is ideal for land use surveys
Standard scale OS maps (1:25,000 Pathfinder series)	A coloured map that shows field boundaries and more detail than the classic Landranger series (see below) making it ideal for walking
Standard scale OS maps (1:50,000 Landranger series)	The most popular OS series accessible to upper KS2 which provides an overview of the features in the local area and surrounding region
Street maps	Useful for tracing local routes and journeys
Road maps	A good way of providing an overview of UK towns and cities and the transport links between them
Historical maps	Useful for studying changes. Availability varies according to area

Atlas maps	Invaluable for developing a knowledge and understanding of key places in the UK, Europe and wider world. Many atlases now contain satellite images and information
Globes	An indispensable aid for all ages and the only way of showing the true proportions of the Earth. Can be used to illustrate planetary motion and to qualify two-dimensional atlas maps
Walls charts	Available at a variety of scales and for different locations. Useful in class displays and for research
Plastic floor maps	Outlines of the world, UK and other areas. Particularly useful with younger pupils
Jig-saw maps	Provide colourful images and symbols for places around the world. Particularly useful for interaction and reflection
Satellite images	Great variations in impact and quality of image
Magnetic compasses	A simple but vital instrument to be used alongside maps or independently in fieldwork
Thermometers	Maximum-minimum thermometers have the advantage of recording variations over a day
Weather recording equipment	Simple recording equipment can be improvised or made by the pupils. Suppliers have a range of instruments for school use including wind vanes and barometers
Measuring height and distance	Long tape measures or trundle wheels for measuring distance and clinometers for angles and heights

Books, photographs and artefacts

Resource	Comment
Textbooks	Bring together a wide range of resources. Often have the advantage of being graded and designed to meet the needs of the national curriculum
Reference books	Can be very valuable in supporting individual or group projects but be careful pupils do not simply copy them
Stories, fairy tales and rhymes	A useful way of developing empathy, motivation and creative mapwork. Very effective for theme or topic work with younger pupils
Locality packs	Provide detailed information, useful for in-depth studies but best used alongside other resources
Teachers' guides and resource books	A useful way of gleaning resources and ideas for self-designed lessons. Often contain activity sheets which are useful if linked to practical enquiries and investigations but rather sterile if used in isolation
CD-Roms	Increasingly available on a range of geographical themes and topics. Ideal for integrating with ICT skills
Photopacks	Collections of photographs on a theme, often with supporting activities. Useful for stimulus displays
Aerial photographs	An excellent resource for amplifying and extending local area work especially if centred on the school. Best used alongside a map of the same scale for comparison
Colour prints and slides	Provide additional information about places and themes. Can be assembled from your own holiday snaps especially if focussed on specific themes
Digital photographs	Increasingly popular and widely used. Can be imported into a Powerpoint presentation or stored or later use
Postcards	A cheap way of adding to a photograph collection
Photographs in brochures, magazines and calendars	Best grouped thematically and added to over time

Artefacts	Can range from tourist souvenirs and craft items to fresh food, plants and flowers from a supermarket or used tickets and labels on paper bags from other countries
Newspapers and brochures	Provide information about current events and issues as well as local language and customs
Rocks, fossil and plant samples	Another cost-free way of collecting geographical 'clues'. Preserve leaves by pressing in a heavy book
Audio tapes, mini discs and CDs	An excellent way of introducing music from around the world as well as songs and nursery rhymes. Your own sound recordings are another approach
Video tapes/DVDs	Shoot your own, virtually cost-free videos or record news items and other relevant programmes
Educational TV programmes	An extremely effective way of bringing images of distant places into a classroom. Usually supported by teachers' notes and activities

Websites

This list identifies some of the better known geography-related websites. A more extensive list of up-to-date web addresses can be found at www.geography.org.uk/phbk.

Resource	Comment
Distant places www.oxfam.org.uk/coolplanet www.globalgang.org.uk www.globaleye.org.uk	Information about children overseas together with some excellent teaching resources Christian Aid site with news, games and gossip from around the world Photographs, teaching activities and games from around the world from Worldaware
Environment www.panda.org www.wastewatch.org.uk	WWF-UK information, teaching ideas and news A national charity dedicated to reducing waste and pollution
Local area www.statistics.gov.uk/neighbourhood	Government data from the 2001 census
Mountains and volcanoes www.peakware.com http://volcano.und.nodak.edu	Explore over 1700 peaks around the world Photographs and images of volcanoes from around the world
Rocks and minerals www.nhm.ac.uk/education www.bgs.ac.uk	Activities and information linked to the Natural History Museum collection British Geological Survey information and education pages
Rivers www.sln.org.uk/geography	Contains an excellent study of the River Eden and other physical geography information
Teaching Ideas www.geography.org.uk www.teachingideas.co.uk www.puzzlemaker.com www.sln.org.uk/geography/	The Geographical Association site covers publications and journals as well as teaching ideas and professional development A wealth of teaching ideas and suggestions assembled from teachers around the country A versatile site which helps pupils with homework as well as having fun with games An award-winning site from Staffordshire geography teachers
UK Maps www.multimap.com www.getmapping.com www.mapzone.co.uk	UK maps and aerial photographs A map site with aerial photographs The Ordnance Survey website for pupils that includes games, competitions and homework help
Weather www.bbc.co.uk/weather www.metoffice.gov.uk	Information about UK and world weather, including weather cams and pollution updates The Meteorological Office sites with live weather warnings, forecasts and educational pages

Index